T0260066

Phenotypic Plasticity

SYNTHESES IN ECOLOGY AND EVOLUTION

SAMUEL M. SCHEINER, SERIES EDITOR

Phenotypic Plasticity

Beyond Nature and Nurture

MASSIMO PIGLIUCCI

THE JOHNS HOPKINS UNIVERSITY PRESS
BALTIMORE AND LONDON

Printed in the United States of America on acid-free paper

9 8 7 6 5 4 3 2 1

The Johns Hopkins University Press
2715 North Charles Street
Baltimore, Maryland 21218-4363
www.press.jhu.edu

Library of Congress Cataloging-in-Publication Data

Pigliucci, Massimo, 1964–
Phenotypic plasticity : beyond nature and nurture / by Massimo Pigliucci.
p. cm.—(Syntheses in Ecology and Evolution)
Includes bibliographical references (p.).
ISBN 0-8018-6788-6 (alk. paper)
1. Phenotype. 2. Genotype-environment interaction. 3. Nature and nurture.
4. Adaptation (Physiology) I. Title.
QH438.5 .P53 2001
575.5′3—dc21 2001029320

A catalog record for this book is available from the British Library

To Giovanni D'Amato,
Guido Barbujani,
Carl Schlichting,
and Annie Schmitt—
all mentors, all friends

I speculated whether a species very liable to repeated and great changes of conditions might not assume a fluctuating condition ready to be adapted to either condition.

—Charles Darwin, letter to Karl Semper (1881)

Contents

— Series Editor's Foreword —

This book is the first in a new series devoted to synthetic works in the fields of ecology and evolution. These fields have entered a period of maturity and synthesis. Recent examples of synthetic activities include the establishment of the National Center for Ecological Analysis and Synthesis at the University of California Santa Barbara and Deep Green, a worldwide collaboration among angiosperm systematists to determine the phylogenetic structure of the flowering plants. In ecology some of the most important research challenges concern global climate change and the loss of diversity. Meeting these challenges requires data and expertise from areas as disparate as geology, climatology, economics, and public policy. In evolution some of the most exciting research is that being performed at the interface between disciplines, such as evolutionary theory, paleontology, developmental biology, and molecular genetics. For example, accounting for macroevolutionary trends has been a challenge since the nineteenth century. By integrating information on genetic architecture, epigenetics, and patterns of natural selection, scientists may now be able to establish rules for macroevolutionary patterns.

This series will provide a unique forum for synthetic and interdisciplinary books. The focus is on the fields of ecology and evolution, but it will reach out to disciplines both within and outside biology. My hope is that this series will play an important role in how new fields develop and in moving established disciplines in new directions.

This book exemplifies these goals. It addresses a subject central and topical in evolutionary biology, *phenotypic plasticity,* and uses it to forge connections between the often disparate disciplines of *genetics, development, ecology,* and *behavior.* I am very pleased that it is the lead-off book in this series, both for its subject matter and for its style. When Dr. Pigliucci approached The Johns Hopkins University Press with his prospectus, I had no hesitation in recommending publication. I had read his papers and his previous book and knew that he could present interesting and provocative ideas in a clear fashion. I was not disappointed. This book provides the type of broad overview needed in a work of synthesis. It will also be somewhat controversial—to my delight. It is thought-provoking and iconoclastic. I do not agree with everything in it, including statements about my own research. But Dr. Pigliucci's ideas are clearly laid out and will undoubtedly spark several research programs.

This debut volume captures the breadth and spirit for which I am striving in this series. I am sure we can look forward to more stimulating works in the years to come.

Samuel M. Scheiner
Arizona State University, Tempe

— Foreword —

Research on phenotypic plasticity has been at the forefront of evolutionary thinking for the last two decades or so. Diversity generated by the ability of any particular genotype to adjust its development and, therefore, phenotype in response to the environment can be dramatic. However, few books on this subject have appeared, which makes this offering a most timely one. Without doubt this book demonstrates why phenotypic plasticity has become so fascinating to many biologists, especially to those who share a bent toward multidisciplinary approaches. Like-minded researchers are most excited by those topics that reflect intersecting areas of biology. In my opinion these are precisely where the opportunities to achieve the most innovative and important insights lie. Phenotypic plasticity must be one of the more striking phenomena in need of integration across different areas of biology. As such it represents both a formidable challenge and an exciting opportunity for research. To study plasticity effectively one needs to be a biologist of many colors: a geneticist, a physiologist, a developmental biologist, an ecologist, and an evolutionary biologist. I think the challenging nature of the topic is also evident in a healthy level of controversy, especially with regard to the nature of genes for plasticity. This book rises admirably to the challenge.

Historically, attempts to integrate ecology and genetics have sought to study the genetic basis of phenotypic variation in natural populations and the ways in which such variability is influenced by natural selection. It can, however, be very difficult to identify the specific targets of selection. Some examples of plasticity can prove excellent tools for exploring natural selection in the wild and addressing the question of to what extent plasticity is the result of adaptive evolution. The other side of the coin—that plasticity can reflect constraints on the ability of organisms to develop optimal phenotypes across environments—is also receiving avid attention.

The ways in which genes map onto phenotypes in natural populations have represented something of a black box in the history of evolutionary ecology. Once again, phenotypic plasticity can provide wonderful material for opening this box. How are developmental pathways within an individual organism modulated to produce the alternative phenotypes? The present opportunities for identifying genes to match up with the developmental and physiological mechanisms are endless. To do this in the context of phenotypes of evolutionary significance in natural populations is an exciting goal indeed. Functional genomics will only provide yet more tools and technologies for further exploration of the developmental and physiological mechanisms underlying the ability to express phenotypic plasticity.

The environment of an individual of a given genotype not only is critical in shaping how it is influenced by natural selection but also may profoundly affect the phenotype it will produce during development. Again the different levels of interaction between genotype, environment, and phenotype are at the heart of thinking about the evolution of phenotypic plasticity, and such complexity is part of its challenge. Those interactions between genotype and environment that are the basis for the evolution of plasticity should provide a cautionary tale to those applying functional genomics: Gene expression may be highly dependent on both external and internal environment.

At the same time that phenotypic plasticity presents such opportunities to the empiricist to explore fundamental issues in biology, the field as a whole has also benefited greatly from contributions from theoretical biology. These have provided both extremely useful frameworks for the empiricists as well as unique insights into how organisms can adapt to spatial or temporal heterogeneity in their environments.

All these issues and many more are explored in this timely and wide-ranging book. It shows what has been achieved but also sets the scene for the many exciting discoveries that the burgeoning molecular world will increasingly open up. Fortunately, the topic of phenotypic plasticity will continue to draw together biologists working with organisms in nature, in the laboratory, and in the test tube.

Paul Brakefield
Leiden University, The Netherlands

Preface

Why write a book on phenotypic plasticity? Plasticity studies happen to hit on one of the fundamental questions that has been asked for a long time: How do *nature* (genetics) and *nurture* (environment) interact to yield the plants, animals, and of course human beings that we see around us? This is certainly the main reason I became fascinated by this field of inquiry—the chance to contribute to the solution of an age-old question that directly affects both our philosophical outlook and our daily lives. A few years ago I co-authored a book on the more inclusive topic of phenotypic evolution (Schlichting and Pigliucci 1998), which included plasticity as one of three main conceptual issues for evolutionary biologists (the other two being allometry and ontogeny). Yet phenotypic evolution and phenotypic plasticity are certainly not one and the same, despite sharing the "phenotypic" adjective. One can think of plasticity as an important, but by no means dominant, component of phenotypic evolution. Students of the latter have to consider a broader range of disciplines, spanning almost the entire gamut of modern evolutionary biology (e.g., no plasticity studies can be done on fossils, but the literature on phenotypic evolution would be sorely lacking if paleontology were not considered). However, plasticity studies have been contributing substantially to our modern understanding of phenotypic evolution in general, so I felt that a closer look at this element of the larger question was justified.

This book is obviously a contribution, not a solution, to the understanding of genotype-environment interactions. Despite almost a century of research, dramatically intensified within the last twenty years, we are still in no position to propose an overarching theory of phenotypic plasticity. What we have is not yet a clear picture, but a giant intricate and fascinating puzzle, with several important pieces positioned in what appear to be the right locations. My aim here is to introduce the reader to the basic pieces of such a puzzle and to provide an interpretation of how researchers in this and closely related fields see the whole picture so far. While I tried to be comprehensive and balanced, the reader will perceive that the book is structured around a particular point of view—my own. After all, a book about ongoing research is neither a textbook nor a review of the subject matter and must interpret developments and suggest new directions. Therefore, while I attempted to cite the most relevant publications on each side of whatever issue I discussed, the reader should have no trouble understanding where I am coming from and why.

An experienced reader in the field of plasticity studies will find that I avoid some of the older questions in order to emphasize other, less well-studied ones. This reflects not just my opinion, but a genuine shift that I see generally among students of phenotypic plasticity. One of the original questions, for example, was how much variation for plasticity can be found in natural populations. We now know the answer: a lot. While papers continue to be published demonstrating new examples of variation for plasticity in plant and animal species, the cutting-edge questions now come from different subfields, such as the ecology of adaptive plasticity and the molecular genetics of plastic responses. This is also the reason why in this book as in some of my papers (Schlichting and Pigliucci 1995; Pigliucci 1996a,b; Pigliucci and Schlichting 1997) I treat the quantitative genetics approach to plasticity as a "classical" field of study. It seems to me that we understand fairly well, at least in a general fashion, how plasticity changes at the population level, provided that there is variation for reaction norms. This is precisely the kind of question that quantitative genetics is so well equipped to answer. But new, and I believe more fundamental, questions remain unanswered and are completely outside the quantitative genetic framework. Where does plasticity come from in the first place? How does natural selection put complex and sophisticated systems of reception of environmental signals in place? What role did gene duplication play in the evolution of genes controlling plastic responses? What are the patterns of evolution of plasticity above the species level? This book cannot provide the answers, but it does offer intriguing clues to problems that will account for most of the action in the next decade or two. My hope is that graduate students, and to some extent colleagues, will find many new ideas for research projects in the following pages.

I start out, in Chapter 1, by addressing the question of what phenotypic plasticity is to begin with, and in Chapter 2 I discuss different types of plasticity. No doubt there will be plenty of room for disagreement in the content of those chapters. Following a brief chapter dedicated to the history of the field (Chapter 3), the remainder of the book traces a logical progression through subdisciplines, from genetics and molecular biology (Chapters 4 and 5), to development (Chapter 6), to ecology, behavior, evolution, and theoretical modeling (Chapters 7, 8, 9, and 10). Each chapter is a discussion of how these broad fields of biological research contribute to our understanding of phenotypic plasticity. The progression culminates with Chapter 11, on phenotypic plasticity and its relationships to the general field of evolutionary biology, and is meant to show that insights into the phenomenon of plasticity can in turn be used to address a panoply of larger questions in biology.

I also feel that I must make a sort of semantic disclaimer at the outset. The reader will find the word *constraint* popping up here and there throughout the book. The literature on constraints in evolution is vast and especially controversial (Antonovics 1976; Gould 1980; Cheverud 1984; Maynard Smith et al. 1985; Cheverud 1988; Gould 1989; Antonovics and van Tienderen 1991; Wake 1991; Arnold 1992; Hall 1992; McKitrick 1993; Oyama 1993; Zelditch et al. 1993; Coleman et al. 1994; Losos and Miles 1994; Price 1994; van Tienderen and Antonovics 1994; Sih and Gleeson 1995; Travisano et al. 1995; Rose and Lauder 1996; Wagner and Altenberg 1996; Baatz and Wagner 1997; Pigliucci et al. 1998; Merila and Bjorklund 1999). Together with Carl Schlichting, I gave the topic my best shot in our earlier book (Schlichting and Pigliucci 1998), and I still maintain that our arguments were, by and large, correct. For the purposes of the present volume, constraints are approached from a quantitative genetic perspective, as genetic correlations or covariances between traits. This is a relatively restrictive definition, but it has the advantage of being easily quantifiable and therefore yielding to empirical study. I am aware of the fact that genetic correlations do not necessarily reflect more fundamental constraints. I also realize that a constraint can actually speed up evolution along a particular direction. For example, if two traits are positively genetically correlated and selection favors an increase in both, the response of the population will be particularly rapid. But the genetic correlation still represents a constraint in the broad sense of limiting the possible range of evolutionary trajectories: Even though a population may be put on a fast evolutionary track by a particular constraint, it will still have to stay mostly on that track while the constraint exists. I also realize that constraints defined in this manner are bound to be local in both space and time. However, I argue that constraints that can temporarily antagonize or be overcome by selection are the most interesting to the evolutionary biologist. More fundamental constraints, such as the impossibility of violating the second principle of thermodynamics, explain very general properties of *all* organisms, but not the differences *among* organisms; and it is the existence of these differences that has justified a whole field of research ever since Darwin (1859).

The epilogue deserves a special explanation since it focuses less on the science of phenotypic plasticity and more on the politics of the nature-nurture debate, especially in humans. While the rest of the book presents the research part of the picture, the last few pages deal with popular as well as technical writings that also represent political and philosophical views. I have included this material because phenotypic plasticity research has fundamental implications for society, and we must not retreat into an aseptic ivory tower and ignore the political and social consequences of science, even though they may be tainted by the ideological agendas of the participants in the nature-nurture debate. My conclusions may surprise some, anger others, and be perceived as the elucidation of the obvious by still others. Nonetheless, given the confusion on the subject in the popular science literature and the media, as well as the tremendous practical consequences of the debate itself, I felt compelled to take a risk and attempt to clarify where the science ends and the ideology begins.

In summary, I offer this book as a review, discussion, and analysis of ongoing research on how genes and environments interact to yield complex organisms capable of sophisticated responses to their surroundings. This is a fascinating field of inquiry that I hope will begin to attract more graduate students and researchers—it has certainly captured the imagination of the general public.

This book would not have been published without the tireless efforts of my science editors, Ginger Berman and Sam Schmidt, and it has greatly benefited from the painstaking work of the series editor, Sam Scheiner, with whom I enjoyed long discussions on phenotypic plasticity. The following people read earlier versions of some of the chapters and tried their best to minimize my errors, factual or conceptual: Chris Boake, Kitty Donohue, Sergey Gavrilets, Laura Hartt, Jonathan Kaplan, Beth Mullin, Ariel Novoplansky, Sue Riechert, and Sonia Sultan. My postdocs—Mark Camara, Hilary Callahan, and Courtney Murren—and several of my graduate students—Anna Maleszyk, Heidi Pollard, and Carolyn Wells—also contributed invaluably. My wife, Melissa Brenneman, convinced me of the importance of a preface and patiently stayed with me throughout the gestation of the book.

—1—

What Is Phenotypic Plasticity?

The human mind is so complex and things are so tangled up with each other that, to explain a blade of straw, one would have to take to pieces an entire universe. . . . A definition is a sack of flour compressed into a thimble.

—Rémy de Gourmont (1858–1915)

CHAPTER OBJECTIVES

To discuss basic concepts related to phenotypic plasticity, starting with the idea of genotype-phenotype mapping function. The concept of reaction norm is introduced and several of its attributes are discussed briefly. A distinction is made between genotypic means and pattern and degree of plasticity. The relationship between plasticity and heritability is discussed, together with experiments showing the environmental variability of the latter. The analysis of variance is introduced and related to attributes of reaction norms and heritability. Plasticity is discussed within the framework of interenvironment genetic correlations and quantitative genetics theory. Multivariate aspects of plasticity such as environmental effects on character correlations and the concept of plasticity integration are introduced.

Phenotypic plasticity is the property of a given genotype to produce different phenotypes in response to distinct environmental conditions. Despite this deceptively simple definition, the concept of phenotypic plasticity has long been a difficult one for evolutionary biologists, and new biology students are still likely to be confused by the apparent overlap of two supposedly distinct entities: the genotype and the environment.

The study of phenotypic plasticity is as old as the idea of genotype itself (Chapter 3), but it received scant attention throughout the first part of the twentieth century. Modern work was conducted initially by a few researchers in the 1970s and early 1980s, with increasing attention from evolutionary biologists in the second part of the 1980s, culminating in a proliferation of papers during the 1990s. (The agricultural literature had been dealing with plasticity for quite some time, so much so that an entire volume would be needed to discuss it, but the interest there was in how to *eliminate* plasticity rather than in understanding where it comes from and how it works.) This intensified focus on genotype-environment interactions has yielded a virtual explosion of empirical papers (e.g., Petrov and Petrosov 1981; Matsuda 1982; Turkington 1983; Scheiner and Goodnight 1984; Via 1984a; Cavicchi et al. 1985; Meyer 1987; Schlichting and Levin 1988; Greene 1989; Weis and Gorman 1990; Hader and Hansel 1991; Schmitt et al. 1992; Andersson and Widen 1993; Day et al. 1994; Cheplick 1995a,b; Winn 1996a,b; Jasienski et al. 1997; Pigliucci and Byrd 1998; Thompson 1999). These in turn have generated a plethora of reviews (Bradshaw 1965; Schlichting 1986; Sultan 1987; West-Eberhard 1989; Thompson 1991; Scheiner 1993a,b; Schlichting and Pigliucci 1995; Via et al. 1995; Pigliucci 1996a). The field has also seen a good number of theoretical papers aimed at modeling the evolution of phenotypic plasticity (e.g., Via and Lande 1985; Gavrilets 1986; Stearns and Koella 1986; Lorenzi et al. 1989; de Jong 1990a,b; van Tienderen 1991; Gomulkiewicz and Kirkpatrick 1992; Gavrilets and Scheiner 1993; Berrigan and Koella 1994; Piepho 1994; van Tienderen and Koelewijn 1994; de Jong 1995; Sibly 1995; Pigliucci 1996b; Zhivotovsky et al. 1996; Scheiner 1998). One of my aims in this book is to assist the interested reader to summarize and make sense of the bewildering number of publications that have accumulated in the field.

So much intellectual effort must have been stimulated by something worthwhile. This something is a century-long quest to understand the relationship between genotype and phenotype (see the epilogue). In this search the environmental factor has, in turn, been ignored, relegated to a passive role, or given center stage. I therefore begin with a discussion of the so-called genotype-phenotype *mapping function* and what we know (and do not know) about it. An understanding of this concept is fundamental if we wish to consider the role of phenotypic plasticity and the idea of reaction norm, which will then be introduced as the basic link relating the three variables that are of interest to us: genotype, environment, and phenotype. I close this chapter with a discussion of the mathematical equivalence and conceptual relationship between reaction norms and a favorite tool of quantitative geneticists—the interenvironment genetic correlation.

Genotype and Phenotype: The Mapping Function Problem

Mendel was the first scientist to grasp, at least intuitively, the distinction between genotype and phenotype, thereby inaugurating the field of genetics as a whole. But it was only after the rediscovery of his work in 1900 (as discussed in Stern and Sherwood 1966) that Johannsen (1911) formalized the separation between the two and coined the terms *genotype* and *phenotype.* Ever since Mendel's time, scientists have been searching for—and occasionally thought they had found—the relationship between genes and phenotypes, what is sometimes referred to as the *genotype-phenotype mapping function* (G→P) (Alberch 1991). For example, the work of Beadle and Tatum on mutants of *Neurospora* in 1941 (which led to a Nobel Prize) elucidated the relationship between genes and proteins, leading to the famous *one gene–one enzyme* hypothesis (recounted in Sturtevant 1965).

The simplest possible G→P mapping function is the one implicitly assumed by Mendel himself, which is also the one so much in vogue with the lay media whenever word of a new "gene for" something is reported (Kaplan and Pigliucci 2001). This hypothesis allows for a direct causal connection between gene action and phenotype. Mendel wrote of a *factor* controlling the color of his pea plants, with each version of this factor determining one of a series of alternative colors (e.g., green or yellow). Similarly, when modern molecular geneticists talk of a "gene controlling obesity," or a "gene causing schizophrenia," they are more or less unconsciously referring to the simple one-to-one mapping function. (I am not suggesting that modern researchers actually *believe* that such a simplistic view of genotype-phenotype relationships would allow us to pinpoint *the* gene "for" complex physiological or behavioral characters, but the language is consistent with a simple scenario of gene action and results in dangerous misrepresentations on the part of the media and the public.) We all know that the one-to-one function is generally correct at the molecular level: One gene produces one protein (sometimes more, in cases in which there are more open reading frames or substantial post-transcriptional splicing [Suzuki et al. 1989]). A scheme of the one-to-one model is shown in Fig. 1.1a.

Very early in the history of genetics, it was realized that most characters do not conform to the Mendelian model (as discussed in Provine 1971). Even though it is possible to demonstrate that they are inherited through the usual mechanisms, they seem to be controlled by many genes. The existence of these so-called quantitative characters—such as the height of a plant or the weight of a human being—opened the whole field of quantitative genetics. Important concepts that anchor the quantitative genetic mapping function are those of *pleiotropy* and *epistasis.* With pleiotropy, one gene can affect several characters simultaneously. (Of course, pleiotropy can also be caused by tight linkage between genes affecting different traits, the physical linkage causing joint inheritance of the genes and, therefore, of the phenotypes.) If there is epistasis, genes affect the actions of other genes, leading to a complex genetic architecture underlying a given trait. As illustrated in Fig. 1.1b, pleiotropy makes it more difficult to trace a given phenotype back to the action of a particular gene (and the same can be said

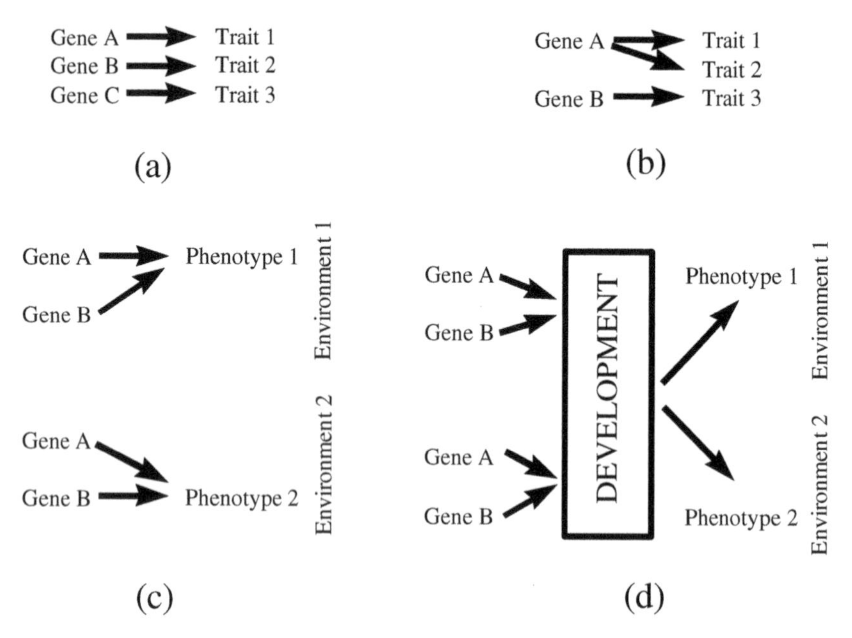

FIG. 1.1. Four schematic hypotheses for the genotype-phenotype mapping function. (a) A simple Mendelian system in which there is a direct relationship between genotype and phenotype. (b) A quantitative genetic model including pleiotropy (other effects such as epistasis are not shown). (c) A model including phenotypic plasticity, in which the same genotype produces different phenotypes in distinct environments. (d) The most realistic model (although still grossly simplified), showing that genotype-environment interactions are mediated indirectly by epigenetic effects during development. (From a concept in Suzuki et al. 1989.)

for epistasis). In the figure, traits 1 and 2 are both influenced by gene A, so it is hard to say that gene A is "for" a particular character. In real cases, one gene can affect many different aspects of the phenotype, either directly or indirectly by interacting with other genes (see, e.g., Zhong et al. 1999).

The two hypotheses of mapping functions discussed so far represented the dominant paradigm in genetics for a long time. As we can see, there was no formal allowance for two other major factors, which have reached center stage in evolutionary research only in the past couple of decades—environmental effects and epigenetics. In Fig. 1.1a and b, the environment is only a source of noise, to be standardized or minimized, either by experimental design (e.g., growing organisms in *common gardens*) or by statistical methods (partitioning of variance components into "genetic" and "environmental," only to focus on the former and discard the latter as noise). In Fig. 1.1c, however, the environment is a player on equal footing with the genotype. Here, the genes respond to the particular environment in which they are expressed and determine the production of two distinct phenotypes, one in each of the two environments. The categorical distinction between genes and environments disappears, to be replaced by a dialectical relationship in which genes do not "do" something in a vacuum, but operate in a particular environmental context (of course, environments by themselves do not do anything either).

Development is the other component of phenotypic determination that has always been ignored; in particular, what happens during the series of processes loosely referred to as *epigenetics* (literally, what happens "beyond the genes" [Waddington 1942, 1961; Gilbert 1991]). I dwell quite a bit in the following chapters on the interaction between epigenetics and plasticity, especially in the framework of modern molecular developmental genetics. For now, let us examine Fig. 1.1d simply to state the problem. In this more sophisticated model, the environment still has a major effect on the phenotype; and this is still somehow mediated by the genes, since we have two distinct phenotypes produced by the same genotype in different environments. However, precisely *how* the genotype-environment interaction occurs is not clear, because the crucial events unfold during development, which accepts genetic and environmental effects as input and produces specific phenotypes as output. Elucidation of this last and far more realistic model of the genotype-phenotype mapping function is one of the field's major challenges, and carries with it the potential for unifying the disciplines of evolutionary, developmental, and molecular biology (Chapter 11).

The Environment and the Concept of Reaction Norm

The fundamental conceptual research tool in phenotypic plasticity is the idea of a *reaction norm,* a term coined by Woltereck (1909) (see also Chapter 3). Once proposed, the idea remained dormant in the remote recesses of evolutionary biology theory for more than half a century, and has only recently claimed center stage. Simply put, a reaction norm is the function that relates the environments to which a particular genotype is exposed and the phenotypes that can be produced by that genotype. In textbooks, reaction norms are often simplified to appear as straight lines, characterized only by their slope and intercept. In reality, genotypic reaction norms may have more complex shapes, but this does not invalidate the idea of describing them as mathematical functions. Reaction norms have several fundamental properties, and a brief discussion of these is in order (Fig. 1.2).

First, let us consider the simplest possible reaction norm—a straight line indicating a linear response of the genotype to a series of environments. This type is found frequently in published reports, mostly as an artifact of studies that, for logistical reasons, are conducted in only two environments. In Fig. 1.2a we see that the reaction norm is typically represented in an

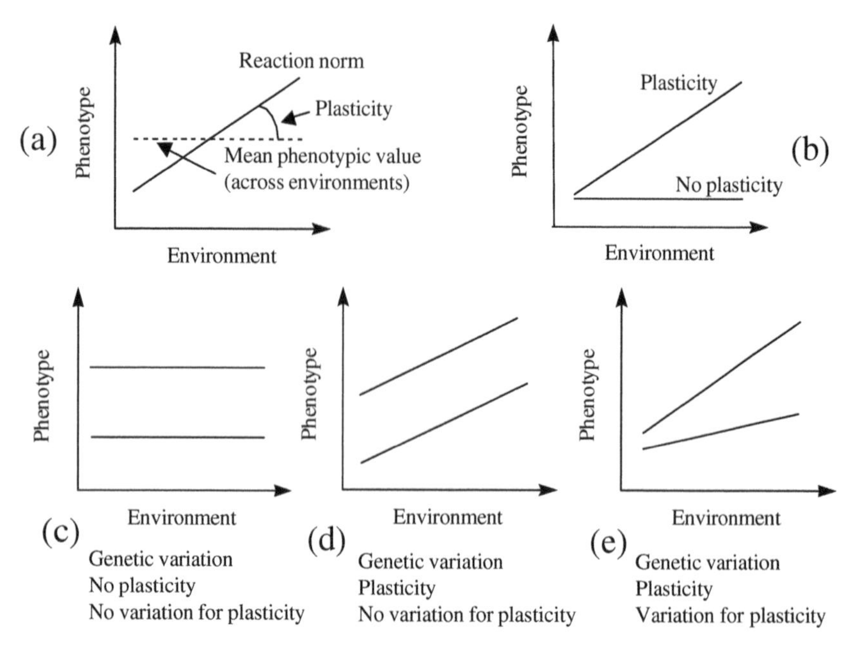

FIG. 1.2. Fundamental properties of (linear) reaction norms—slope (plasticity) and height (across environment mean): (a–b) relationship between plasticity and reaction norm, showing that the two are not synonymous; (c–e) independence of the concepts of genetic variation, plasticity, and genetic variation for plasticity.

environment-phenotype plot in which the environmental parameter (biotic or abiotic) is on the abscissa and some measure of the phenotype (behavior or another characteristic of the organism) is on the ordinate. Only one reaction norm is traced in the diagram, the solid line. The dashed line represents the average phenotypic value of that genotype *across environments.* This is usually referred to as the *genotypic mean* of the trait and is by definition *mathematically* (but not necessarily biologically) independent of the *plasticity* displayed by the same reaction norm. On the other hand, genotypic means, as well as plasticities, depend *on the range of environments used in the experiment.* If one subjects a genotype to a different set of environments, recalculates its reaction norm, and therefore its *mean value,* this second mean is likely to be different from the one calculated using the first set of environments. All this is of course a consequence of the fact that no genotype produces anything in the absence of a specific environmental context.

There has long been a debate in the field of plasticity studies about the meaning of a possible genetic correlation (see below) between the plasticity and the character mean of a given reaction norm. The absence of this kind of correlation has been taken as prima facie evidence that character means and character plasticities are genetically independent of each other (Bradshaw 1965). In fact, biologically, these two aspects of the reaction norm might or might not be interrelated, depending on the genetic mechanisms that control the expression of the trait. Other authors have argued that the across-environment character mean is irrelevant, and that the appropriate comparison is between plasticity and the means within each environment (Via 1993). There are, however, two problems with this latter approach. First, plasticity is a property of a genotype that makes sense *only* across a series of environments, so it is not clear what biological information one would gain from comparing an across-environment trait with a series of within-environment ones. Second, plasticity is calculated from the reaction norm, which itself is determined empirically by estimating a series of within-environment means. Therefore, the two measures are bound to be mathematically interdependent, and any test of their biological independence would be biased (Yamada 1962).

Another crucial distinction is the one between reaction norm and plasticity. The two terms are sometimes used as synonyms, but they are not synonymous. Strictly speaking, a reaction norm is a function, of whatever shape, that describes the genotype-specific relationship between environments and phenotypes. Plasticity is an attribute of the reaction norm, distinct from other attributes such as the across-environment mean. One can easily see that a genotype can be plastic or not plastic and have a well-defined reaction norm in either case (Fig. 1.2b).

Groups of reaction norms, of course, make up properties of populations, not of individual genotypes. Figure 1.2c, d, and e shows the distinctions among genetic variation for trait means, plasticity, and genetic variation for plasticity. I illustrate this first by referring to the graphs and then discuss the mathematical modeling. Figure 1.2c displays two genotypes characterized by different reaction norms. Both are flat with respect to the environmental axis, and therefore neither genotype is plastic—by definition. Since there is no plasticity for any genotype, obviously there is no genetic variation for plasticity. However, the two genotypes do differ in their across-environment mean, sometimes referred to as the overall mean or "height" of the reaction norm. Therefore, there is genetic variation within the population for one of the two fundamental parameters of the reaction norm. The situation illustrated in Fig. 1.2d is slightly different. Now the reaction norms still differ in the across-environment means, but they are also plastic. However, they are plastic in the same fashion (they are parallel along the environmental continuum considered), so there is genetic variation and plasticity but no genetic variation for plasticity. In Fig. 1.2e the two genotypes have reaction norms that differ not only in their height, but also in their degree of plasticity. In this case, therefore, we have genetic variation for the trait mean, plasticity, *and* genetic variation for plasticity.

From an evolutionary perspective, the cases in Fig. 1.2c and d show a population in which the trait mean can evolve, but the plasticity cannot. The case in Fig. 1.2e, on the other hand, presents a population in which *both* the trait mean and plasticity can evolve, because they are both variable across genotypes. Of course, these combinations can lead to many other scenarios as well.

Figure 1.2 also pinpoints the important difference between the *degree* (or amount) of plasticity, and the *pattern* of plasticity. All the reaction norms in the figure have the same pattern of plasticity in the sense that the trait value increases with an increase in the environmental value. The slope of each reaction norm, on the other hand, directly quantifies the degree of plasticity. Incidentally, this is one of the two main reasons why plasticity studies conducted in only two environments are very popular, the other reason being that they are logistically much simpler. In the case of two environments, measuring both the pattern and degree of plasticity is straightforward as the degree is indicated either by the slope or by the difference between the values in the two environments, and the pattern is given by the sign of that difference or by the sign of the slope. Proceeding when there are three or more environments is much less intuitive. Gavrilets and Scheiner (1993) have proposed the use of polynomial functions to represent reaction norms, a solution similar to one advocated by de Jong (1990b). However, a biological interpretation of quadratic and cubic coefficients is much less intuitive than the interpretation of a slope. David's group proposed an interesting alternative in terms of transformations of the basic polynomial function, which provides a good solution to the problem (Morin et al. 1997; Gibert et al. 1998). However, we must take into account that a large data set is necessary for a reliable fit of a higher-level curve to an environmental series. It should be noted as well that in the case of two environments the researcher does not actually perform a regression analysis: The two points are simply connected graphically and either the angle or the difference is measured. Gomulkiewicz and Kirkpatrick (1992), using the so-called *infinite-dimensional model* made yet another attempt to quantify the properties of reaction norms (Chapter 10). To date, there is no widely accepted and biologically intuitive method for quantifying the degree and pattern of phenotypic plasticity for more than two environments.

Figure 1.2e can also be used to elucidate a powerful concept first articulated by Lewontin (1974): The heritability of a trait depends on—among other things—the environment in which it is measured. (It also depends, e.g., on the particular set of gene frequencies, which means that it can be different from population to population, or for the same population at different times—as long as genetic drift, migration, selection, mutation, and assortative mating are relevant factors in the evolution of that population.) Heritability, contrary to the mechanistic connotation implied by the term, is simply a measure of how much genetic variation there is in a population for a particular trait relative to the total phenotypic variation. Therefore, it measures how quickly that trait will respond to selection. The generalized mathematical definition of heritability is

$$h^2 = \frac{\sigma_G^2}{\sigma_P^2} \tag{1.1}$$

where σ_G^2 is the genetic variance for the trait (which can be broad or narrow in sense, depending on how the genetic variance is calculated, which in turn depends on the experimental design), and σ_P^2 is the phenotypic variance for the same trait (including any possible source of variation, e.g., genetic, environmental, developmental, and error). The observed reaction norms should be considered as estimates of the true reaction norms, as is the case for any quantity in statistics, so they are surrounded by confidence intervals. Whenever the confidence intervals of the reaction norms of a population overlap, no statistically significant phenotypic variation can be attributed to differences of genotype, and the heritability is zero. If we consider the extreme left of the environmental gradient in Fig. 1.2e, we see that the reaction norms almost converge toward the same phenotypic value. Even small confidence intervals around each bar in that environmental range would cause an overlap and an estimate of zero heritability. (One has to remember that the power of these analyses, as with any analysis in statistics, depends on the sample size, and that the magnitude of effects per se is not a reliable guide to statistical significance.) However, if we suppose that the confidence intervals remain of roughly the same magnitude on any point along the reaction norms, then at the right end of the environmental gradient the two reaction norms would probably not overlap, and we would conclude that the heritability of that trait is large.

An important consequence of the above is that selection in one range of environments would be ineffective, but it would work perfectly well in another range. This principle was demonstrated empirically by Neyfakh and Hartl (whereas it has been known in crop science at least since Vela-Cardenas and Frey 1972, 1993). These authors noted that the relationship between the rate of embryonic development and temperature in *Drosophila melanogaster* is linear for most of the natural environmental range (from about 17° to 27°C). After that (in the 27° to 32°C range), however, the reaction norm becomes strongly nonlinear and can be fit more adequately by a cubic function. Neyfakh and Hartl then conjectured that embryonic development is less constrained toward the upper extreme of the reaction norm (i.e., at higher temperatures), which might lead to an increase in the expressed genetic variation, that is, to a higher heritability in that range. Selection for shorter developmental rates is usually ineffective in *D. melanogaster,* but these authors were able to apply selection for faster development in two populations: one an artificial mix of four inbred lines and the other a collection of flies from a natural site. In both cases, Neyfakh and Hartl were successful in altering developmental rates in the desired direction after only three generations of selection, with realized heritabilities of 5.7 and 8.7% in the two respective populations. It is interesting that the shape of the developmental rate/temperature cumulative distribution remained unaltered and simply shifted laterally. The authors could not determine whether the response to selection was due to multiple genes reacting simultaneously or the result of pleiotropy at a few loci with major effects.

Using a set of garden transplant experiments, Mazer and Schick (1991a,b) analyzed variation in heritability as a function of the environment for several characters simultaneously. They obtained maternal and paternal half-sib families of wild radish (*Raphanus sativus*) and grew them in three densities. They then measured survivorship as well as several life history traits (days to germination and to flowering), morphological traits (petal area), and reproductive characters (ovule number per flower, pollen production per flower, and modal pollen grain volume). They were able to show that heritability for these traits changes as a function of the density of planting. However, each trait behaves differently, so that it is not possible to draw generalizations about which density would elicit the highest heritability (and therefore the strongest response to selection [Fig. 1.3]). As a consequence, Mazer and Schick predicted that both the direct response of a character and the correlated response of other traits are environment-dependent parameters, which would make it extremely difficult to model phenotypic evolution in this system.

Several studies of this type have been conducted on natural populations in the field (as opposed to laboratory or greenhouse, as in the two preceding instances). Bennington and McGraw (1996), for example, performed a reciprocal transplant experiment of two ecotypes of *Impatiens pallida.* One population was collected from a floodplain and the other from a hillside characterized by large temperature extremes and able to support only scattered patches of individuals of this species. They again found significant differences in heritability between the sites, in that both populations displayed a reduced heritability in the stressful environment. This kind of result implies that colonization of novel or extreme environments might be hampered by an environment-dependent reduction in the genetic variance available to selection. This is an entirely new mechanism for explaining the difficulty of colonizing extreme habitats even for species that show a high degree of genetic variation in most of their range (i.e., under "normal" conditions).

On the other hand, Parsons (1987) showed, and Ward (1994) corroborated mathematically, that narrow-sense heritability may actually *increase* in extreme environments. Ward pointed out that the usual interpretation of

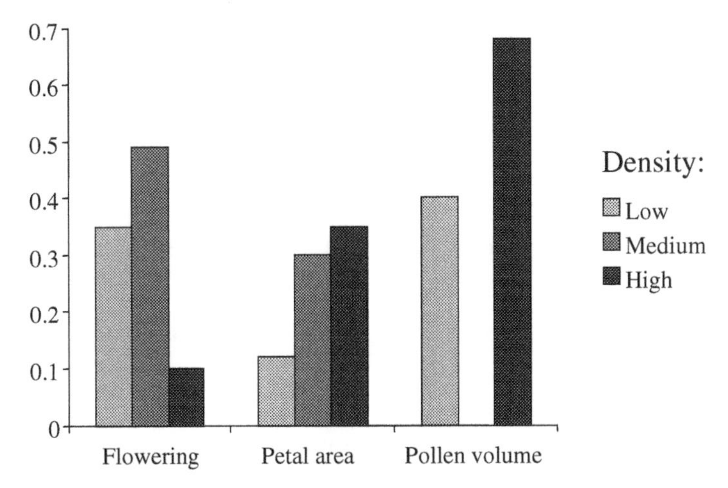

FIG. 1.3. Changes in heritability of different traits with the environment in wild radish. Note how the heritability of the same trait (flowering time, petal area, or pollen volume) depends dramatically on the environment (low, medium, or high density of conspecifics). Also, the pattern of variation of heritability with the environment is not the same for different traits. (From data in Mazer and Schick 1991a,b.)

this trend is that new genes, or specific *stress-genes,* are expressed in response to novel environmental conditions, thereby actually altering the genetic basis of a given trait (the explanation favored by Parsons). However, he questioned such an interpretation on the basis that the statistical definition of quantitative genetic parameters such as variances and heritabilities does not allow direct inferences about the biological basis of the variation in any trait. He therefore constructed a model of quantitative genetic parameters based on a specific underlying mechanism of gene expression, informed by current knowledge of biochemical pathways. The results indicated that an increase in heritability under stress can be achieved even if exactly the same genes act on the phenotype under study.

Hoffman and Merila (1999) have summarized the panoply of hypotheses proposed so far to explain the relationship between the quality of the environment and measures of heritability in that environment. They recognized the following categories:

Unfavorable conditions increase heritability because:

1. Stress alters rates of mutation or recombination.
2. Selection seldom occurs in novel or rare (and presumably stressful) environments, thereby leaving more allelic variation available to be expressed in those environments.

3. Conversely, selection favors particular phenotypes under common (presumably favorable) conditions, thereby reducing allelic variation in those environments.
4. Resource limitation magnifies phenotypic differences.

Unfavorable conditions decrease heritability because:

1. The environmental component of variance increases under stress.
2. Organisms do not reach their genetic potential under poor conditions.

Or, we cannot predict the relationship between environment and heritability because:

1. Measurement error is affected in an unpredictable way.
2. Genotype-environment interactions influence estimates of heritability.

As the reader can see, there is a scenario with an accompanying rationalization for every taste.

Of course, plasticity per se can also have a degree of heritability, as does any trait that can be reliably measured and that is variable within natural populations. However, both the statistical and biological meanings of this heritability have been foci of much discussion. One simple way of estimating the heritability of phenotypic plasticity is to use the parent-regression analysis so common for other traits in quantitative genetics. Jain (1978) did exactly that in two populations of *Bromus mollis,* a grass. He measured plasticity of flowering time in the parental generation and in their offspring. The slope of the resulting regression is an estimate of the heritability for that character. (If and when plasticity can be legitimately considered a character is discussed in Chapter 9.) The results were clear-cut. Both populations showed a significant regression slope, and the population with the higher genetic variance for flowering time also turned out to have the highest heritability for the plasticity of flowering time.

Scheiner and Lyman (1989) proposed a formalism for the definition of the heritability of plasticity based on the analysis of variance typical of many plasticity experimental designs (Chapter 2). They proposed two equations; the most straightforward, which is analogous to the general equation for heritability (1.1), is

$$h_{pl}^2 = \frac{\sigma_{G\times E}^2}{\sigma_P^2} \tag{1.2}$$

where $\sigma_{G\times E}^2$ is the variance of the genotype-environment interaction in an analysis of variance, and σ_P^2 is the usual total phenotypic variance. These same authors, however, pointed out that there are some problems in using this particular definition, because the associated least-square standard errors have not been derived mathematically—although they can in principle be computed by resampling techniques (Becher 1993). Sheiner and Lyman calculated the heritability of phenotypic plasticity in response to temperature of thorax size in *Drosophila melanogaster* and found significant heritability of plasticity.

The concept of reaction norm and its graphic representation in an environment-phenotype diagram are tightly linked with the analysis of variance (ANOVA) as a major method of statistical analysis of plasticity experiments (Lewontin 1974; Westcott 1986). The link is appealing because there is a simple, direct correspondence between attributes of reaction norm diagrams and the interpretation of main and interaction effects in an analysis of variance. A typical ANOVA model can be written out in the following fashion:

$$\sigma_P^2 = \sigma_G^2 + \sigma_E^2 + \sigma_{G \times E}^2 + \sigma_{err}^2 \quad (1.3)$$

where σ_P^2 is the total phenotypic variance; σ_G^2 is the portion of that variance that is attributable to differences among genotypes, *irrespective of the environment* (these first two parameters enter into the calculation of broad-sense heritability, as we have seen); σ_E^2 is the variance associated with changes in the environment, *independently of the genotype;* $\sigma_{G \times E}^2$ is the so-called *interaction variance,* that is, the variation due to the fact that some genotypes might respond to the environment in a different way than others; and σ_{err}^2 is the residual or *error variance,* which includes actual experimental error, microenvironmental variation not due to the experimental treatments, developmental noise, and anything else that cannot be accounted for by the main and interaction effects in the model. An actual plasticity experiment (Chapter 2) can be much more complex, with the inclusion of nested effects and three-way interactions. (For example, the genetic variation can be partitioned between species, among populations nested within species, and among genotypes nested within populations. Each of these nested effects would imply a corresponding "$G \times E$" term for species, populations, and genotypes.)

The correspondence between ANOVA effects and reaction norms is very simple; σ_G^2 quantifies the average spread of the height of reaction norms across environments. As long as reaction norms diverge in *any* environment(s), we will get a significant σ_G^2 term in the analysis of variance. Even if the reaction norms are perfectly parallel to the environmental axis (no plasticity) and/or perfectly parallel among themselves (no variation for plasticity), this term can still be significant (Fig. 1.2c). The σ_E^2 source of variation is found to be significant whenever reaction norms show an average trend, so that genotypes tend to express different phenotypes in different environments. This is so even if all genotypes show the same trend (no variation for plasticity) or do not differ from one another within a given environment (no average genetic variation). $\sigma^2_{G\times E}$ is the most important component for our purposes, in that it is a direct quantification of the amount of genetic variation for plasticity. A significant $\sigma^2_{G\times E}$ occurs any time that different reaction norms have distinct slopes, *regardless of the significance* of the other two effects. This last statement may generate some confusion. The best way to visualize such an occurrence is to consider the extreme case of several genotypes that display perfectly symmetrical responses to the environment (Fig. 1.4, lower right panel). In this case, a group of genotypes increases its phenotypic value (of whatever character one is considering) with an increase in the environmental value; a second group decreases its phenotypic value by roughly the same amount. Such an extreme "crossing" of reaction norms would yield no net genetic variation (i.e., $\sigma^2_G = 0$) because of the symmetry of the spread across the environmental gradient, so the average differences among genotypes cancel each other. There also would be no net plasticity (i.e., $\sigma^2_E = 0$), because there would be no net trend toward an increase or decrease of phenotypic value with an increase in environmental value. On the other hand, there would be an extremely significant genotype by environment interaction because some genotypes would be responding to the environment in a dramatically different fashion than others.

One problem related to the visualization of phenotypic plasticity as a property of a reaction norm is the fact that—strictly speaking—a reaction norm can be traced along an environmental gradient. For example, let us consider the case of the variation of a given trait in response to increasing levels of light, nutrients, water, density of conspecifics, or any other variable that can be ordered in a nonarbitrary way (see, e.g., Sultan and Bazzaz 1993). What we then obtain is not just a reaction norm that is a convenient visual summary of the genotype-phenotype relationship, but one that also immediately lends itself to a biological interpretation in terms of how changes in the phenotype are a continuous function of changes in the environment. Some plasticity studies, however, "mix" different factors, such as a control condition and a variety of distinct stresses (see, e.g., Schlichting and Levin 1990; Lois and Buchanan 1994). In these cases data can still be visualized as a reaction norm, which may help the investigator to spot meaningful patterns, but the order in which the environments are plotted is arbitrary. This arbitrariness can mislead a reader, and there is no unique or self-evident biological interpretation of the resulting "reaction norm." On the other hand, it can be argued that these "discontinuous" reaction norms are actually "slices" of a multidimensional reaction norm, in which each environmental factor has been sampled only at one extreme of a gradient. In general, I would simply suggest retaining the term *reaction norm* for all these plots, but for authors to make clear if—and for what reason—they have ranked the environments in any particular fashion.

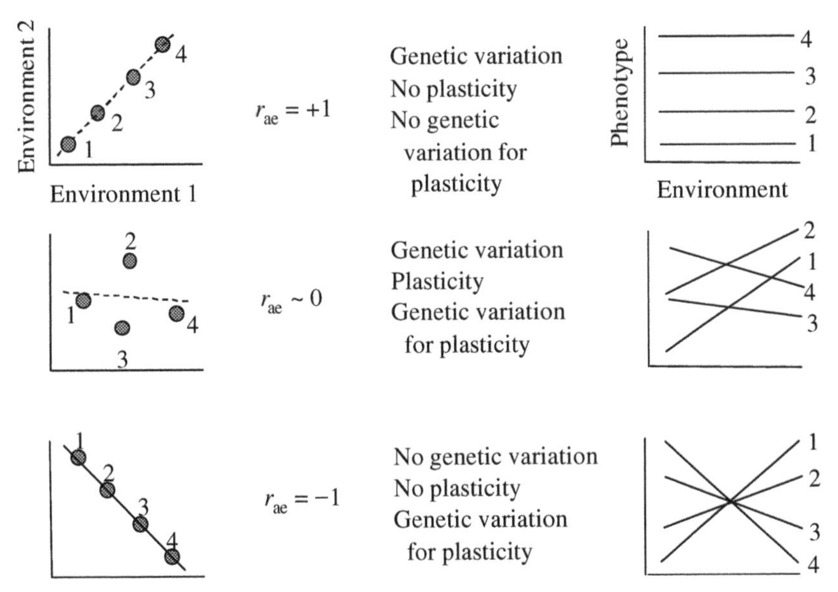

FIG. 1.4. A direct comparison between two views of plasticity data: the one based on interenvironment genetic correlations (left) and the one based on reaction norm plots and analysis of variance (right). The values of the interenvironment genetic correlations and the interpretation of the results of an analysis of variance are detailed in the central columns. Genetic variation here refers to the across-environment means only. (After Via 1987.)

The problem of ranking environments has also been tackled in an entirely different manner. An approach originally proposed by Yates and Cochran (1938) has been used extensively in the applied literature (see, e.g., Falconer 1990). Basically, the goal is to rate treatments not in accordance to what we, as humans, think are "different," "good," or "bad" environments, but to adopt the point of view of the organism. To do so, the researcher can use some measure of overall size or fitness to rank the environments. For example, one could calculate the mean biomass of all plants (across genotypes) in each of a series of predetermined treatments. Then, one would rank the environments in terms of increasing or decreasing *mean biomass,* the idea being that the average allocation to biomass is a direct reflection of the quality of that environment *as perceived by the organism.* There are some obvious advantages in adopting this strategy. For one, there is no longer a difference between ordered and unordered environments: Any combination of treatments could be ordered, and reaction norms would always be plotted along a biologically relevant gradient of increasing environmental quality. The second advantage is that one is freed from making dangerous assumptions about what is and is not stressful for the organism under study. However, there are also several problems intrinsic in this approach. First, if one is interested in studying reaction norms of size or fitness, one ends up essentially plotting the same data on both axes. Second, if a statistical analysis is attempted using the resulting ranking of environments, it turns out that the assumption of independence common to regression analysis and analysis of variance is violated. This translates into the impossibility of carrying out standard statistical procedures for that data set. The problem becomes less serious for an increasing number of genotypes and can be entirely sidestepped by using resampling techniques to estimate the relevant statistics. Third, if different genotypes perceive the quality of the environment differently, then the actual ranking of environments based on genotypic means is an artifact of which reaction norms happen to be more frequently represented in that particular population. The biological implications of genotype by environment interaction would be overlooked in favor of a simplifying search for "mean" trends.

The Concept of Genetic Correlation and Its Application to Plasticity Studies

There is a very different way of thinking about phenotypic plasticity data apart from utilizing reaction norms. One could consider the expression of the same trait in two (or more) environments as different traits *related by a certain degree of genetic correlation.* Referred to as the *character-state approach* (Via et al. 1995), it was advanced by Via (1984b) within the context of evolutionary quantitative genetics, but the concept is much older. Falconer (1952)—adopting a standard quantitative genetics approach—was the first to point out the "problem" posed by environmentally induced variation in a character under selection. Following his initial suggestion, a series of studies improved on the mathematical treatment of the multiple expressions of one trait in several environments. Robertson (1959a,b) presented the theoretical arguments substantiating Falconer's intuition. Dickerson (1962) pointed out the equivalence between the genetic correlation approach and the classical analysis of variance/reaction norms framework (see above), a point recently extended by van Tienderen and Koelewijn (1994) and de Jong (1995) to the multivariate case. Yamada (1962) discussed the type of statistical model (random or mixed) that should be used to extract the variance components necessary to calculate the genetic correlations, as well as the statistical assumptions underlying the character-state approach (such as equal heritabilities or homogeneity of variance between environments, a condition seldom met by real plasticity data sets). Several other authors further contributed to the extension of these statistical methods (Eisen and Saxton 1983; Fernando et al. 1984; Robert et al. 1995a,b), but the fundamental idea remains the one proposed by Falconer.

Visualizing reaction norms as interenvironment genetic correlations has problems of its own. First, the number of parameters that have to be estimated increases quadratically with the number of environments. Therefore, one rapidly loses precision in the description of multienvironment reaction norms, and it soon becomes impossible to obtain meaningful estimates of the parameters even for a small number of environments. Second, unlike the case of the polynomial approach described above, there is no way to extrapolate results to other environments because a genetic correlation is a single number relating two specific environments, not a continuous function. The so-called *infinite* dimensional model (Gomulkiewicz and Kirkpatrick 1992), a conceptual extension of the genetic correlation approach, does not suffer from this limitation (see Chapter 10).

As much debate and misunderstanding has plagued comparisons between the two approaches of reaction norm–analysis of variance versus character state–genetic correlation, I will discuss briefly what a genetic correlation essentially is, what (if any) differences there are between the two methods in mathematical and biological terms, and why we should care about one or the other.

First of all, a genetic correlation is the standardized genetic covariance between two traits and can be calculated as

$$r_G = \frac{\mathrm{cov}_{XY}}{\sqrt{\sigma_X^2 \sigma_Y^2}} \tag{1.4}$$

where cov_{XY} is the genetic covariance between traits X and Y; σ_X^2 is the genetic variance of trait X, and σ_Y^2 is the genetic variance of trait Y. In the case of the same trait expressed in two environments, the formulation is analogous:

$$r_{ae} = \frac{\mathrm{cov}_{1,2}}{\sqrt{\sigma_1^2 \sigma_2^2}} \tag{1.5}$$

where r_{ae} denotes the interenvironment genetic correlation, and the subscripts 1 and 2 stand for a measure of the same trait in treatments 1 and 2, respectively.

From a biological perspective, there are two possible causes for a genetic correlation: pleiotropy and linkage. In the case of pleiotropy, there is a significant genetic correlation between two traits if both are under the control of the same gene(s). In fact, the degree of genetic correlation is assumed to be proportional to the degree of genetic overlap in the control of the two traits. A positive genetic correlation would then imply that the same genes cause an increase or a decrease in the phenotypic value of both traits. A negative genetic correlation would suggest that—presumably due to some trade-off—the same genes would increase the value of one trait while simultaneously decreasing the value of the other one. This has direct implications for the future evolution of the two traits: If selection favors a simultaneous increase in their value and there is a positive genetic correlation, the population will respond very quickly. However, if there is a negative genetic correlation, the population might literally be stuck in the middle of two opposing forces (if these are approximately equal in strength). A correlation around zero would give the two traits the maximum latitude for independent evolution, so that selection could, for example, increase or decrease the mean of one trait while keeping the other one invariant.

The second causal mechanism that can be responsible for a genetic correlation is physical linkage. In this situation different genes might control each of the two traits, but if the two (or more) loci are close to each other on the same chromosome, then recombination between them will seldom occur. For instance, the population starts with a particular combination of alleles at those loci, say, the allele at the first locus, which causes an increase in trait 1 is always coupled with the allele at the second locus, which causes a decrease in trait 2. Such a combination will probably be maintained for a long time, until a rare crossing over causes the segregation of non-parental combinations. From an evolutionary perspective, the short-term consequences are the same as in the case of pleiotropy. The long-term consequences, however, are very different, in that we expect recombination to eventually break down linkage, but only evolution of the genetic system itself (e.g., by mutation) would be able to decouple traits connected by pleiotropy (Wagner 1995; Mezey et al. 2000).

Figure 1.4 shows two series of diagrams illustrating the representation of the same data (genetic variation and plasticity among four genotypes) as either character states (left) or reaction norms (right). The two methods of quantifying variation for plasticity would respectively lead one to calculate an interenvironment genetic correlation or to carry out an analysis of variance. The first row in the figure represents a case in which the reaction norms are parallel to each other and to the environmental axis. From a character-state standpoint, the four genotypes fall on a straight line with a positive slope in an environment 1 versus environment 2 diagram. Mathematically, the across-environment genetic correlation would be $r_{ae} = +1$, and an analysis of variance would detect genetic variation for across-environment means, but not plasticity or variation for plasticity. The second row displays a situation in which there is some, but not extensive, crossing of the reaction norms. This gives a rather random scatter in the environment 1 versus environment 2 diagram. The resulting genetic correlation would be close to zero, while an analysis of variance would detect genetic variation, phenotypic plasticity, and genetic variation for plasticity. Finally, the bottom row depicts the most extensive possible pattern of reaction norm crossing. Here, different genotypes react strongly to the environment, and in diametrically opposite ways. In an environment 1 versus environment 2 diagram, the genotypes would again align along a straight line, this time with a negative slope. Consequently, $r_{ae} = -1$, and an analysis of variance would detect only genetic variation for plasticity, and not genetic variation for the overall mean or the presence of overall plasticity, as discussed above. It is clear from the two parallel series of diagrams in Fig. 1.4 that there is a decoupling between the genetic correlation and the amount of genotype by environment interaction. While the genetic correlation goes from positive to zero to negative from the top to the bottom of the figure, the amount of genotype by environment interaction steadily increases. This is because the genetic correlation does not measure genetic variation for plasticity, but rather a mixture of variation for plasticity and for character means. This is an important point, since the rationale for the introduction of the concept in evolutionary quantitative genetics was that the genotype by environment variance does not have a direct evolutionary meaning (Via and Lande 1985). On the other hand, the G by E has a perfectly clear meaning (it measures the available genetic variation for phenotypic plasticity), but it is not mathematically suitable for plugging directly into the standard equations describing phenotypic evolution derived by Lande and Arnold (1983). However, it is indirectly usable for that purpose, through appropriate transformations, as was pointed out independently by Yamada, Eisen and Saxton, van Tienderen and Koelewijn, and de Jong.

Roff (1994) attempted to further extend the idea of interenvironment genetic correlation to dimorphic traits such as the presence or absence of wings in insects. This extension is possible if we assume that the dimorphic, or two-state appearance of the character, is actually due to an underlying continuous and quantitative series of genetic effects. However, whereas we generally simply do not know anything about the genetic basis of dimorphic traits, in some cases we do know at least that they are controlled by one or two genes with major effects (Tauber and Tauber 1992).

One further problem with r_{ae} is that its biological interpretation is not as straightforward as the within-environment r_G. Two other phenomena can cause changes in the interenvironment genetic correlation, but they usually do not enter into the makeup of standard genetic correlations: allelic sensitivity and environment-dependent regulatory switches (Schlichting and Pigliucci 1993, 1995). They are also ignored in the reaction norm depiction of plasticity data, since that is a description of patterns, not of causal mechanisms. Allelic sensitivity refers to the fact that the same set of alleles controlling a given character can trigger a distinct response of that character to different environments. For example, an enzyme coded by a certain gene will probably show some kind of kinetic curve describing its ability to perform its metabolic function at different temperatures or pH. Therefore, any character affected by the functioning of that enzyme is likely to exhibit changes that are proportional to either the temperature or the pH in the environment. If alleles at loci producing enzymes involved in a plastic response code for molecules with certain kinetic curves, it may lead to an interenvironment genetic correlation of less than 1, *without any change whatsoever in the genetic basis of the trait.* This is not equivalent to changes in gene expression due to, for example, epistasis, which would have a superficially similar effect. Therefore, a genetic correlation of less than 1, normally interpreted as evidence for the existence of alterations of the genetic architecture, can in fact be the outcome of a simple biochemical phenomenon, which tells us nothing about the mechanistic basis for that trait.

Environmental-dependent regulatory switches are gene products that sense an environmental change and trigger a switch between two or more alternative developmental pathways (Pigliucci 1996a). Several examples of this phenomenon are discussed in Chapters 5, 6, and 11, and they include phytochrome-mediated shade avoidance in plants, temperature-dependent sex determination in reptiles, and photoperiod-induced changes between winged and apterous forms in insects. One can think of environment-dependent regulatory switches as a form of epistasis, in that a regulatory gene turns on or off a set of other genes, whose function thereby depends on the epistatic effects of the regulatory gene(s). Epistasis can also affect within-environment genetic correlations by effectively decoupling characters (yielding a zero correlation), even though their genetic bases are, from a functional standpoint, overlapping.

Schlichting and Pigliucci (1995) argued that both allelic sensitivity and regulatory switches can dramatically alter the interenvironment genetic correlation in a variety of complex ways, so that there is no link whatsoever between the value of the correlation coefficient and the genetic basis of the traits under study. Both Houle (1991) and Gromko (1995) made this point in a more general context.

Houle elegantly demonstrated that the actual value of a genetic correlation depends on the details of the genetic architecture of the traits being considered. For example, the existence of a regulatory gene that allocates resources differentially to two traits may well result in a measurable negative genetic correlation between the traits, thereby highlighting a trade-off. But if there are many more loci involved in acquiring resources, the apparent genetic correlation might be positive, simply because most of the genetic variation comes from loci that tend to increase the value of both traits: The existing trade-off would be completely masked.

Following a different line of reasoning, Gromko concentrated on the effects of pleiotropy. He simulated selection on populations characterized by different combinations of pleiotropic effects and concluded that many combinations of pleiotropy can yield the same value of the genetic correlation, thereby breaking the causal link necessary to use genetic correlations as inferential tools. While Houle had demonstrated that genetic correlations are not necessarily constraints, Gromko extended the findings to show that the converse is also true: Response to selection can be constrained even though the apparent (measurable) genetic correlation is near zero.

Yet another insightful contribution to our understanding of the mechanisms underlying interenvironment genetic correlations has come from a completely different field of research: molecular physiology. Markwell and Osterman (1992) performed a mutagenic screening for chlorophyll-deficient genotypes in *Arabidopsis thaliana.* Figure 1.5 shows the scatter of the M_3 mutants in environment versus environment phenotypic space for two traits: chlorophyll content (left) and chlorophyll b/a ratio (right). The two environments were growth chambers at 20° or 26°C. It is interesting that the scatter of the mutants away from the control is very different for the two traits. In the case of chlorophyll content, there is a clear interenvironment *genetic* correlation (actually a *mutagenic* correlation): Most mutants produce more or less chlorophyll than the wild type, but *do so independently of temperature* (i.e., there is a high interenvironment genetic correlation). On the other hand, the graph for the chlorophyll b/a ratio shows a pattern of concentration of the mutants in the immediate phenotypic neighborhood of the wild type, with few noticeable exceptions falling outside the main diagonal. This suggests that the ratio is more difficult to alter genetically, but the exceptions show that it can be changed very dramatically in one environment *while the value in the second environment remains the same.* (Compare the relative positions on both axes of the extreme mutants and the wild type.) This kind of detailed discussion of genetic constraints would not be possible if we were examining an analogous plot from a natural population (see, e.g., Fig. 2.1), because we would not know if the genotypes differed by single mutations or at many loci simultaneously.

Given all of the above, the continued use of genetic correlations to predict evolutionary trajectories for phenotypic plasticity is at least questionable and requires further discussion. Despite their mathematical equivalency (van Tienderen and Koelewijn 1994; de Jong 1995), ultimately both interenvironment genetic correlations and reaction norms are simply heuristic descriptions of variation within populations, and the associated statistics are in no way to be understood as indicators of underlying mechanisms. Unfortunately, long-term predictions of evolutionary trajectories *are not meaningful* if one ignores the details of the genetic machinery. This is so because several markedly different combinations of alleles can yield the same genetic correlation of patterns of reaction norms, yet another form of the classic pattern/process problem in evolutionary biology (and in any other historical science). The response to selection depends on the alleles actually present in the population, not simply on the readily observable

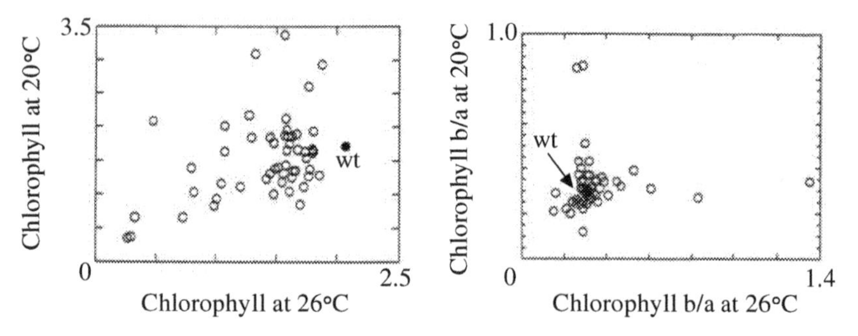

FIG. 1.5. Effect of mutations on genetic correlations. A wild-type *A. thaliana* (solid circle in both graphs) was mutagenized, and the M_3 generation (open circles) was grown under two temperatures. The mutants showed a very different across-environment correlation for two traits (chlorophyll content and chlorophyll b/a ratio) and a different scatter away from the wild type. (From Markwell and Osterman 1992.)

phenotypic patterns (Pigliucci 1996b). Further, basic population genetics (Hartl and Clark 1989) has amply demonstrated that specific combinations of alleles at several loci can lead to completely different stable equilibria for a population, taking very different routes to get there. Quantitative genetics has worked so nicely in practice (Roff 1997) because plant and animal breeders (and some evolutionary biologists) have applied it to short-term predictions of response to selection, for which the details of the underlying genetics are less crucial. Therefore, even though quantitative genetics provides us with a valuable exploratory tool to assess the current variation for quantitative characters and reaction norms in natural populations, its value as a theoretical framework for the description of long-term evolutionary trends is more uncertain (Pigliucci and Schlichting 1997).

A final note about the difference between the reaction norm and character-state approaches to visualize and analyze plasticity data. The two methods are mathematically equivalent and simply represent alternative summaries of the current state of a given population, but they also reflect different ways of thinking about phenotypic plasticity. The character-state approach implies—in a biological, not a mathematical, sense—that the observed phenotypic plasticity is the by-product of within-environment variation (Via 1993). In this view, the lines that connect environment-specific means in a reaction norm plot are artifacts that can be used as aids to pinpoint patterns, but are not biological realities. Consequently, plasticity is seen to be the result of selection within environments and is not itself regarded as the target of selection. In some situations, this is undoubtedly true. For example, Via (1984a) studied two ecotypes of phytophagous insects that normally feed on distinct hosts. When she manipulated the system so that each ecotype was also forced to feed on the "wrong" host, she found strong plasticity for fitness. However, this is not surprising and indeed can only be the by-product of selection within environments: Clearly *plasticity for fitness* is not adaptive, since ideally an organism should maintain high fitness in the greatest possible number of circumstances. These ecotypes had probably been under selection on a single host for enough time to show a maladaptive response when exposed to a relatively novel environment (Service and Rose 1985; Holloway et al. 1990; Joshi and Thompson 1996).

On the other hand, there are many instances in which one can select for phenotypic plasticity per se. This argument was clearly articulated by Bradshaw (1965) and recently expanded by Scheiner (1993b) and by Schlichting and Pigliucci (1995) (see Agrawal et al. 1999 for a couple of empirical examples). I discuss the ecological conditions that favor selection on plasticity per se in Chapter 7. Here, let me just note that there are many well-documented instances in which a plastic response *could not have evolved if it had not been the direct target of selection.* These cases include any instance in which there are specific environmental receptors (such as light receptors in both plants and animals) that determine one of a series of alternative developmental pathways for the organism (see Chapters 5 and 11). The very existence of an environment-dependent switch does not make any sense if we think of evolution as occurring independently within each of the two (or more) environments. All in all, the most reasonable answer to Via's (1993) question, whether adaptive phenotypic plasticity is the *target* or the *by-product* of selection in a variable environment, is: It depends.

It should be clear, then, that the confusion characterizing the debate on the genetic basis of plasticity is due to the fact that three conceptually distinct problems come into the mix (Fig. 1.6). We could discuss the best way to visualize and analyze the data (a statistical question); or we might be interested in probing the mechanistic basis of plastic responses (a genetic question); or, finally, we might want to know if the plastic response originated because of direct selection for environmental sensitivity or as a by-product of selection within distinct environments (an evolutionary question). The fact that the researchers involved in this debate (including myself) are interested in all three questions has not contributed to a clear conceptual separation among them. It is to be hoped that we can now appreciate these differences and move on.

Environmental Effects on the Multivariate Phenotype: Plasticity of Correlations and Correlations of Plasticities

Most of our discussion so far has been framed in terms of univariate reaction norms, that is, focusing on one trait at a time. Several authors, however, have investigated environmental effects on the multivariate, whole-

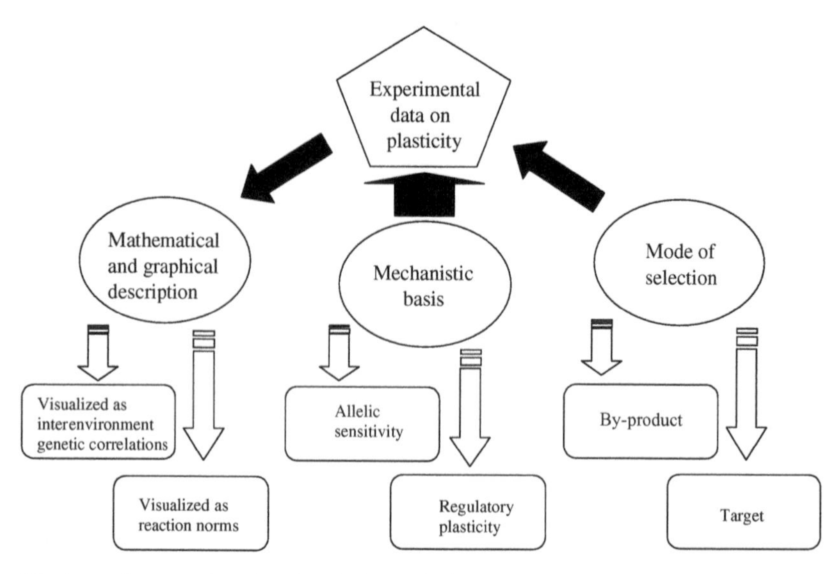

FIG. 1.6. Conceptual explanation of the confusion generated by the debate on the genetic basis of plasticity. Raw data from plasticity experiments can start discussions in three distinct and, for all effective purposes, independent directions: (1) how to analyze and visualize the data (as interenvironment genetic correlations or as reaction norms—ANOVAs); (2) what are the mechanistic bases of the observed plasticity (allelic sensitivity, regulatory plasticity, or a mixture of the two); (3) what mode of selection generated the currently observable patterns of plasticity (selection within environments or selection on plasticity).

organism phenotype. Pivotal in this field has been the work of Schlichting (1986, 1989a), who focused on the twin concepts of phenotypic and plastic integration. *Phenotypic integration* refers to the network of character correlations quantifying the relationships among life history, allocation, and reproductive traits making up the phenotype of an organism. Character correlations have long been at the center of empirical research (Berg 1960; Clausen and Hiesey 1960; Antonovics 1976), and have been the focus of much theoretical debate (Cheverud et al. 1983; Service and Rose 1985; Via 1987; Cheverud 1988; Houle 1991; Gromko 1995). Character correlations are informative on two counts. First they tell us something about the *functional* relationships among traits. Why is it that some characters are correlated and others are not? What role did selection play in shaping the current network of correlations among traits? Second, they hint at *constraints* on future evolution (in the specific sense defined in the preface). If a phenotypic correlation is the reflection of an underlying genetic or developmental constraint, then the population may be precluded from following certain evolutionary trajectories. The additional contribution of the environmental component certainly increases the complexity of the problem. Schlichting showed this clearly by studying the correlation network for several characters in *Phlox drummondii* exposed to optimal conditions, low nutrients, low water, leaf removal, and small-pot stresses (Schlichting 1989a,b). Some correlations (e.g., root/shoot versus days to flower) disappear when going from one stress (low nutrients) to the control, while others (e.g., root/shoot versus shoot weight) appear only under optimal conditions (Fig. 1.7). Clearly then, phenotypic integration can be plastic, since it can be altered by environmental circumstances. While Schlichting's work focused solely on phenotypic correlations, later authors demonstrated environmental effects on genetic correlations as well (Mazer and Schick 1991a,b; Stearns et al. 1991; Windig 1994; Bennington and McGraw 1996). Stearns and collaborators (1991) offered a theoretical framework within which to discuss the effects of phenotypic plasticity on genetic correlations and detailed the biological circumstances under which we should expect changes in genetic correlations. They identify two levels of analysis. At the phenotypic level, a sign change in the correlation between two traits can be brought about by the particular pattern of reaction norms characterizing a population. Indeed, if reaction norms cross along an environmental gradient, then the rank of the genotypic means changes along the same gradient. If this alteration occurs for one trait but not for another one, the correlation between those traits is also altered, and it becomes either positive or negative, depending on which sign it had before the crossing point.

The second level of investigation is physiological and is based on the *Y-model* of van Noordwijk and de Jong (1986), later elaborated upon by Houle (1991). Here the genetic correlation is a result of the fact that a certain amount of resources is distributed (allocated) to two different traits, thereby generating a genetic trade-off. This *structural pleiotropy,* however, is again affected by the environment, since the amount of resources available is a function of environmental influences, as well as of the capability of the genotype to gather those resources efficiently. If the environment is favorable, the underlying genetic trade-off will not translate into a phenotypic one, since both characters will benefit from an increased allocation of resources. The work of Stearns et al. therefore directly links phenotypic plasticity to a major topic of research in evolutionary biology: the impact and origin of constraints.

An idea closely related to the plasticity of correlations is that of *plasticity integration* (Chapter 7). It follows from the fact that plasticity is nothing but a trait in its own right: If characters (in the standard sense of character *means*) can be correlated, then the plasticities of those same characters can be correlated as well. A classic example is provided by Schlichting's (1989a) reanalysis of data in *Sesbania macrocarpa* and *S. vesicaria* after a study by Marshall et al. (1986). He plotted the relationships among the *plasticities* (not the trait means) of several traits in these two species raised in

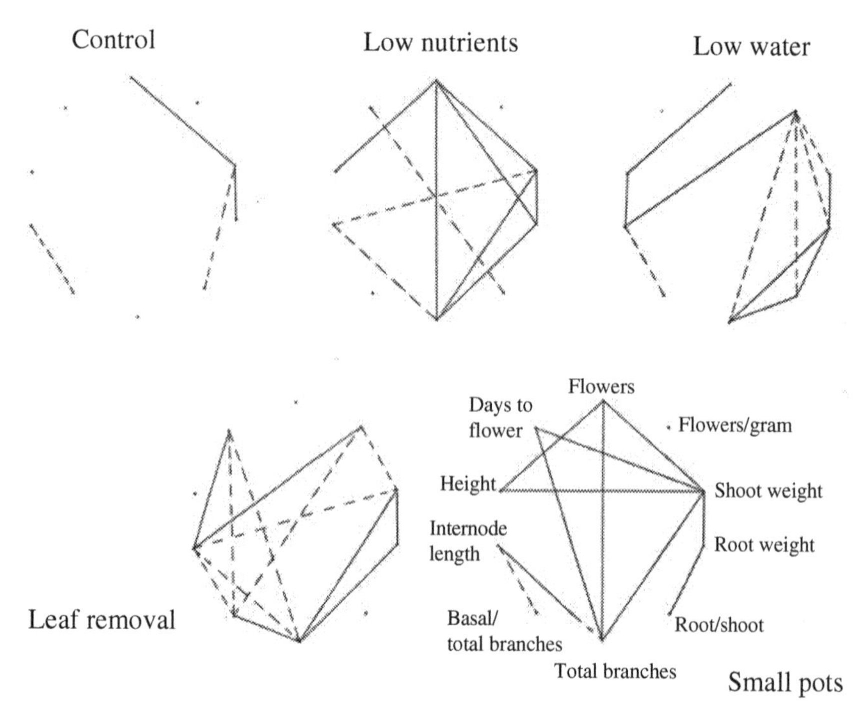

FIG. 1.7. Network diagrams showing the change in character correlations when plants of *Phlox drummondii* are exposed to a variety of stresses and compared to a control under benign environmental conditions. (From Schlichting 1989a.)

ten environmental treatments. *S. vesicaria* was shown to display a more integrated plastic phenotype than *S. macrocarpa,* as measured by the number of significant correlations among plasticities (Fig. 1.8). Are these differences the result of historical accidents, or is there selection on the degree of interdependence among plasticities as a function of the particular ecological situations encountered by each species? This is just one of many as yet open questions about multivariate phenotypic plasticity.

Phenotypic and plastic integration have been repeatedly confused in the past, and there have been very few studies to date focusing on either one (e.g., Pollard et al. 2001). This lack of attention is probably due to two very distinct causes. First, such an explicitly whole-phenotype level of analysis runs contrary to the extremely successful but conceptually more limited mechanistic framework that characterized much of twentieth-century biology (Sultan 1992). Second, given the large numbers of organisms, environments, and traits involved, it simply is very cumbersome to carry out meaningful experiments on plasticity integration. It is clear, however, that integration studies are pivotal to an understanding of the evolution of

Sesbania macrocarpa

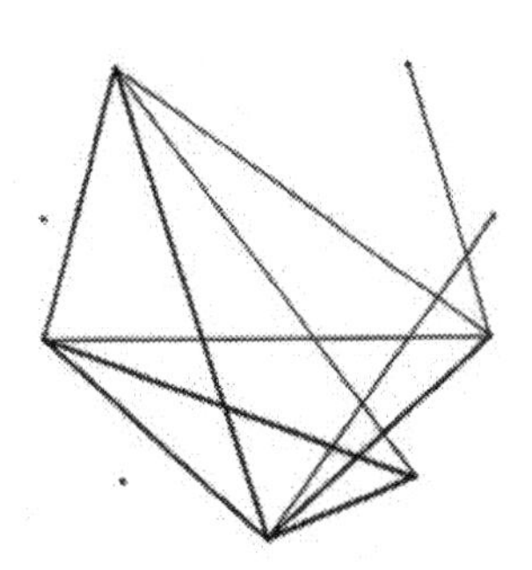

Sesbania vesicaria

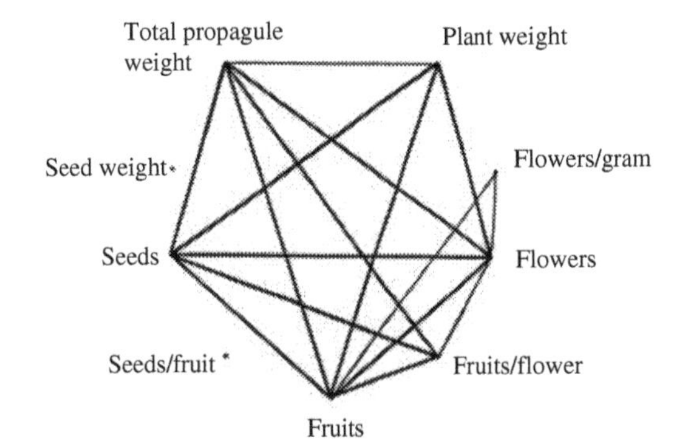

FIG. 1.8. Network diagrams showing the different patterns of correlations among the plasticities of several traits in two species of *Sesbania* exposed to ten environmental treatments. (From Schlichting 1989a after data in Marshall et al. 1986.)

characters themselves (Wagner 1996; Wagner and Altenberg 1996), and I think they will play an increasingly important role in modern evolutionary biology.

Conceptual Summary

- Phenotypic plasticity is the property of a given genotype to produce different phenotypes in response to distinct environmental conditions.
- Plasticity is one of several components—together with epigenetics, pleiotropy, epistasis, and others—that enter into our understanding of the so-called genotype-phenotype mapping function.
- Plasticity is one of the attributes of a reaction norm (the other being its height in environment-phenotype space) and can be characterized by its pattern and degree.
- The environment can affect heritability, the ratio between genetic and phenotypic variance, because given genotypes may be more or less phenotypically differentiated from each other, depending on the environmental conditions.
- The relationship between environmental stress and genetic variation is very much under discussion, with all possible patterns having been predicted and described.
- It is possible to calculate the heritability of plasticity itself, since the latter can be considered a phenotypic trait as any other.

- There is a close correspondence between the statistical technique of analysis of variance and the visual description of genotype-environment interactions in reaction norm diagrams.
- Genotype-environment interactions can also be treated as inter-environment genetic correlations. This has the advantage of providing a straightforward incorporation of plasticity in standard quantitative genetics equations. However, the approach suffers from a variety of limitations, both computational and conceptual.
- Multivariate plasticity has two distinct aspects. On the one hand, character correlations can be affected by the environment, so that the degree and pattern of phenotypic integration may change significantly in response to the external conditions. On the other hand, the correlations among the plasticities of traits (plasticity integration) can also be very different, depending on which environments and taxa one is considering.

2

Studying and Understanding Plasticity

For a man to attain to an eminent degree in learning costs him time, watching, hunger, nakedness, dizziness in the head, weakness in the stomach, and other inconveniences.

—Miguel de Cervantes (1547–1616)

CHAPTER OBJECTIVES

To discuss what is generally meant by phenotypic plasticity and the scope of plasticity studies. A series of empirical approaches to the study of plasticity is presented briefly, including environmental manipulations, transplant experiments, phenotypic manipulations, artificial selection, and quantitative trait loci analyses. The chapter ends with a discussion of the relationships between classical (morphological/developmental) plasticity studies and other types of organismal responses to environmental heterogeneity, such as physiological adjustment and behavior.

Understanding the way in which plasticity studies are actually done will be valuable, so this chapter begins by summarizing several of the research methods that have been used to date, highlighting the advantages and limitations of each. This is followed by a discussion of the differences and similarities among the different kinds of plasticity.

Empirical Approaches to Studying Phenotypic Plasticity

There are several distinct experimental approaches to the study of phenotypic plasticity, with some elements common to all. I will provide a few examples for each category to help identify similarities and differences among the approaches. Those examples that rely heavily on molecular data will be treated more extensively in Chapter 5. This is by no means meant to be a representative survey of the field, as the number of experimental papers on plasticity would require a book of this size just to list the references.

Environmental Manipulations

The standard or "classical" plasticity experiment is one of many possible variations or subsets of a rather straightforward complete factorial design, although sometimes the factorial is not complete or split-plot, Latin-squares, or other types of designs are used because of logistic difficulties in setting up a full factorial (Zolman 1993). In the case of a single factor, the organisms are exposed to two or more levels of a particular environmental variable, biotic or abiotic. The factors most frequently investigated in plasticity studies are temperature and food availability for animals and water, light, and nutrients for plants. The effects of some abiotic variables are investigated less frequently, usually because they entail logistical problems in setting up the experiment (e.g., photoperiod or light quality). Biotic factors, such as density of conspecifics, interspecific competition, and predation/herbivory, are also studied, though less commonly than abiotic parameters, and, again, I think this reflects the ease or difficulty of experimental manipulation rather than any fundamental assumption about the relative importance of one factor over another.

In a typical experiment, several genotypes sampled from different populations and—more rarely—from different species, are studied. This of course requires a minimum number of replicates per species/population/genotype within each treatment. The result is that plasticity studies tend to be characterized by a very large number of experimental subjects and the need for extensive blocking to account for the inevitable variation owing to microenvironmental effects. This in and of itself largely explains why plasticity studies have never been very popular among ecological geneticists: If one carries out *common garden* experiments (i.e., in a single environment) one suddenly halves, or even reduces to a third or a quarter, the number of experimental units.[1]

As discussed in Chapter 1, the results of a classical plasticity experiment can be presented either in terms of reaction norms or in environment versus environment plots, and they can be statistically analyzed either by an analysis of variance or by computing interenvironment genetic correlations. Andersson and Shaw (1994), for example, used the latter approach to study plasticity of *Crepis tectorum* to light availability. They started with two contrasting ecotypes, a *weed* population and an *outcrop* population. Weed ecotypes of *C. tectorum* occur in a wide range of agrestal and ruderal habitats; outcrop ecotypes tend to be found on exposed bedrocks. Furthermore, in Sweden, weed ecotypes occur throughout the range of the species, whereas outcrop ecotypes are found only in the Baltic lowland. Thus, the two ecotypes selected by Andersson and Shaw represented habitats with very different light regimens, and one can argue that these differences are to some extent reflected in the differential biogeographical distribution of the two types. They then obtained F_3 and F_4 inbred families from a cross between the two parental ecotypes and exposed them to full sunshine and 50% shade in a greenhouse. They measured several phenotypic traits ranging from leaf number, size, and shape to flowering time, total size, and branching, as well as reproductive fitness (e.g., number of heads per plant and number of flowers per head). They found a range of interenvironmental genetic correlations from 0.06 to 0.84, with most of them being statistically significant. Curiously, for the three traits for which they had data on both the F_3 and the F_4 populations, the correlations increased in magnitude across generations, going from not significant to significant in one case, with the other two being significant in both generations. (It should be remembered that genetic correlations are, among other things, sensitive to the actual allelic frequencies within a population, which can in turn be altered by inbreeding.) Perhaps more interestingly, the characters measured during the experiment could be divided into two broad categories (Fig. 2.1). In one case (six traits: leaf dissection, leaf number, flowering date, plant height, branch length, and head width), the progeny fell well within the two extremes represented by the parents, *along an almost straight line.* This indicates that the plasticity of those traits is highly constrained, in that a higher value for the trait under full sunshine would be accompanied by a proportionally higher value for the same trait under partial shade. Also, the fact that the progeny fell within the parental values suggests no disruption of major epistatic effects regulating the system. On the other hand, five more traits (leaf length, leaf width, seed length, heads per plant, and flowers per head)—including all measures of fitness—showed a very different pattern: The progeny scattered in a much more random fashion around the parental means, *and several inbred families lay outside the parental range.* This implies a much weaker genetic constraint and perhaps the presence of major epistatic effects underlying the plasticity of these characters to light availability.

An example of a similar study carried out in *reaction norm* fashion is Gupta and Lewontin's (1982) classic work on the response of *Drosophila pseudoobscura* to temperature and density. They raised thirty-two laboratory strains (isochromosomal for the second chromosome), originally taken from three geographic locations, under three temperatures (14°, 21°, and 26°C) and at two densities of conspecifics. They also kept track of male and female reaction norms, given the common gender-based differences in phenotypes among animals and in *Drosophila* in particular. There were ten replicates, which yielded a total of 1,536 phenotypic measurements per trait.

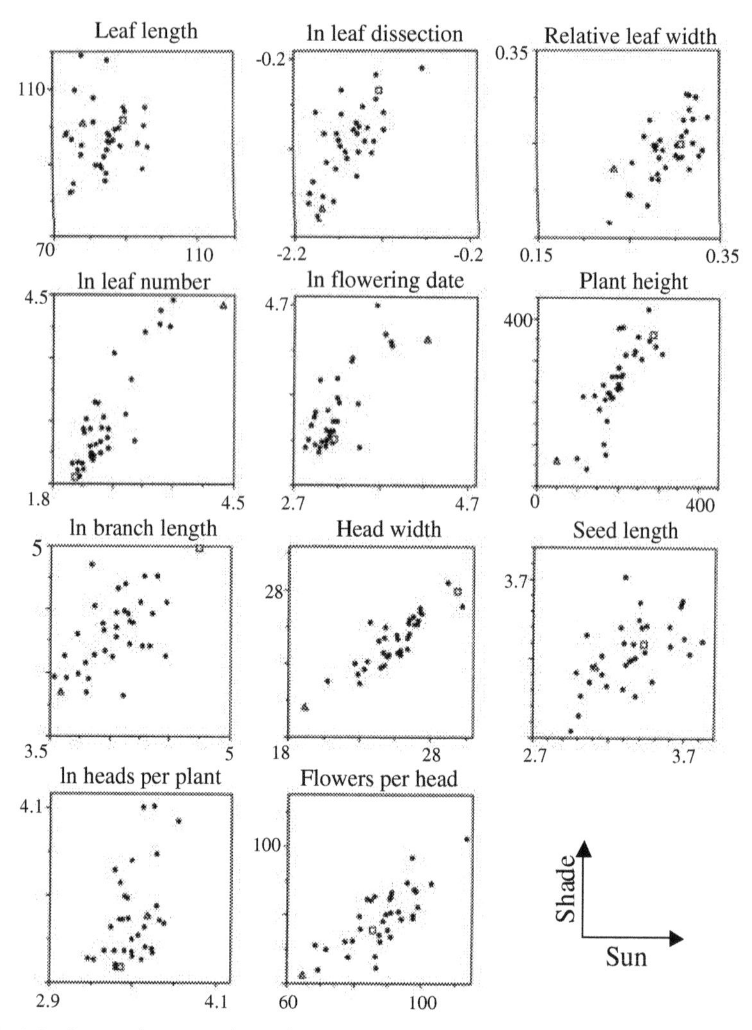

FIG. 2.1. Interenvironmental genetic correlations for plasticity to light availability in *Crepis tectorum*. Each graph represents the behavior under normal light (abscissa) and low light (ordinate) for several individual F_3 families. Note how some correlations are strong, whereas others show a wide scatter of the points in environment versus environment space. Also, note that the empty symbols identify the parental lines: For some traits, most of the progeny lies between the parents, but for other characters several progeny families are scattered outside the parental range, indicating strong epistatic effects. (From Andersson and Shaw 1994.)

Gupta and Lewontin did not indicate whether or not the temperature range chosen was representative of the natural habitat of this species, but they did note that there was no estimate of density in natural populations, so that it was impossible to determine if either of the two densities was stressful for the flies. Perhaps their most striking result was the comparison of the reaction norms of bristle number and development time (Fig. 2.2). Bristle number is an important character in *Drosophila,* but it is probably under much less intense selection than a fundamental life history trait such as development time. Gupta and Lewontin observed a wide range of variation in reaction norms to temperature in their lines, although all the lines tended to show fewer bristles at the highest temperature. On the other hand, reaction norms for development time in response to temperature formed a very tight bundle, with development time decreasing steadily with temper-

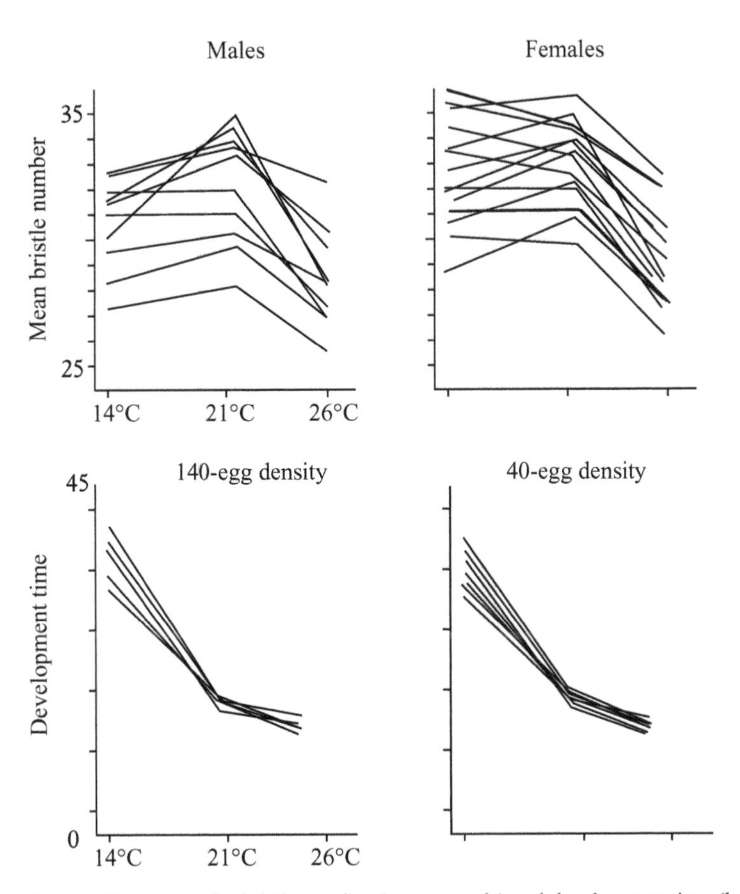

FIG. 2.2. Reaction norms for bristle number (upper panels) and development time (lower panels) for males (left) and females (right) *Drosophila pseudoobscura* collected within a population in California. Note how the reaction norms for bristle number are highly variable within the population, whereas those for development time form a tight bundle. (Redrawn from Gupta and Lewontin 1982.)

ature and showing very little variation among lines. According to the fundamental theorem of natural selection (Fisher 1930), this pattern can be interpreted as the outcome of very intense selection on the plasticity of development time and either relaxed or variable selection for plasticity of bristle number. When this type of contrast can be made we gain a glimpse into possible selective pressures on phenotypic plasticity and thus onto its ecological relevance (see Chapter 7).

Transplant Experiments

An equally venerable approach to the study of plasticity is to transplant organisms (usually plants, but not necessarily) from one field site to another. The idea was originally born out of the concept of *ecotype* (Turesson 1922), which is in some sense the exact opposite of adaptive phenotypic plasticity. An ecotype is a genetically specialized population that has evolved specific adaptations to cope with a particular—and usually narrow—environmental range. Alpine plants provide typical examples of ecotypes, with their morphologies particularly suitable for high winds and low temperatures. From the point of view of ecological theory, this is a *specialist.* Adaptive plasticity, on the other hand, by definition provides the plant or animal with a more flexible spectrum of opportunities, in that the phenotype can be adjusted to the specific environment, while retaining the ability to survive in a less restricted range of conditions. Such a genotype would then be considered a *generalist.* While ecotypes can indeed be plastic, such plasticity is likely to be either incidental or the residual of previous history. (I will return to the relationship between ecotypes and plasticity in Chapter 7.) It is therefore interesting that a technique used to demonstrate the existence of ecotypes has been successfully adapted to study the very distinct phenomenon of phenotypic plasticity.

In a reciprocal transplant experiment a series of genotypes from several locations (at least two) is collected, replicated under controlled conditions (either by cloning or through an appropriate breeding design), and then put back into the field at all locations, including the one of provenance. The general expectation is that each population/genotype will do better in the location from which it was sampled and worse anywhere else, especially if the other locations are edaphically or otherwise biologically different. This expectation corresponds to the hypothesis that each local population is in fact an ecotype, or presents ecotype-like characteristics of adaptation to the local conditions. I have to point out, however, that the prediction applies only to measures of fitness and not to other phenotypic traits: If a population is an ecotype we would expect to see plasticity for fitness because it would not do well in the alien or novel environment. Other phenotypic traits might show no plasticity across locations (e.g., the same number of bristles or leaves might be produced). It is very important to realize that this lack of plasticity in some traits might actually be the determinant of the undesirable fitness plasticity. On the other hand, one might find a population that is very plastic for morphological or life history traits, and that, *as a consequence,* can maintain a relatively high fitness (reduced or no fitness plasticity) in several locations, a possibility first described by Bradshaw (1965; see also Bazzaz and Sultan 1987; Sultan 1987). This is a consequence of another interesting phenomenon: *plasticity integration* (Schlichting 1986, 1989a; see Chapter 1). Both concepts, which I discuss further in Chapter 7, were present in embryonic form in Adams (1967) within the context of crop yield.

A definitive example of reciprocal transplant experiments is found in the long sequence of papers published by Clausen, Keck, and Hiesey over the span of twenty years (Clausen et al. 1940; Clausen and Hiesey 1960). The central question in their work was, "what are the mechanisms that cause living things to differentiate into recognizable species . . . rather than to evolve into a flow of continuous variation?" (Clausen and Hiesey 1960). To answer it they carried out extensive experiments on the genera *Achillea, Potentilla,* and *Viola,* which are now classic and still outstanding examples in ecological genetics. One of their major (and by and large underappreciated) results was that "coherence is as much a part of evolution as is variation" (Clausen and Hiesey 1960). That is, they found extensive variation in reaction norms between coastal and alpine races of their species (Clausen et al. 1940), while at the same time demonstrating extensive covariation of characters, so that an "alpine" or "coastal" syndrome could easily be described. In the view of Clausen et al. it was this coherence of the phenotype that eventually would allow interspecific differentiation and evolution.

Transplant experiments continue to be used today, and they are an invaluable complement to controlled condition experiments in the lab or greenhouse. In fact, I would argue that neither of them makes sense without the other. In a laboratory, while one is obviously able to distinguish among the effects of distinct aspects of the environment and study their interactions at leisure, it is very difficult to match the range and especially the complexity of real field conditions. On the other hand, transplant experiments capture the essence of a complex and multifaceted environment, but make it very difficult to analyze the causes of phenotypic variation. A research program that relies on only one of these two approaches is bound to be incomplete.

Phenotypic Manipulation: Phenocopies and Mutants

A class of experiments broadly defined as *phenotypic manipulation* represents a relatively recent experimental approach to the study of the genetics and evolution of phenotypic plasticity (Schmitt et al. 1999). These experiments can be carried out in two different ways, but the basic idea is that the animal or plant can be "manipulated" so as to test specific adaptive hypotheses concerning the presence or absence of a given plastic response. One approach relies on "tricking" the organism into displaying the response even though the conditions are not appropriate. For example, *Daphnia* can alter its morphology by producing a prominent helmet as a defense against predators (Dodson 1989; De Meester 1993; Tollrian 1995; Burks et al. 2000). But it is possible to induce the formation of the helmet even without the predator, because its developmental pathway is actually triggered by chemical compounds released by the predator in the water rather than by the presence of the predator itself (De Meester 1993). These cases can be considered examples of phenocopies[2] and can be used to test hypotheses about the cost of producing the plastic phenotype, as well as about the appropriateness of the response itself (i.e., to test if the plasticity is indeed adaptive). A second approach is based on single-gene mutations that alter the phenotype in very specific ways, eliminating the plastic response but otherwise not affecting the normal development of the organism. It then becomes possible to test the adaptive plasticity hypothesis, as well as to gain some insight into how many and what kind of genes are involved in the plastic response.

One example of the first type of manipulation is the study of shade avoidance plasticity in *Impatiens capensis* by Dudley and Schmitt (1996). Some plants can alter their growth habit and flowering schedule if they are growing under a canopy of inter- or intraspecific competitors (Schmitt and Wulff 1993). (I return to the molecular basis and other aspects of this plastic response in more detail in Chapter 11.) Dudley and Schmitt literally "tricked" the plants into producing the elongated stem phenotype typical of a canopy, even though they were not growing under vegetation shade, by altering the ratio between the red and the far-red portion of the solar spectrum (R:FR) with the use of appropriate filters. The R:FR ratio is the actual cue that signals the presence of neighbors, because neighboring plants will absorb the red—which is photosynthetically active—and reflect the far-red, which is not. Elongated and nonelongated plants were then either subjected to the presence of a dense stand of conspecifics or grown at low density. The elongated phenotype should be adaptive at high density because it gives the plant the ability to reach the top of the canopy and gain access to light. On the other hand, the nonelongated phenotype should be favored at low density, not only because there is no need to overtop its neighbors, but also because solitary long stems are more prone to mechanical damage (they cannot "rest" on their neighbors, which is what happens in dense stands). The results were simple and clear-cut (see Fig. 7.6): Under high density, elongated plants had a significantly higher fitness than nonelongated ones; at low density, exactly the opposite pattern emerged. This is strong confirmation of the adaptive plasticity hypothesis.

An example of the second approach is my own work with Johanna Schmitt (Pigliucci and Schmitt 1999), again on shade avoidance, but this time in the weedy mustard *Arabidopsis thaliana*. Molecular biologists have identified several mutations in *A. thaliana* that seem to specifically affect the shade avoidance response. Schmitt and I used these to investigate the effects of partial or total lack of plasticity to canopy shade in this species. We compared an inbred wild-type line from which the mutants were originally derived with several types of known one-gene mutants. I will discuss three of them here: Two mutants have no functional phytochromes (the molecule responsible for the perception of the R:FR ratio), which means that all five phytochrome genes known in *A. thaliana* are inoperative. This results from a single mutation that impedes the synthesis of the chromophore, the actual photoreceiving part of the molecule. The two mutants defective in all phytochromes differ because the two mutations map in distinct locations in the genome, that is, they represent the effects of two different genes. The third mutant considered here is a less severe one, in that among phytochromes A, B, C, D and E, only B is nonfunctional. Figure 2.3 shows some of the results we obtained by exposing the mutants and the wild type to normal light and to a canopy shade, created either by planting grass around each *A. thaliana* individual or by using selective filters (the results were the same with both methods).

Let me discuss the behavior of the wild type first. The graph on the left shows the production of leaves before the plant switches from the vegetative to the reproductive phase of its life cycle. This is usually a measure of meristem allocation to vegetative growth, but in the case of *A. thaliana* it

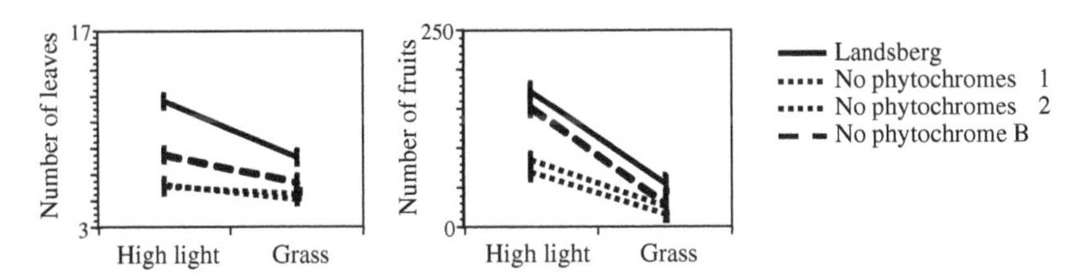

FIG. 2.3. Effects of single-gene mutations on the shade avoidance response in *Arabidopsis thaliana*. Landsberg is a wild-type inbred line from which several mutations were isolated. Two of the mutants have no functional phytochromes, while the third mutant lacks only phytochrome B. The left graph shows that the number of meristems being allocated to leaf production under normal light is reduced when one or all of the phytochromes are nonfunctional (reduced plasticity). This lack of plasticity translates into a significant reduction in fitness relative to the wild type (right graphs), *only in the environment in which the phytochrome affects the phenotype.* (Full data in Pigliucci and Schmitt 1999.)

also represents a developmental measure of flowering time because the plant flowers only after the rosette has produced a certain number of leaves; the threshold depends on the genotype and on the particular environment. The wild-type flowers develop earlier under shade than under full sunlight, in accordance with the shade avoidance hypothesis, which suggests that it is advantageous to accelerate the life cycle under competition. The result in terms of reproductive fitness (right graph) is that the wild type indeed does not do as well under competition as without competition, *but at least produces seeds* even under the adverse condition. We also demonstrated that plants that flower later under a real canopy are unable to produce *any* progeny. But is this plasticity adaptive? If so, we would expect that eliminating or reducing the ability to respond plastically would entail a corresponding reduction in reproductive fitness—which is exactly what we observed with the mutants. Mutants that lack all functional phytochromes are basically insensitive to changes in R:FR and always flower early (left in Fig. 2.3). Correspondingly, their fitness is much lower than the wild type *in the environment in which the plastic response occurs.* The fitnesses of the two mutants are not different from the wild type under the grass canopy, where plants flower as early as they possibly can: It is the *lack of a delay in flowering* under high light that puts the mutants at a disadvantage. A similar, but less dramatic, scenario holds for the phytochrome B mutant. In this case the reduction in fitness is barely noticeable, which suggests that at least one other phytochrome (probably C or D; see Aukerman et al. 1997 and Devlin et al. 1998) is also involved in the shade avoidance syndrome, and compensates to some extent for the lack of Phy B (an example of genetic redundancy [see Pickett and Meeks-Wagner 1995]).

Several other phenotypic manipulations have been carried out in both animals and plants (Newman 1992; De Meester 1993; Landry et al. 1995; Schmitt et al. 1995; Sinervo and Doughty 1996), and Sinervo and Svensson (1998) presented a critical discussion of the advantages of this approach as well as of laboratory selection experiments (below) for the study of the evolution of plasticity.

Artificial Selection

Selection to alter the population mean of a trait, thereby demonstrating that it has a genetic basis (or, more correctly, demonstrating the existence of genetic variation for that trait), has also been used for plasticity studies, albeit more rarely than for other types of characters (Brumpton et al. 1977; Falconer 1990; Hillesheim and Stearns 1991; Huey et al. 1991; Scheiner and Lyman 1991; Leroi et al. 1994; Holloway and Brakefield 1995; Loeschcke and Krebs 1996; Bell 1997b).

A classical study in this respect is the one by Brumpton et al. (1977) on *Nicotiana rustica.* They reasoned that if plasticity (*sensitivity* in their article) is genetically independent of the across-environment character mean (*mean performance* in their terminology), one would expect to be able to select for all possible combinations of high/low mean and high/low plasticity. To test this, they performed two cycles of selection on the progeny of a cross between two varieties of *N. rustica* and monitored changes in plasticity and overall mean for two traits: flowering time and plant height. From a qualitative point of view, they found three of the four expected combinations of means and plasticities, but these differed according to the trait examined. In the case of flowering time, it was impossible to obtain the combination of low mean and high plasticity. For plant height, it was not possible to obtain families characterized by high mean and low plasticity. From a quantitative perspective, all selection responses were lower than would have been expected if mean and plasticity were completely genetically independent, indicating some level of genetic covariance between the two. Further, the degree of genetic variation for means and plasticities was different, with that for the latter being about half of that for the former.

Another way of studying plasticity via selection experiments is by comparing the reaction norms of different lines obtained in response to selection under constant (but distinct) environmental conditions (Roper et al. 1996; Bell 1997b; Chippindale et al. 1997; Bubli et al. 1998), thereby addressing the evolution of plasticity as a correlated response. An example is provided by the study of life history evolution in *Drosophila melanogaster* undertaken by Chippindale et al. (1997). Their premise was that the relationship between genetic change and phenotypic plasticity is not generally well understood. (Of course, alterations in phenotypic plasticity are a form of genetic change.) These authors therefore used two sets of divergent lines, selected respectively for short and long generation times, and exposed them to two types of environmental manipulations (yeast and mate availability). They found parallels and differences in the life histories of their flies in terms of both plasticity and genetic differentiation induced by the selection process. For example, Fig. 2.4 shows that when the flies are raised on a low yeast diet their fecundity is uniformly low, regardless of mate availability or selective history. At high food levels, however, the two selected lines diverge dramatically in the plasticity of their fecundity: Short-generation flies have high fecundity even without mate availability, whereas long-generation flies go from almost zero to about 60 eggs per female per day when exposed to mates.

We see, then, that selection experiments have been used to study plasticity directly, to assess the effects that classical selection protocols can have on plasticity, and to address the potentially complex interactions be-

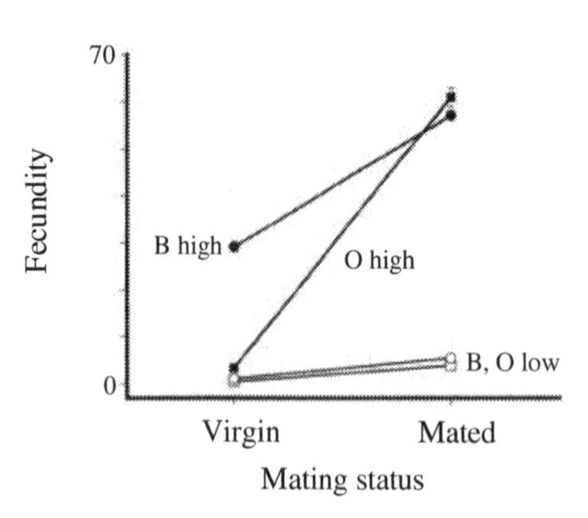

FIG. 2.4. Interaction between plasticity and response to selection in *Drosophila melanogaster.* Circles indicate lines characterized by short generation time, and squares mark lines with long generation time. High and low refer to the presence of yeast in the diet of the flies. Virgin and mated refers to their reproductive status. Note how B females have high fecundity when fed an abundant diet, regardless of their mating status. O females, on the other hand, require *both* mating and a high diet for their fecundity to be as high. (From Chippindale et al. 1997.)

tween selection histories and current environmental variation. All of these are of course highly relevant to different facets of our understanding of phenotypic evolution in heterogeneous environments.

QTL Mapping

QTL (quantitative trait loci) mapping is an emergent, yet very old, technique for the study of the genetic basis of phenotypic variation in natural or artificial populations. The basic principle, which is very simple, can be traced back to Sturtevant's idea of mapping a genetic locus by estimating its recombination rate with other known markers (Suzuki et al. 1989). QTL mapping can be done with visible markers or isozymes, but it is usually carried out using RFLPs (restriction fragment length polymorphisms), or other DNA-level markers such as RAPDs (randomly amplified polymorphic DNAs) and AFLPs (amplified fragment length polymorphisms), simply because then there are enough markers scattered throughout the genome to make a relatively precise localization of the QTLs feasible and statistically significant. The process begins with two lines, artificially obtained or sampled from natural populations, that show a distinct phenotypic difference of interest (e.g., different flowering schedules). The two are then crossed, and the F_1, F_2, and sometimes later generations are screened for the frequency of recombination between the phenotype under study and the available markers. If the phenotype shows up in statistically significant association with a series of markers on the same chromosome, it is logical to conclude that at least one gene affecting that trait is physically linked with those markers, and is therefore to be found on that chromosome. QTL mapping enables the researcher—given an adequate quantity of markers and of progenies and, ideally, a small genome—to pinpoint the candidate gene with relative accuracy.

QTL studies are rapidly becoming more common in ecological genetics (Mitchell-Olds 1995; Wu 1998), and are offering a view of the genotype-phenotype mapping function that involves the action of several genes of major effect, often in a complex pleiotropic and epistatic fashion. To date, most QTL studies are still focusing on crop plants (see, e.g., Zhikang et al. 1995; Wu et al. 1997), and few of them have addressed phenotypic plasticity directly (Paterson et al. 1991; Wu 1998; Leips and Mackay 2000), but this is changing rapidly. A typical example is work on genotype by environment interactions in an intervarietal cross of almond, as reported by Asins et al. (1994). They studied fifteen quantitative traits in a large progeny group, and followed their effects across several "environments," represented by different growing seasons. They identified a minimum of seventeen QTLs, only three of which behaved homogeneously over the years. This points to extensive changes in gene expression as a function of the particular environment in which the organism finds itself. It is also worth noting that these authors found that most of the environmentally sensitive QTLs were associated with traits *whose heritabilities changed dramatically* from year to year (see Chapter 1).

QTL studies provide the evolutionary biologist with a powerful tool for exploring the genetics of observable variation in natural populations, but they do have some drawbacks. First of all, most QTL studies highlight the presence of genes with major effects. I am a staunch supporter of the idea that genes with major effects are much more important in evolution than they have been given credit for after Goldschmidt's (1940) infamous *hopeful monsters;* therefore, I view this kind of evidence very favorably. Nevertheless, it is also clear that QTL mapping intrinsically favors the detection of genes with major effects, since the statistical power (and hence the sample sizes) necessary to reveal smaller effects is usually beyond the logistics of a typical experiment of this type. Further, in QTL studies we are able to cross only a limited number of inbred lines originating from natural populations. This introduces a large sampling error, and we should be aware of the fact that we are still ignoring *most* of the variation that is out there (and *its* genetic basis). To assume that most of the variation present in natural populations can be attributed to the QTLs that we find as a result of one or a few crosses between extremely divergent lines would be an unforgivable mistake. These objections notwithstanding, I expect to see a dramatic increase in QTL studies within the next few years, and look forward to the new insights they might lead to concerning the genotype-phenotype relationship. More specific examples of QTL studies of genes affecting plasticity are discussed in Chapter 5.

How Many Types of Plasticity?

Now that I have discussed the basic ideas underlying the theory (Chapter 1) and empirical research (above) on phenotypic plasticity, it may be time to pause and ask: What *is* plasticity, and what *is not* plasticity? How many different *kinds* of plasticity can we expect to deal with? There are two ways of approaching these questions, one based on the phenomenology and temporal dynamics of plasticity and the other on a consideration of natural selection and the developmental-genetic machinery that makes plastic responses possible.

First, let us consider the phenomenology. Several researchers have noted similarities between classically defined phenotypic plasticity and two other phenomena that allow organisms to change in response to environmental challenges: behavior (see Chapter 8 for discussion and references) and physiological reactions (Maresca et al. 1988; Scharloo 1989; Herrera et al. 1991; Spencer and Wetzel 1993; Gehring and Monson 1994; Lang et al. 1994; Maxwell et al. 1994; Rakitina et al. 1994; Jagtap and Bhargava 1995; Mantyla et al. 1995; Alpha et al. 1996; Fink et al. 1997; Lotscher and Hay 1997; Wang et al. 1997). Strictly speaking, these are not examples of *morphological* plasticity, which is what most classic studies of phenotypic plasticity address, but behavior and physiological acclimation are aspects of the phenotype nevertheless, albeit characterized by their own mechanistic basis and specific patterns of variation.

The question is one of trade-off. If we include all these phenomena under the umbrella of phenotypic plasticity, on one hand we create a huge field of inquiry, with ramifications throughout every branch of modern ecology and evolutionary theory. On the other hand, we may be confounding the proverbial apples and oranges, mistaking superficial phenomenological similarities for deep biological correspondences. The decision, of course, should be made on the basis of the likelihood and type of such correspondences. It is clear that morphological plasticity, physiological acclimation, and behavior are similar strategies, in that they allow the organism to change (its morphology, physiological status, or behavior) in order to cope better with an environmental challenge. However, very different time scales and mechanisms characterize these three phenomena. Most behavioral changes and physiological acclimations can be measured in seconds or minutes, whereas morphological plasticity is a developmental process that takes days to months. Furthermore, learning and seasonal acclimation are exceptions that tend to muddle a distinction based only on temporal considerations. Also, the mechanistic bases of each phenomenon can be quite distinct. Behavioral alterations are mediated by hormonal reactions; physiological adjustments may be due to hormones or to rapid alterations in gene transcription and translation levels; morphological phenotypic plasticity may require a complex cascade of genetic switches and epigenetic effects, although part of this cascade is mediated by hormones. My suggestion therefore is to keep in mind that these three realms of biological investigation are to some (perhaps large) extent related, but to distinguish among them by applying the proper qualifiers (as in developmental or physiological plasticity).

The second type of taxonomy of plastic responses—based on considerations of developmental biology and natural selection—has a long history, which starts with Schmalhausen's (1949) distinction between *dependent morphogenesis* and *autoregulatory morphogenesis.* A similar dichotomy was proposed by Smith-Gill (1983) in her paper on *phenotypic modulation* versus *developmental conversion.* Schlichting and I (1995) suggested an analogous conceptual separation when we defined *allelic sensitivity* and *regulatory plasticity.* These names, however, refer to different aspects (developmental for the first two and genetic for the latter) of distinct classes of environmentally induced responses. Once again, the relationships among classes are not straightforward, which is a recipe both for confusion and for stimulating new empirical research and theoretical advances.

Let us consider the genetic distinction between allelic sensitivity and regulatory plasticity and if and how this corresponds to the developmental difference between phenotypic modulation and developmental conversion. Distinct aspects of reactions to temperature can typify both categories. It is well known that growth rates, for example, are altered in both plants and animals by a decrease or increase in temperature, with an "optimal" range of temperatures well within the limits of the thermal niche of the organism. It is also intuitive to think that most of these differences may be a direct result of alterations in the kinetic properties of enzymes involved in key metabolic functions, since most enzyme activities peak at certain combinations of physical parameters, such as temperature or pH. As discussed in Chapter 1, this is allelic sensitivity (literally, in the sense that the product of each allele can be "sensitive" to environmental changes in a different fashion). One might advance the argument that allelic sensitivity at the genetic level is likely to result in a *modulation* of the phenotype throughout development, that is, in a continuous and proportional response of the genotype to the environment (because the kinetic curves are a continuous function of temperature or pH).

On the other hand, we also know that changes in temperature can trigger very specific alterations in gene expression that protect the organism from the ensuing environmental stress—a syndrome known as heat-shock response (Maresca et al. 1988; Rickey and Belknap 1991; Komeda 1993; Harrington et al. 1994; Loeschcke and Krebs 1994; Prandl et al. 1995; Loeschcke and Krebs 1996; Krebs and Feder 1997b; Rutherford and Lindquist 1998). Genetically, this is a case of regulatory plasticity, in that specific regulatory genes are involved. It can trigger a precise cascade of cellular and higher-level changes of the phenotype, therefore qualifying as *developmental conversion* (in the sense that the phenotype "converts" between two or more possible discrete states: induced and noninduced).

Should we conclude from this that regulatory plasticity is always associated with developmental conversion and that allelic sensitivity implies phenotypic modulation? Not necessarily (Fig. 2.5), since it is easy to imagine a continuous molecular response giving rise to a discontinuous phenotypic outcome (as in the case of threshold traits [Roff et al. 1997]). Analogously, it is conceivable that a molecular switch will actually result in a gradation of phenotypes because of intermediate levels of regulation that can enhance or attenuate the original molecular signal. The situation is similar to the one in which students of heterochrony find themselves (McKinney and McNamara 1991; Raff 1996). *Heterochrony* is defined as an alteration in the relative timing of developmental events. Several authors have looked for the genetic and molecular basis of such alterations (Wray and McClay 1989; Ambros and Moss 1994; Schichnes and Freeling 1998). While these can of course be found, the simple idea that molecular heterochrony causes developmental heterochrony is invalid. It can easily be demonstrated that a change in the rate of a molecular process does not necessarily cause developmental heterochrony, whereas some developmental heterochronies can be caused by alterations in gene action that have nothing to do with timing (Nijhout et al. 1986; Cubo et al. 2000).

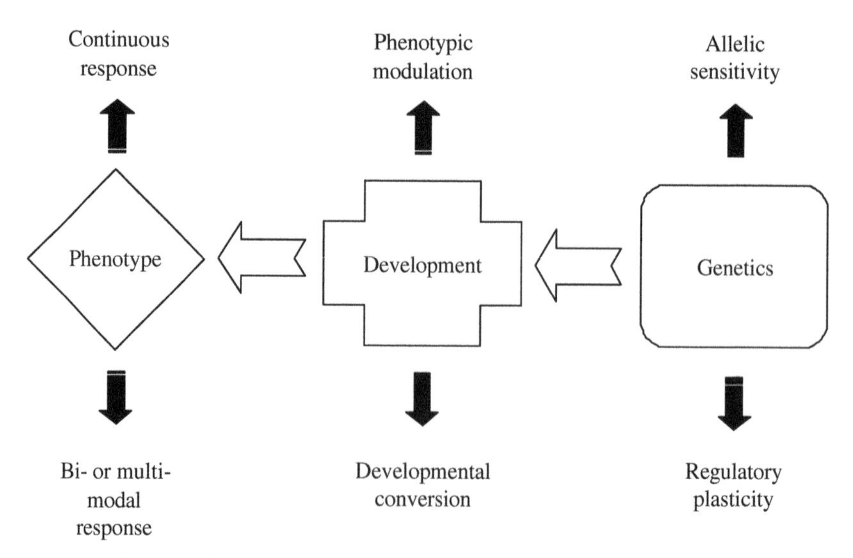

FIG. 2.5. Conceptual relationships between the genetic basis and developmental phenomena underlying observable plastic responses. While the top and bottom rows may be read as a logical progression, the boxes actually indicate that either allelic sensitivity or regulatory plasticity can translate into phenotypic modulation or developmental conversion. The latter, in turn, may result in visible continuous or discontinuous phenotypic classes.

Schlichting and I (1993, 1995) have suggested that one generalization can be drawn from studies of regulatory plasticity: It certainly is the result of a historical process of adaptation, though it may not be currently adaptive. It is difficult to imagine any process other than natural selection putting an environmental sensor in place that triggers a developmental cascade. Allelic sensitivity, on the other hand, could be currently adaptive—and might even have been selected for in the past—but it does not necessarily have to be so. Unfortunately, as I just argued, a purely phenotypic study does not provide sufficient evidence for regulatory plasticity, and one has to go all the way to the identification of environmentally sensitive regulatory switches to make a convincing case.

To summarize, phenotypic plasticity can encompass a variety of biological phenomena, and there can be legitimate disagreement among researchers about the proper scope of this field of inquiry. Rather than getting entangled in discussions centering on elusive definitions, however, we should simply take note of the relationships between morphological/developmental plasticity and a variety of other levels of organismal responsiveness to environmental conditions and make these the subjects of studies on genotype-environment interaction in the broad sense of the term.

Conceptual Summary

- Phenotypic plasticity can be studied empirically in a variety of ways. The classical approach is by manipulating selected environmental conditions in a laboratory or in the field while exposing replicates of several genotypes to each set of conditions.
- A second approach is represented by reciprocal transplant experiments, which have a long history in evolutionary ecology. Whereas these can essentially be seen as plasticity experiments under natural—hence more realistic—conditions, the experimenter has no control over the simultaneous variation of several environmental parameters.
- A more recent method involves phenotypic manipulations, which can be carried out in two ways: either by altering an environmental cue to "trick" the organism into displaying the "wrong" kind of phenotype, or by using single- and multiple-gene mutants impaired in the reception of the environmental cue of interest. In both cases, these studies are particularly useful to address hypotheses concerning functional morphology and costs of plasticity.
- Plasticity can also be studied by artificial selection, in the same way as any other trait with a genetic basis. Selection can be directly on plasticity or on some other aspect of the phenotype; in the latter case, changes in plasticity are of interest as correlated responses to selection.

- Quantitative trait loci mapping has been used successfully to identify genes involved in genotype-environment interactions in natural populations. While this technique has some intrinsic limitations (e.g., the bias toward detecting genes of large effect and the limited number of natural populations that can be sampled), its use will probably steadily increase over the next few years.
- Morphological/developmental phenotypic plasticity is closely related to other ways in which organisms respond to changes in environmental conditions, such as physiological adaptation and learning behavior. All of these can be studied within the general framework of genotype-environment interactions.
- It is possible to differentiate between kinds of plasticity at the genetic level (allelic sensitivity versus regulatory plasticity), at the developmental level (phenotypic modulation versus developmental conversion), and at the phenotypic level (continuous versus threshold responses to the environment). However, the relationships among levels are not necessarily straightforward.

3

A Brief (Conceptual) History of Phenotypic Plasticity

The best history is but like the art of Rembrandt; it casts a vivid light on certain selected causes, on those which were best and greatest; it leaves all the rest in shadow and unseen.

—Walter Bagehot (1826–1877)

CHAPTER OBJECTIVES

To highlight and discuss some of the major milestones of past research on phenotypic plasticity, starting with Johannsen's distinction between genotype and phenotype. The contributions of Schmalhausen during the Neo-Darwinian synthesis and the pivotal work of Bradshaw are summarized, with particular reference to their thinking on the evolution of plasticity. A classic publication by Lewontin is discussed as far as the relationship between analysis of variance and analyses of causes is concerned. The more recent contributions of Via and Lande on theoretical treatments of reaction norm evolution provide the bridge to contemporary literature in the field.

The history of the field of phenotypic plasticity is a very long one, and it has two virtually separate threads. On the one hand, there is the basic research motif, in which plasticity has been the sometimes controversial, often misunderstood, concept underlying the nature versus nurture debate that has accompanied humankind through several centuries of philosophical diatribe (see the epilogue). On the other hand, there is the practical aspect of phenotypic plasticity—or rather, the *impractical* one, in that plant and animal breeders have always considered the fact that an organism can change its "yield" according to the environment—a very unnerving and economically disturbing fact. These two approaches evolved almost independently of one another, featured very different groups of researchers with distinctly different backgrounds, and even developed distinct terminologies and statistical approaches. Falconer (1990) gives a nice example of the breeder's perspective on studying phenotypic plasticity (or, as he calls it, environmental sensitivity). References to this line of research can be found in his paper, as well as in classic works by Jinks and collaborators (Perkins and Jinks 1966; Brumpton et al. 1977). I focus here on the conceptual issues involved in the history of the field as seen by an evolutionary biologist, emphasizing the impact (or lack thereof) of plasticity studies on the development of modern evolutionary thinking. A treatment of the history of research in the broader area of phenotypic evolution studies, in which some of the players were of course the same, is provided in Chapter 2 of Schlichting and Pigliucci (1998).

In the Beginning: Johannsen and Woltereck

Before there could be phenotypic plasticity, there had to be a distinction between genotype and phenotype. Johannsen introduced such a distinction at the beginning of the twentieth century (Johannsen 1911; the concept was probably grasped earlier by August Weissman with his theory of the separation of germ and somatic lines proposed in 1883), thereby marking the birth of modern genetics at least as clearly as the rediscovery of Mendel's work by Correns, De Vries, and von Tschermak (see Sturtevant 1965 for a historical perspective). Johannsen soon realized that there was no direct correspondence between the "factors" hypothesized by Mendel and the traits that a biometrician could easily measure in any plant or animal. Mendel had established a *link* between the two, so that one could predict the trait's appearance from a knowledge of the combination of factors that was present in a particular organism. But there were two disturbing phenomena that did not quite fit with a simple one-to-one correspondence between genotype and phenotype. First, the character state seemed to depend not just on the factors present but also on their particular combination. Second, even a completely homozygote line would show some variation, usually distributed as a bell curve, around the character mean.

Recognition of the first problem—that is, of the fact that the genotype-phenotype mapping function (Chapter 1) is not a linear, one-to-one correspondence—opened the way to the study of two classes of gene-gene interactions, termed *dominance* and *epistasis* by Bateson during the first decade of the century (King and Stansfield 1990). It immediately became clear that, within a given locus, the two alleles in a diploid organism could show all sorts of relationships in terms of dominance, from complete recessivity of one allele to perfect additivity of effects (co-dominance).

The second problem, the manifest variability of genetically homogeneous lines, simultaneously opened the way to the study of quantitative genetics and—more pertinent for us—to investigations of genotype-environment interactions. Johannsen correctly reasoned that, if a homozygote could still show variation around a mean, there had to be another, nongenetic, source for this variation. Some kind of environmental influence was the only possible candidate, because although experimenters could completely control and homogenize the genetic background of an organism, they could never succeed in reducing microenvironmental fluctuations ("noise") to zero. The phenotype would thereby result from the interaction between the genetic signal and the environmental one, which would yield a quantitative distribution of character states around the genotypic mean. Hence the concept of a genotype distinct from the phenotype.

Johannsen's next step was to recognize that the macroenvironment would also affect the phenotype produced by a given genotype, and to realize that this effect would be much more profound than any catalyzed by minor fluctuations of environmental conditions during controlled experiments. This was the basic idea that led to the formulation of the concept of *reaction norm.* But Johannsen was not the one who formally introduced the term. That was the work of Woltereck (1909), who described a peculiar phenomenon in *Daphnia,* which is now considered to be the first study of adaptive phenotypic plasticity clearly recognized as such at the time of publication (Fig. 3.1).

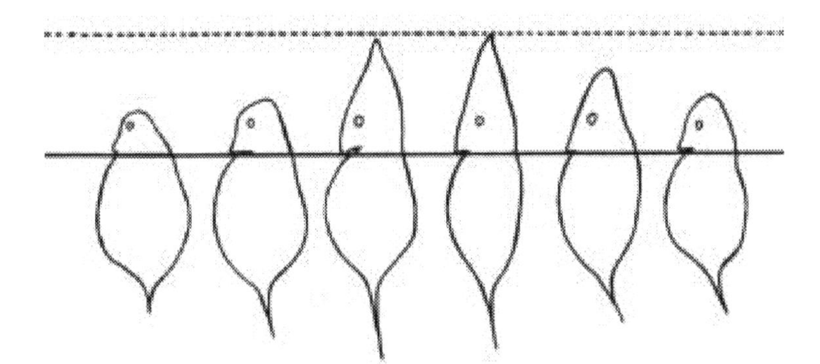

FIG. 3.1. Cyclomorphosis in *Daphnia,* the first published example of adaptive phenotypic plasticity. The development of the "helmet" or crest (middle individuals) is catalyzed by the presence of a predator, and it can be induced simply by the release of specific chemicals in the water, even without the actual physical presence of the predator. The helmet reduces the predation rate but probably carries a cost, either metabolic or in terms of swimming efficiency, which is presumably why *Daphnia* that are not exposed to the cue do not develop the helmet. (From Dodson 1989, adapted from Woltereck's original.)

Woltereck, however, completely misunderstood the significance of his findings, thereby opening the way to long-standing confusion about the meaning and significance of phenotypic plasticity, which lasted until the mid-1980s and, to a lesser extent, still haunts plasticity research today. Woltereck's reasoning was simple. Johannsen had suggested that the genotype was distinct from the phenotype and that the genotype represented some sort of "fixed" signal underlying any phenotypic trait. However, the case of *Daphnia* clearly demonstrated that under some environmental circumstances different genotypes could display exactly the same phenotype. Therefore, there is no unique relationship between genotype and phenotype, and the idea of an independent, underlying genetic factor should be abandoned.

Johannsen's reply was polite, but devastating. He argued that phenotypic convergence in one or a few environments did not invalidate the idea of a fundamental role of genes in the shaping of organismal appearances. On the contrary, he wrote, it represented the best demonstration that each genotype has a unique way of responding to environmental circumstances. If we limit our focus to one or a few environments, we might not be able to differentiate some genotypes from others. But if we broaden the study to all biologically relevant environments, then each genotype (or at least most of them) will be characterized by its own reaction norm. Accordingly, the study of phenotypic plasticity should be central to the whole endeavor of genetics, implying that the solution of the nature-nurture debate was finally at hand. However, most geneticists bought more or less consciously into Woltereck's argument (although probably not directly from him, since his paper remained obscure until the renaissance of plasticity research in the mid-1980s). They retained the idea of genes but relegated their environment-specific effects to unwanted "noise," to be minimized experimentally and altogether avoided from a theoretical standpoint. It took the better part of the twentieth century and the efforts of many researchers to overcome this initial mistake and finally put phenotypic plasticity at the forefront of modern evolutionary biology.

Plasticity during the Neo-Darwinian Synthesis: Schmalhausen

Researchers had to wait four decades after the Johannsen-Woltereck debate for the next major conceptual advance in the field of phenotypic plasticity. However, the pause was well worth it, since the work of the Russian geneticist Schmalhausen (1949) included a full-blown framework that would eventually lead (after *another* four decades) to the currently ongoing synthesis of plasticity research with the Neo-Darwinian paradigm from which it was initially excluded.

Schmalhausen's book, *Factors of Evolution: The Theory of Stabilizing Selection,* was first published in Russian in 1947 and then translated into English by his friend Theodosius Dobzhansky in 1949. The book was very well received and widely cited immediately after publication. However, it somehow got lost during the period of "hardening" of the Neo-Darwinian synthesis, when extreme positions such as Simpson's quantum evolution (which he himself toned down considerably in later years) gave way to a more moderate and traditional gradualistic Darwinism (Mayr and Provine 1980).

Several important conceptual points that Schmalhausen made in *Factors* are keys to modern ideas about the evolution of phenotypic plasticity, and in fact of phenotypes in general. First, at the beginning of the book he clearly distinguished between what he called the external and internal factors in the origin of change, assigning equal importance to both. Many biologists throughout the twentieth century have also considered this distinction, but have attempted to elevate one or the other to the status of *the* relevant factor directing evolutionary change. For example, before the modern evolutionary synthesis, many paleontologists were rejecting Darwin's ideas precisely because they believed that some sort of internal force, independent of natural selection, was responsible for all the major evolutionary trends seen in the fossil record. (Similar ideas have been advanced by the so-called "process structuralists" [Resnik 1994] and by some researchers in the field of complexity theory [Kauffman 1993].) Indeed, a great part of G. G. Simpson's contribution to the Neo-Darwinian synthesis consisted of showing how population genetics and natural selection could account for such trends without invoking some sort of orthogenetic principle. However, the triumph of the Neo-Darwinists was, perhaps, *too* complete. For decades developmental biology and the role played by internal factors on phenotypic evolution were completely ignored. Debates on developmental constraints, epigenesis, and internal forces are now an integral part of evolutionary research (Maynard Smith et al. 1985; Nijhout et al. 1986; Cheverud 1988; Gould 1989; Wagner 1989; Atchley and Hall 1991; Vogl and Rienesl 1991; Hall 1992a,b; Thomas and Reif 1993; Raff 1996; Pigliucci and Kaplan 2000). All of this, however, was very clear in Schmalhausen's book. He considered internal and external environments as two distinct yet similar factors limiting or channeling the evolution of phenotypes. It seems to me that modern evolutionary biologists have begun to agree on a similar qualitative conclusion only within the past fifteen years or so, albeit from the vantage point offered by quantitative and molecular genetics (Wagner 1988).

Immediately after making the distinction between internal and external factors in his book, Schmalhausen proceeded to introduce the concept of reaction norm, essentially the same idea stumbled upon by Woltereck and correctly framed by Johannsen. Merging the developmental and ecological components, Schmalhausen came close to the more recent idea of the *developmental reaction norm* (DRN) (Pigliucci and Schlichting 1995; Pigliucci et al. 1996; Schlichting and Pigliucci 1998). The DRN concept is really quite simple, and yet it embodies the three major fields of research that affected evolutionary biology during the twentieth century: *allometry,* the study of relationships among characters at the whole-organism level; *ontogeny,* focusing on the "internal" changes emphasized by Schmalhausen; and *plasticity,* the realm of the "external" factors. Schlichting and Pigliucci (1998) suggested that—even though it might be difficult to consider all three aspects of the DRN in individual empirical studies—its neglect from a theoretical standpoint has hindered the achievement of a true evolutionary synthesis.

The third major concept introduced in *Factors* is one of two meanings of the term *stabilizing selection.* Schmalhausen uses the expression in two different contexts, one corresponding to the modern idea of selection curtailing the extremes of a phenotypic distribution. (This is not an original contribution of the Russian geneticist; it goes back to the work of W. Bateson, R. A. Fisher, and J. B. S. Haldane.) The innovative alternative meaning refers to a process of stabilizing the reaction norm itself. Schmalhausen correctly surmised that organisms are exposed to a variety of environments, which can occur at different frequencies. In a sense, then, distinct portions of the potential reaction norm of a genotype are "exposed" to natural selection with frequencies proportional to those of the different environments. This in and of itself can bring about the evolution of reaction norms. But the process of stabilizing selection described in *Factors* is more far-reaching, and it has the potential to create a substantial bridge between micro- and macroevolution.[1] In essence, Schmalhausen considered what would happen to the reaction norm following a change in the environment, such as a climate shift, or simply the migration of organisms to a novel environment, thereby defining a new niche. Portions of the old reaction norm will no longer be exposed to selection and will therefore evolve by drift and mutation accumulation; other portions of the norm will be exposed to and molded by natural selection. This process will gradually alter the potential reaction norm of a given genotype, and the new norm will be *stabilized;* that is, it will produce the new normal (we would say wild-type) phenotype. Of course, new mutations might also occur, either during the process of evolution of the new norm or during the stabilizing interval. Schmalhausen proposed that the effect of new mutations—initially usually disruptive—is "absorbed" by the existing genetic background. This idea is very similar to Waddington's (1942) genetic assimilation, and it has evolved into the modern concept of evolution by selection on gene modifiers.

The theory of stabilizing selection (in the latter sense) and the absorption of new mutational effects constitute the basis for Schmalhausen's discussion of adaptogenesis, or the origin of adaptations. Here a fourth major conceptual issue is brought to bear: the idea of integration. The organismal perspective that permeates *Factors* is nowhere more manifest than in this section of the book. Organisms are made of different—more or less recognizable (Wagner 2001)—parts. What makes these parts work together in unison? Schmalhausen's answer is: integration through common regulation. Research on regulatory genes and *plasticity integration* (Chapter 1) was far into the future at this point, but Schmalhausen realized that a significant component of adaptive evolutionary change has to occur at the level of alterations in the elements that regulate and coordinate the function of different traits. This idea was also being championed experimentally by Clausen, Keck, and Hiesey (Clausen et al. 1940; Clausen and Hiesey 1960), as well as by Berg (1960). It was part of Wright's (1932) concepts of pervasive epistasis and adaptive gene complexes, and it has recently become the centerpiece of a new wave of studies on the evolution of pleiotropy and character correlations (Sharrock and Quail 1989; Dorweiler et al. 1993; Carroll et al. 1995; Schlichting and Pigliucci 1995; Wagner 1995; Wagner and Altenberg 1996; Baatz and Wagner 1997; Cheverud et al. 1997; Pollard et al. 2001).

Despite the energetic promotion of Schmalhausen's work by his compatriot Dobzhansky, a combination of factors caused it to fall temporarily into oblivion. First, the book (or at least the English translation) is written in a rather obscure and redundant style. Second, Schmalhausen soon found himself in the position of not being able to continue and disseminate his work because of Lysenko's all-out attack against "capitalist" genetics. Third, the Neo-Darwinian synthesis itself evolved toward a conservative center, with little room for more creative and complex ideas that were not immediately understandable in population genetic and systematic terms.[2] Fourth, biology was soon to be dominated by the rapid development of molecular biology, and the emphasis would therefore shift toward mechanistic rather than organismal approaches for decades to come. We consequently had to wait another sixteen years for the next major conceptual contribution to the field of phenotypic plasticity—Bradshaw's fundamental review of 1965.[3]

Bradshaw and the Reinvention of Plasticity Studies

Bradshaw's (1965) review, entitled "Evolutionary Significance of Phenotypic Plasticity in Plants," is an excellent example of a single work credited with the revitalization of an entire field of research (although, interestingly, it took at least ten years for its effects to actually be felt). Despite the fact that empirical studies of plasticity had continued among plant breeders since Schmalhausen's time, few evolutionary biologists were paying any attention to the phenomenon. On the contrary, most evolutionists considered plasticity a source of measurement error, to be reduced as much as possible either by experimental design (common garden experiments) or by statistical analysis (removal of "environmental" and "nongenetic" components of variance). It is symptomatic that Falconer (an animal breeder) published a paper in *The American Naturalist* (an evolutionary journal) on the topic, entitled "The *Problem* of Environment and Selection" (Falconer 1952, my italics), and as we have seen it led to the idea of treating reaction norms as across-environment correlations (Chapter 1). Bradshaw, a plant breeder, also had strong evolutionary interests, and unlike Falconer he saw plasticity as a fascinating alternative way for organisms, and especially plants, to adapt to changing environmental conditions. It is probably regrettable that he used the word *plants* in the title of his review, since his concepts apply across kingdoms. The unfortunate consequence of that title is that researchers studying animals tend to disregard the importance of Bradshaw's pioneering ideas, attributing them to much more recent authors.

A cursory reading of Bradshaw's table of contents reveals the breadth of his approach. We find everything from the genetic control of plasticity to a discussion of its adaptive significance, from an analysis of what circumstances favor the evolution of plasticity to a classification of different types of plastic responses. These are the very themes that have been the focus of much more recent theoretical and empirical research, and they constitute the underlying framework of this book. I attempt to summarize Bradshaw's main points as he saw them, in the hope of clarifying the history of current debates and their intellectual underpinnings.

Genetic Basis

Bradshaw acknowledged that the interaction between environmental effects and genes was so complex and the state of mechanistic biology (at the time) so primitive that it would be a long time before we would actually be able to uncover the details of the genetic machinery underlying phenotypic plasticity. However, he did emphasize that development would have to play a central role in such a mechanistic understanding. He pointed out that there must be a relationship between the rapidity of development of a structure and the degree to which this structure can be affected by environmental modifications. This principle has considerable explanatory power. For example, plants displaying indeterminate (open-ended) growth show plasticity in the number of parts formed (because meristems are not all committed to a fate early in development). By contrast, species that have determinate growth, like most but not all animals, are more likely to be plastic in the *size* of those parts. At the same time, Bradshaw also cautioned against explanations of patterns of plasticity rooted entirely in developmental pathways. Since developmental mechanisms are clearly more conserved across species than plasticity, there must be some degree of freedom for the latter to evolve independently of developmental constraints.

After a brief summary of the findings of genetic variation for plasticity across and within species, Bradshaw concluded that "Such differences are difficult to explain unless it is assumed that the plasticity of a character is an independent property of that character and is under its own specific genetic control" (p. 119). This quote had profound influence on the debate concerning the genetic basis and evolution of phenotypic plasticity (which started with Scheiner 1993b, Schlichting and Pigliucci 1993, and Via 1993, and continues to this day). One of the best examples of specific genetic control of plasticity cited by Bradshaw is the work on *Potamogeton.* Many species in this genus have heterophylly; that is, they are capable of producing different types of leaves in response to water availability. There are nineteen species in the subgenus *Potamogeton,* many of which produce natural hybrids. If we look at the distribution of heterophylly in the parents and hybrids, we clearly see that it is not random: Only hybrids that have at least one heterophyllous parental ancestor display heterophylly (Fig. 3.2). It

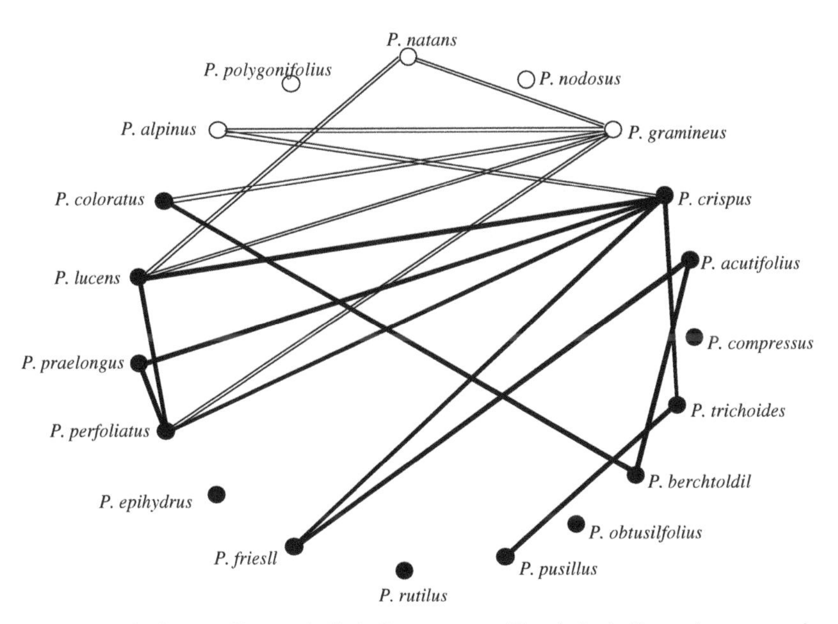

FIG. 3.2. Inheritance of heterophylly in *Potamogeton.* The circles indicate nineteen species of the subgenus *Potamogeton,* and the lines mark their natural hybrids. Black is no heterophylly; white is heterophylly. It is clear that all hybrids between heterophyllous and non-heterophyllous species display the plasticity. (From Bradshaw 1965.)

would be extremely interesting to use modern comparative methods to assess how many times heterophylly evolved independently in this group (i.e., outside of hybridization events), as well as to carry out further genetic dissection to ascertain the number and type of genes involved.

Bradshaw also pointed out that the plasticity of a given trait can be completely independent of the plasticity of another trait *for the same genotype.* The general conclusion—which still seems to be counterintuitive for students—is that it does not make any sense to talk about more or less plastic genotypes. A given genotype can be more plastic than another for a certain trait, but less so for a different trait. Hence the plasticities of different traits must be under at least partially distinct genetic control. On the other hand, development must play a part here: Plasticities of developmentally related traits do tend to be correlated. Thus the emerging picture is one of a constellation of plasticities, some of which are related to each other and others of which can evolve independently. The pattern of plasticity also varies with the particular environment one has chosen to study. Accordingly, "plasticity must be considered in relation to specific environmental effects at specific stages of development" (p. 122). This sentence and the examples cited in the review eventually gave rise to a host of studies specifically focused on the developmental aspect of phenotypic plasticity (Diggle 1993, 1994; Martin-Mora and James 1995; Pigliucci and Schlichting 1995; Brakefield et al. 1996; Leimar 1996; Imasheva et al. 1997; Pigliucci et al. 1997; Blanckenhorn 1998).

Bradshaw then tackled the long-standing debate over the relationship (or lack thereof) between heterozygosity and plasticity. As early as 1965 he had concluded that the evidence goes both ways, and that "it is clear . . . that stability may be determined by gene systems unrelated to heterozygosity as such" (p. 123). It is uncanny that this debate is still with us, despite the fact that plenty of new evidence points in the same direction as the conclusion that Bradshaw came to more than three decades ago (Chapter 4).

Finally, Bradshaw examined some of Waddington's classic (but then current) experiments on genetic assimilation to conclude that plasticity can indeed react to selection at a very rapid rate. This was the first explicit link between genetic assimilation and plasticity from an evolutionary point of view.

Fitness and Plasticity

Bradshaw readily recognized that some plasticity is not adaptive, and in fact is the result of a reduced growth rate under stressful environmental conditions. He also pointed out that plasticity of fitness *must* be selected against, since the ideal condition for an organism is to maintain as high a level of fitness as possible regardless of environmental fluctuations. But some kinds of plasticity can be adaptive, and plasticity for a given trait can result in fitness homeostasis. Bradshaw reckoned that this is more likely for plants, or for sedentary organisms in general, than for animals. This is because a plant is literally rooted to its spot and must endure environmental heterogeneity without the option of behavioral flexibility (Chapter 8) that would allow escape to seek a more favorable microhabitat. If plasticity can be adaptive, then it follows that we must ask under what circumstances we might (or might not) expect phenotypic plasticity to evolve as a character in its own right (i.e., not simply as a by-product of selection within distinct environments).

Bradshaw considered the evolution of plasticity under each of the three standard modes of natural selection: disruptive, directional, and stabilizing. In the case of disruptive selection, he argued that plasticity would be favored whenever the time scale of environmental fluctuation is of the same order or smaller than the generation time. In this case, evolution by alteration of gene frequencies (leading to the formation of ecotypes or to genetic polymorphisms) would be very inefficient. In fact, the population would always lag behind the environmental change, having started to adapt genetically to a situation that has already changed and that will be altered again before the inertia of selection on allelic frequencies allows it to catch up with the new situation. He even quoted Darwin in what is probably his only reference to plasticity (from a letter to Karl Semper in 1881): "I speculated whether a species very liable to repeated and great changes of conditions might not assume a fluctuating condition ready to be adapted to either condition" (p. 127).

Bradshaw discussed several pertinent examples, reaching the general conclusion that adaptive plasticity can evolve very rapidly, since closely related species, or even geographical populations of the same species, can display a particular pattern of plasticity or lack of it. For instance, northern European populations of *Lolium perenne* are characterized by a marked plasticity of the rate of leaf expansion in response to temperature. Southern populations of the same species, however, lack plasticity to temperature, presumably because the environment in which they live never reaches temperatures low enough to endanger the survival of the individual. Of course, the grain of environmental heterogeneity can be small on a spatial, rather than a temporal, scale, but similar considerations would apply.

Directional selection would seem to be a situation in which plasticity has little to contribute to the evolutionary process. Bradshaw pointed out, however, that if genetic variation in the population is limited, plasticity can afford persistence of the population for a certain period of time. This concept was elaborated upon much later by West-Eberhard (1989) and by Schlichting and Pigliucci (1995), and it relates plasticity to Waddington's (1942) concept of genetic assimilation. Essentially, the idea is that plasticity can facilitate natural selection and a directional shift in the population mean if some genotype shows plasticity in the same direction favored by selection. These genotypes will then constitute "bridges" from one generation to the next, allowing the time for favorable genetic variants to appear by mutation or recombination.

According to Bradshaw, stabilizing selection (in the ordinary sense, not in Schmalhausen's) is a case in which plasticity can play a very unusual role. Instead of allowing a single genotype to assume different phenotypes, it may cause different genotypes to produce the same phenotype. The best examples of this phenomenon are cases in which individuals supposed to be part of a single *ecotype* when studied in their natural habitat reveal instead variation for quantitative characters if brought into a different environment, such as a greenhouse. This *genetic redundancy,* which was subsequently addressed from a more mathematical perspective (Goldstein and Holsinger 1992), also implies the ability of plasticity to "buffer" genetic variation against selection (Gillespie and Turelli 1989).

Finally, Bradshaw considered conditions that *do not* yield the evolution of phenotypic plasticity, such as intense and prolonged stabilizing selection. Even under disruptive selection, plasticity may not be favored, especially if there are *costs* to the maintenance of a plastic response. This is the first explicit consideration of costs, another topic recently at the forefront of research in this field (Chapter 7). Plasticity would also not evolve if the grain of environmental change is too coarse or if the conditions are altered too suddenly and unpredictably (see, e.g., Levins 1963; van Tienderen 1991; Sibly 1995).

Bradshaw's contribution, as I mentioned earlier, did not have an immediate impact, since readers had to wait almost another decade for the next theoretical advance and almost two decades for the flourishing of new empirical studies within an evolutionary framework. But his work was and continues to be read, and the lag time between its publication and the renaissance of plasticity research is a clear tribute to his foresight and his ability to anticipate new directions of research in the field.

Plasticity, IQ, and Heritability: The Contribution of Lewontin

As noted previously, the study of phenotypic plasticity can be thought of as the modern incarnation of the nature versus nurture debate and, in fact, its partial solution in terms of a continuous dialectical interaction between nature and nurture (more on this in the epilogue). It is therefore significant that one of the landmark papers in the field was published in 1974 by Richard Lewontin, not as a study in its own right but as a rebuttal to another work dealing with the always delicate issue of the inheritance of intelligent quotient (IQ) in humans.

That issue of the *American Journal of Human Genetics* had two papers by the Newton Morton group (Morton 1974; Rao et al. 1974) that dealt with estimating heritabilities and components of variance for human phenotypes. Lewontin agreed with the cautionary approach of Morton and his colleagues, who pointed out that, given the complexity of human social classes and racial divisions, environmental effects can confound genetic analyses. But Lewontin went much further. His position was that the problem is not just that psychologists do not know much about genetics, but that the real issue is that human geneticists themselves keep using the wrong tool to pursue analysis of causes, namely, the analysis of variance. This was an all-out attack on the misuse of one of the most venerable of statistical techniques, and I think it deserves to be thoroughly discussed—not because it succeeded in its most radical objectives, but because it sparked a debate and much research on the limits of quantitative genetics as a discipline (see, e.g., Lewontin 1984; Mitchell-Olds and Rutledge 1986; Turelli 1988; Barton and Turelli 1989; Turelli and Barton 1994; Shaw et al. 1995; Pigliucci and Schlichting 1997). It should also be noted that the following discussion applies mostly (but not solely) to human genetics, because of the impossibility of carrying out appropriate experimental manipulations that can be combined with an analysis of variance. Several examples of such approaches in other systems are discussed throughout this book and represent a significant part of the empirical effort to study phenotypic plasticity.

Lewontin's basic argument is simple and straightforward: Analysis of variance is not synonymous with analysis of causes. Few scientists would disagree with this statement, and one of the first things that every student of statistics is taught is that even a very significant correlation does not imply anything at all about the causal relationship between two variables. The debate is about *how much* we can infer about causes by studying variation. This is a crucial question, because if it turns out that we can deduce little about causality by studying patterns of variation, then quantitative genetics has no explanatory power and is reduced to a descriptive science. Lewontin was very conscious of this, since he ended the introduction to his paper with the following statement: "I will come thereby to some very annoying conclusions."

Lewontin started by distinguishing two meanings of causal analysis. In the first sense, we are interested in discriminating between two mutually exclusive hypotheses about what may have generated a given pattern. For example, we might want to know if a certain phenotypic character is "caused" by a few genes of major effects or by multiple genes of small effects. The solution to this problem belongs to the field of segregation analysis, and it is obviously important in terms of the mechanistic understanding of any phenotype. The second form of causal analysis comes into play when we want to disentangle phenomena that interact to generate the observable phenotype, for example, when we know that a certain character is influenced by both the genetic constitution of the individual and by the environment in which that individual lives. For most of the twentieth century, geneticists assumed that a phenotype could be caused by *either* genes *or* environments, but not by both. This conceptual error is embodied in the common definition of *phenocopy,* wherein it is supposed to represent an environmentally induced phenotype that mimics a mutation. But in both cases (phenocopy and mutation) the phenotype is actually the result of the interaction between environments and genes; the only difference is that the phenocopy is probably the result of environmental influences on many genes with small effects, whereas the "mutant" arises from the environment interacting with a single gene of major effects.

No serious geneticist would disagree with the statement that it is ridiculous to attempt to quantify the separate contributions of genetics and environment to any phenotypic trait in a given individual. For example, nobody would say that if you are six feet tall, three and a half of those feet are due to your genes and the rest to what you ate while you were growing up. But this impossibility, Lewontin argued, is a major blow to the Cartesian-reductionist program of looking at the world in terms of its minimal components in order to understand the way it works. So quantitative geneticists have substituted a different target for the one that is unachievable. Instead of studying causes, we study variation. A quantitative geneticist is therefore interested not in what proportion of a single individual's height is due to genes or environments, but rather in what deviation from the average height is caused by a certain deviation in the average genotype or in the average environment.

However, Lewontin submitted, our tool for carrying out such an analysis of variation, the formal statistical technique of analysis of variance, is in some sense tautological. The mathematical model relates the phenotypic variance to the genetic variance, the environmental variance, and their interaction. But all terms are in fact expressed in units of phenotypic variation, there is no actual quantification of environments or genotypes, and the two sides of the equation must balance by definition. A first consequence of this is that we lose the connection between variation and causality. Let us assume that we find little "environmental" variance for a phenotypic trait. We cannot tell if this is due to reduced variation in the environment itself or to the fact that the genotypes are not sensitive to environmental changes, no matter how large these might be. These two alternatives are profoundly different from a biological standpoint. A second consequence is that the environmental and genetic "means" used in the analysis of variance are in fact *local* values, and they can change dramatically if the environment or the genetic constitution of a population changes. This implies not only that any analysis of variance is limited in space and time, but that we cannot derive from it any *functional* statement: Functional relationships among genes (and between genes and environments) do not change in a parallel fashion with gene frequencies or environmental heterogeneity.

Lewontin then contended that the real object of study is the reaction norm, since this can be thought of as a table of correspondence between phenotype and the genotype-environment combination. Such a table embodies a many-to-many, not a simple one-to-one, equivalence, and it implies that both genotype and environment have to be the (equal) focus of our efforts as they are both causes (in a mechanistic sense) of the phenotype. Lewontin then used this new focus on the reaction norm to illustrate two fallacies of the analysis of variance, once again contending that the latter only gives the *illusion* of separating causes. First, a change in the environments that we focus on can alter the genotypic variance, since the same genotypes will respond differently to a new set of environments. Hence, an alteration in the environment can bring about a change in the apparent genetic variance (this is related to the discussion of plasticity of heritability in Chapter 1; see also Roff and Bradford 2000)! Analogously, the amount of environmental variance explained by the model can change if the frequency of the genotypes in the population is altered, because the distribution of reaction norms will be different. Therefore, a change in the genetic constitution of a population can alter the measured environmental variance!

Second, the use of an analysis of variance model to "remove" the effect of a given source of variation is conceptually erroneous. For example, we can "fix" the environment to highlight genetic effects. But *which* environment should we fix? If we choose an environment in which reaction norms are highly divergent (i.e., genotypes produce very different phenotypes) we can end up with an *increase* in the total variance by "eliminating" one source of variation. A similar reasoning applies if we wish to "fix" the genotype. By choosing a particularly plastic genotype, we may end up again increasing the total phenotypic variance. Lewontin concluded by pointing out that the observable distribution of phenotypic variances is actually the complex result of three distinct levels of causality: (1) the functional relationships between genotype and environment that result in a reaction norm; (2) the historically determined distribution of allelic frequencies in the population; and (3) the actual distribution of environments. Much more than simple descriptive statistics is necessary to untangle these causal forces.

Lewontin then proceeded to dissect the reasons for one of the fundamental assumptions of analysis of variance: additivity. Additivity is important because an analysis of variance is actually able to separate genetic and environmental causes *only if* they are completely (or almost completely) additive. That being the case, differences between genotypes would remain the same regardless of the environment, thereby allowing a true partitioning of variation into causal components (genetic and environmental). In other words, interactions would be far less important than the main effects. Lewontin saw four philosophical underpinnings for the assumption of additivity. The first was Occam's razor, that is, the belief that the simplest hypothesis is that there are no interaction effects, which is the fundamental assumption of Cartesian science. The problem, of course, is that there is no actual basis for believing that to be true of the real world (and, in fact, the plasticity literature accumulated since Lewontin's article clearly shows this not to be the case). Second, the suggestion is usually made that the assumption of negligible interaction is a "first approximation," to be subsequently refined, similar to a mathematical analysis of a complex system based on Taylor's expansion series. But Taylor's series is a local analysis, which is carried out precisely when one is interested in local effects. That is also what the analysis of variance does, although in this case the stated purpose is the search for *general* patterns. Third, many empirical biologists claim that interactions turn out to be small when the analysis is carried out, and they can therefore be justifiably neglected. The problem is that interactions are found mathematically by accounting for as much variation as possible in terms of main effects and then attributing the leftover variance (except for the error component) to the generic interaction term. It then becomes rather tautological to claim that therefore interactions are demonstrably small. Also, the relative importance of interactions depends on the genotypes being analyzed and on the shape of their reaction norms, which makes it to some extent a problem of sampling. Fourth, Lewontin argued that additivity sometimes simply fits well with social prejudices. Social standings and political pressures heavily influence most human genetics research on the heritability of IQ. Thus the assumption of additivity made in many such studies conveniently leads to the conclusion that, for example, educational programs are irrelevant since a "superior" genotype will continue to be superior regardless of the (educational) environment. As we shall see in the epilogue, there is not an iota of evidence to support either such a contention or most other conclusions about the shape of human reaction norms.

Lewontin's devastating attack did not catalyze the downfall of human quantitative genetics, nor did it preclude the publication of further questionable studies on the inheritance of human IQ (Bouchard et al. 1990). But its broad conceptual framework made it a milestone in the field of phenotypic plasticity, and it eventually spurred theoretical and empirical studies on the limits of quantitative genetics and the restricted value of such crucial parameters as genetic variances and covariances (Mazer and Schick 1991a,b; Willis and Orr 1993; Simons and Roff 1994; Ward 1994; Bennington and McGraw 1996; Merila 1997; Hoffmann and Schiffer 1998, 1999; Sgrò and Hoffmann 1998).

Modeling Plasticity: Via and Lande

The final major contribution I mention here—though it is so recent that it barely qualifies as "historical"—is Via and Lande's (1985) proposal of a theoretical framework to model the microevolution of phenotypic plasticity. In considering this paper, it must be remembered that most of the more recent mathematical work in this area either is framed in reference to Via and Lande or is an attempt to improve or supersede it (Chapter 10).

I discussed Via and Lande's basic idea and its historical roots in Chapter 1: By using the quantitative genetic concept of genetic correlations between characters and coupling it with Falconer's (1952) suggestion that a reaction norm can also be thought of as the genetic correlation between expressions of the same trait in different environments, they achieved two fundamental results. First, they gained a powerful tool with which to relate empirical data and mathematical modeling. Second, they could use the theoretical arsenal of quantitative genetics to describe the evolution of plasticity as a variation on the already well-established theory of quantitative phenotypic evolution.

The simulations described in the 1985 paper were designed to explore the evolution of character means under both soft and hard selection (Fig. 3.3), provided that the environment varies in a coarse-grained fashion. We will see that coarse-grained variation is actually the least likely condition for the evolution of adaptive phenotypic plasticity (Chapter 7), which explains some of the discussions that followed upon Via's work in particular. Via and Lande concluded that a panmictic population subjected to a bivariate fitness function (i.e., experiencing two environments) will eventually reach the optimum phenotype in each environment. This is true unless one of the following exceptions occurs:

1. There is a cost to plasticity (van Tienderen 1991, see Chapter 7).
2. There is no residual genetic variation.
3. The across-environment genetic correlation is ±1—and therefore the system is entirely constrained to a particular evolutionary trajectory and most of the phenotypic space is not reachable.

Via and Lande found further restrictions on the evolution of reaction norms. Even if the absolute genetic correlation is not 1, very positive or very negative correlations will not only slow down the course of evolution, but in fact force the population through temporary maladaptation. Further, and more intuitively, the pace of evolution within a given environment is proportional to the frequency of occurrence of that environment, so that rare environments weigh less than common ones under both soft and hard selection. Finally, these authors found differences between the details of the evolutionary trajectories when comparing hard and soft selection, with en-

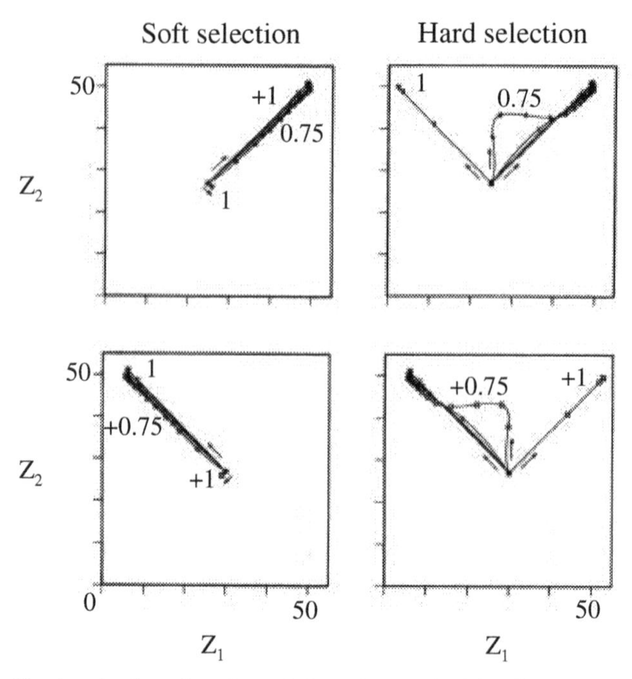

FIG. 3.3. Simulated trajectories of populations characterized by different reaction norms (quantified as interenvironment genetic correlations) under soft and hard selection. Numbers on the trajectories indicate the genetic correlations, and Z_1 and Z_2 are the expressions of the same trait in two environments. The optimum is always located at (5,50) and the initial conditions and direction of evolution (arrows) are shown. The two environments occur with equal frequency. Note how soft selection always leads to one outcome (at or away from the optimum, depending on the initial conditions). Hard selection, on the other hand, can lead to very different outcomes, as a function not only of initial conditions but also of the value of the genetic correlation. (From Via and Lande 1985.)

vironments characterized by the highest mean fitness contributing the most to the gene pool under hard but not soft selection. The very fabric of the adaptive landscape was different under the two regimes, with adaptive topographies showing a different number and distribution of peaks in the two cases and consequently distinct evolutionary trajectories (Fig. 3.3).

In general, the fact that Via and Lande's work generated so much follow-up and controversy (including the one in which I found myself entangled—see Chapter 1) is exactly what one would expect from a milestone in scientific research. Many empirical and theoretical papers on plasticity have been published in the fifteen years since Via and Lande's work, and several reviews on phenotypic plasticity have appeared to summarize them (Schlichting 1986; Sultan 1987; West-Eberhard 1989; Thompson 1991; Scheiner 1993a; Gotthard and Nylin 1995; Schlichting and Pigliucci

1995; Via et al. 1995; Pigliucci 1996a; Nylin and Gotthard 1997). I discuss some of what I consider the best examples and the most important conceptual advances published in the primary literature throughout the rest of this book, but these reviews remain an invaluable source of information and references for the serious student of genotype-environment interactions.

I think it is clear from this brief (and admittedly a bit idiosyncratic) walk through the history of plasticity studies that this field of research is both old and vital. Understanding where certain ideas came from and in what context they developed is not just an exercise in the history of science. It allows us to better appreciate the intellectual complexity of studies on genotype-environment interactions as well as the scope of the effort that has gone into a century of theoretical and empirical investigations. Many new students are being attracted to the enterprise of understanding how organisms cope with environmental changes, and appreciating the controversies in which others have become entangled should help them to avoid repeating past mistakes as well as point us all toward new horizons.

Conceptual Summary

- Johannsen first clearly realized the distinction between genotype and phenotype.
- Woltereck carried out the first study on phenotypic plasticity using *Daphnia* and coined the term *reaction norm*. However, he evidently did not understand the concept of genotype-environment interaction, although it was clear to Johannsen.
- Schmalhausen contributed to the conceptualization of the evolution of plasticity during the Neo-Darwinian synthesis, proposing two meanings of *stabilizing selection,* one of which refers to the evolution of a novel reaction norm under new environmental conditions. His work (and possibly his chances of influencing the synthesis) was cut short by Lysenko's rise to prominence in the Soviet Union.
- In the 1960s Bradshaw essentially invented the field of plasticity studies as it is understood today. Among his many important contributions are the ideas that plasticity has a genetic basis and can therefore evolve as any other character and that the same genotype may be more or less plastic depending on the environments and traits that are being considered.
- Lewontin, in a seminal paper published in 1974, made it clear that the statistical tool of analysis of variance is not equivalent to a study of causality in quantitative genetics. In the process, he clarified the relationship between ANOVA and reaction norms, being the first to suggest that heritabilities can change depending on the environment.

- In the mid-1980s Via and Lande, developing an original idea by Falconer, provided a framework for the incorporation of phenotypic plasticity into the general theory of evolutionary quantitative genetics. Their work is still at the basis of modern theoretical studies of reaction norm evolution.

4

The Genetics of Phenotypic Plasticity

When a thing ceases to be a subject of controversy, it ceases to be a subject of interest.

—William Hazlitt (1778–1830)

CHAPTER OBJECTIVES

To discuss various genetic models of phenotypic plasticity as well as the evidence for genetic variation for plasticity within and among natural populations. The idea of genetic constraints on plasticity is introduced on both univariate and multivariate aspects of reaction norms. The relationships (or lack thereof) between plasticity and heterozygosity, developmental noise, canalization, and homeostasis are discussed. The chapter ends with a section on the theoretical modeling and empirical evidence relating plasticity to the maintenance of genetic variation in natural populations.

The genetics of phenotypic plasticity is probably the single most controversial subject of study in the field. One often gets the impression that biologists not directly involved in plasticity studies still view with disbelief the idea that plasticity is genetically based and not just "environmental." I have three objectives in this chapter. First, I discuss the ample evidence that plasticity is indeed under some kind of genetic control, regardless of what *sorts* of genes underlie a plastic response (a more comprehensive discussion of this topic is deferred to Chapter 5, where we examine the molecular genetics of plasticity). I review some of the evidence for the presence of natural variation for plastic responses at the within-population, among-population, and among-species levels. This is accompanied by a discussion of the kind of insights that emerge from the study of variation at these distinct levels of the biological hierarchy. Second, I focus on genetic constraints on phenotypic plasticity, elucidating how they have been measured and discussing if and how they can be released (again, I am using the word *constraint* here in the general sense discussed in the preface). Third, I consider the relationships (or lack thereof) between plasticity and a host of classical biological phenomena, such as heterozygosity, developmental noise, and canalization, with which plasticity has been historically associated in one way or another. This leads us into a brief discussion of a possible link between plasticity and one of the oldest questions in populations genetics: How is genetic variation maintained in natural populations?

Is Plasticity Genetically Controlled?

As I noted in Chapter 3, Bradshaw (1965) reportedly was the first to clearly state that phenotypic plasticity is under independent genetic control, although the same was implied in Schmalhausen (1949). His assertion, however, was simply the next step on the long path leading to the development of the idea. As we have seen (Chapter 1), the discovery of reaction norms by Woltereck was immediately and correctly interpreted by Johannsen (but not by Woltereck himself) as meaning that genotypes have a specific way of responding to environmental conditions. Since a reaction norm is by definition a property of a genotype, its main attributes—the amount and pattern of phenotypic plasticity—must be under genetic control.

On the other hand, practically any character, regardless of how artificially it is defined, will probably turn out to be "under genetic control," if by that we simply mean that there is genetic variation for the expression of that trait in natural populations. For example, we really do not know what intelligence is or how to measure it (Feldman and Lewontin 1990), but this has not stopped us from trying to calculate its heritability through the study of variation in that raw—and clearly artificial—measure of intellectual skills known as the *intelligence quotient* (Bouchard et al. 1990). Analogously, plant and animal breeders do not really know if the *characters* they are interested in from an economic viewpoint are actually *biological* traits of the organisms on which they impose a selective pressure. Certainly, yield in corn or the commercial qualities of cow's milk are only indirectly related to the natural biology of plants or mammals. However, it is equally obvious that changes in the genetic makeup of these organisms will have *some* consequences for corn yield or milk productivity, even though natural selection might have acted on different components of the phenotype or in a different fashion than the artificial breeder (for an in-depth discussion on the concept of character, see Wagner 2001).

This leads us to consider the problem of the genetics of phenotypic plasticity from two distinct standpoints. First, is plasticity a *natural* character?

In other words, does selection act on phenotypic plasticity to shape it in response to the specific environments that a particular organism happens to encounter during its own lifetime or the lifetime of its progeny? This is an ecological question. I advanced arguments in favor of a positive answer under certain circumstances in Chapter 1, and we examine the ecology of phenotypic plasticity in more detail in Chapter 7. Second, is there genetic variation for plasticity available to respond to natural selection? This question deals more directly with the genetics of plasticity in natural populations, which is the subject of the first part of this chapter.

Before examining some of the experimental evidence, I have to discuss Scheiner and Lyman's (1991) classification of the genetic bases of plasticity, since it bears directly on the topics covered in this chapter as well as on our examination of the molecular biology of plastic responses in Chapter 5. These authors suggested that we can think of the genetic bases of plastic responses as falling into three distinct categories:

1. *Overdominance.* There is an inverse relationship between plasticity and heterozygosity: The more homozygous a genotype is, the more plastic its reaction norm will be.[1]
2. *Pleiotropy.* Plasticity depends on the expression of the same set of alleles in different environments, analogous to the classical quantitative genetic case of several traits sharing a common genetic control through pleiotropic gene effects.
3. *Epistasis.* (At least) two distinct classes of genes control the two fundamental characteristics of a reaction norm: its plasticity and its across-environment average value (sometimes referred to as its *height*).

The overdominance model basically assumes that plasticity is an accident resulting from lost or reduced homeostasis in a genotype that has become too highly homozygous, but there is not much evidence to support it, despite some theoretical explorations of its mathematical consequences (Gillespie and Turelli 1989). As regards the other two models, we are tempted to see a close correspondence, on the one hand, among pleiotropy, allelic sensitivity, and phenotypic modulation and, on the other, among epistasis, regulatory plasticity, and developmental conversion. However, we discussed in Chapter 2 (see also de Jong 1995) the reasons why such correspondence is far from established and can in fact be doubted on the grounds that the same genetic mechanism can provide the basis for either developmental mode.

The major limitation of Scheiner and Lyman's classification is that it is entirely phenomenological, that is, it is based on the observation of patterns rather than on actual investigation of the underlying causes. For example, the pleiotropic and the epistatic models can be studied entirely in a quantitative genetics (i.e., statistical) way, without knowing what the genes are doing. The prediction of the epistatic model, in particular, is very simple: The genetic correlation between the height and the slope of a reaction norm should be less than 1, thereby indicating at least partially distinct gene complexes acting on the two aspects of the reaction norm. While this is an excellent preliminary approach, and it certainly is the only available venue for most experimental systems to date, it is not entirely satisfying. It is fairly easy to imagine situations in which the genetic control of plasticity can be very specific and based on biochemical epistasis, while a mutation in the same gene might still alter both the height and the slope of the reaction norm (no statistical epistasis). As an actual example, we can refer to the case of the phytochrome mutants affecting the shade avoidance response discussed in Chapters 2 and 11. It is clear from Fig. 2.3 that a mutation in any of the genes coding for the photoreceptors alters both the plasticity and the average response of the wild type. Nevertheless, since we know the true mechanistic basis of this response, we can readily classify shade avoidance as a case of plasticity controlled by regulatory (epistatic) interactions among a very reduced number of genetic elements. If we were considering exactly the same data from a collection of natural genotypes, we would have concluded that the pleiotropic model would (statistically) fit the available evidence better. This cautionary tale should be kept in mind any time we attempt to infer more than is legitimate from quantitative genetic analyses (Pigliucci and Schlichting 1997).

Genetic Variation for Plasticity in Nature: Genotypes, Populations, and Species

As with any other biologically relevant trait, the study of natural variation for phenotypic plasticity can be conducted on at least three distinct levels: among genotypes within a population, among populations within a species, and among closely related species. Literally thousands of examples of this kind of research accumulated in the literature throughout the twentieth century. As attempting even a cursory review or a summary of the most relevant ones would require a multivolume opus, I have chosen some specific examples from the classical and recent literature to illustrate both the insights and limits of these kinds of studies.

Within-Population Variation

The point of investigating variation for plasticity among genotypes within a natural population is basically to estimate the amount of genetic variation available within that population to respond to future selection. In other words, we want to know the heritability of plasticity (see Equation 1.2). The examples that I examine briefly and countless more demonstrate the following points: (1) There is abundant genetic variation for plasticity within natural populations (Sultan 1987; Scheiner 1993a), which implies that plasticity might respond to selection (either natural or artificial). (2) Genetic variation for plasticity can be found in both plants and animals, as well as in practically any other organism (from bacteria to fungi) that has ever been studied. (3) The same population can harbor genetic variation for the plasticity of one trait, while being invariant for the plasticity of another trait *to the same environmental variable.* (4) The genotypes of a certain population can show variation for plasticity of a trait to one set of environments, but not to a different set. This leads to the question of why there are such contrasting patterns of natural variability, with heritabilities for plasticity spanning a wide range. (5) There do not seem to be any genotypes characterized by patterns of plasticity that confer the highest fitness across the entire range of biologically relevant environments for that species (i.e., there are no *Darwinian monsters*). (6) It is possible to detect negative across-environment correlations for the expressions of the same trait in multiple contexts, which may indicate a trade-off and therefore constraints on the evolution of plasticity.

The first example comes from a classic work by Dobzhansky and Spassky (1944), who analyzed the variation in reaction norms of development rate and viability in response to temperature and density of conspecifics in *Drosophila pseudoobscura.* Their general conclusion (rather novel at the time) was that "natural populations contain a tremendous variety of genotypes with different reaction norms" (p. 290). Some of the specifics of that paper are also very interesting and representative of similar investigations. The authors studied the reaction norms to three temperatures (ranging from 16.5° to 25.5°C) and to a variety of population densities of homozygote and heterozygote genotypes from the Mount San Jacinto area in California. They found that the reaction norms for development rate were less plastic overall than the ones for viability. Plasticity for viability showed peaks at low and, less frequently, intermediate temperature, but never at high temperature (Fig. 4.1). On the other hand, some genotypes had highest viability at low, others at intermediate, and still others at high density (i.e., throughout the range of densities tested). Most important, no genotype proved to be superior to the others under all conditions, a fact that has immediate implications for the maintenance of the observed variation in the face of selection on viability and development rate.

Windig (1994) investigated reaction norms to four temperature levels in the tropical butterfly *Bicyclus anynana.* These insects exist in two seasonal forms, one adapted to dry conditions and the other more suitable for

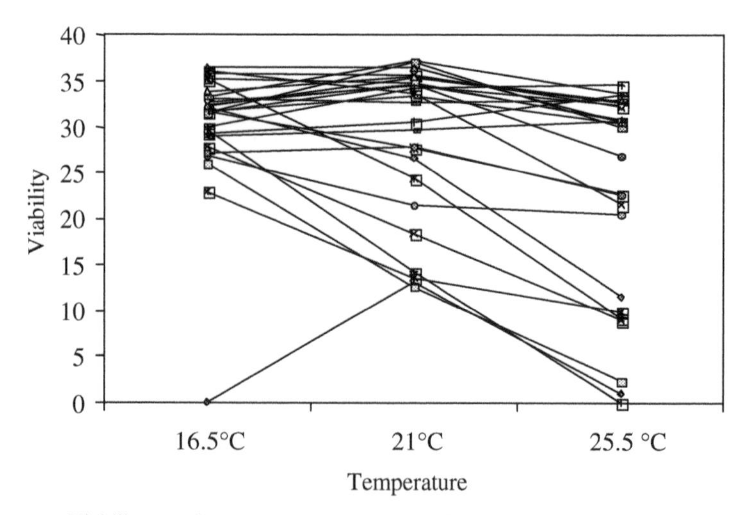

FIG. 4.1. Viability reaction norms to temperature of twenty-six genotypes of *Drosophila pseudoobscura* homozygous for different kinds of second chromosomes. Note how most genotypes are less viable at the highest temperature, while a few peak at intermediate temperature. (From data in Dobzhansky and Spassky 1944.)

wet seasons. Windig studied the reaction to temperature of larval development time, pupal weight, and two traits related to wing pattern characteristics (which he termed *seasonal form* and *thermal form;* Fig. 4.2). He found that development time and seasonal form were more plastic than the other two characters in twenty-one full-sib families. The reaction norms for development time were the only ones to actually cross along the environmental gradient (close to the intermediate temperature). The crossing of the norms translated into a negative across-environment genetic correlation, simultaneously showing a high degree of variation for plasticity and the presence of a constraint, because selection for fast development at one extreme of the gradient would presumably imply a correlated response at the other extreme, which would result in slower development.

Delasalle and Blum (1994) studied germination and survival in response to four levels of salinity and two of temperature in the freshwater perennial *Sagittaria latifolia.* They found genetic variation for plasticity among families in response to salinity, but not in their reaction to temperature. Apparently, the response to the two environmental gradients was independent, as indicated by the lack of significance of the temperature-by-salinity effect in their analysis of variance. Again, variability for one kind of plasticity does not imply variability for another kind, even for the same traits.

As a final example of variation within populations, let us consider a study of reaction norm variation to four levels of nutrients in *Arabidopsis thaliana* (Pigliucci 1997). Selfed families from three populations were com-

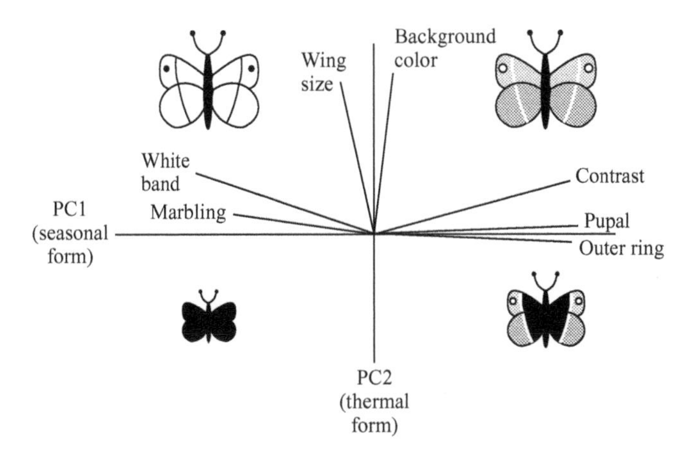

FIG. 4.2. Phenotypic variation induced by genotype-environment interaction in butterflies of the species *Bicyclus anynana.* The variation is summarized by principal components axes, the horizontal one differentiating the seasonal forms and the vertical ones separating the thermal forms. (Redrawn from Windig 1994.)

pared for parameters describing the life history of the plant (onset of sexual reproduction, offset of sexual reproduction, and growth rate of the main stem during the reproductive phase). Plots for the early flowering ecotype (Fig. 4.3) clearly show that most of the variation among reaction norms is actually due to a few genotypes that tend to flower later and for a longer period of time, whereas they elongate at a slower rate. Even though most of the reaction norms in these genotypes were flat (little plasticity), the analysis of variance still detected significant genetic variation for reaction norms for onset of flowering and growth rate, but not for offset of reproduction.

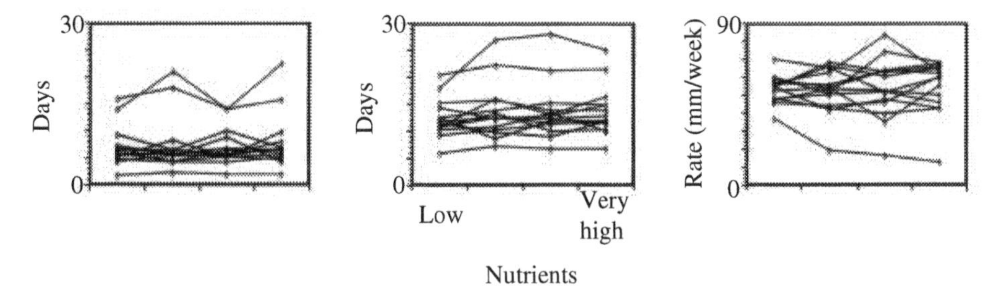

FIG. 4.3. Reaction norms of genotypes of an early flowering population of *Arabidopsis thaliana* along a nutrient gradient. Left to right: onset of flowering; offset of reproductive growth; rate of growth of the main stem during reproduction. (From Pigliucci 1997.)

Among-Population Variation

Most studies of interpopulation differences in plasticity are a by-product of investigations of within-population variation carried out in several populations simultaneously. Before we proceed any further, let me clear the field by stating unambiguously that there is nothing wrong with considering *mean* plasticities as an attribute of a population or species. While it is true that by definition a reaction norm is a genotypic property, we have no qualms whatsoever in calculating population- or species-level means for all sorts of morphometric traits, and I do not see why plasticity should be an exception. For instance, even though the size of a mandible is clearly an *individual* attribute (determined by genotype and environment), we speak freely of mean mandible size in different populations, and we can even differentiate species based on this and similar characters. Of course, the more variation there is for *any* trait within one hierarchical level, the less informative it is to abstract that information at the next level in the form of a mean value. But this problem applies to anything we might wish to measure, including behavioral traits and, of course, phenotypic plasticity.

The available studies on plasticity at the across-population level lead to the following general conclusions, as we shall see for some specific cases below: (1) There is ample variation for plasticity among populations of the same species. (2) Ecotypic (i.e., mean genetic) differentiation does not preclude the existence of plastic reaction norms, with each population retaining a substantial capability to alter its phenotype in response to alternative environmental conditions. (3) It is possible to suggest a link between the environmental regime and/or provenance of a given population and its pattern of plasticity, but such links are subjected to substantial variation owing to historical events and the idiosyncratic evolutionary path of each population. (4) The phenotypic correlation matrices relating different aspects of the multivariate phenotype of populations are prone to both genetic differentiation and marked phenotypic plasticity, as is the case for single traits. (5) Marked differences among populations in parts of the reaction norm cannot necessarily be attributed to the result of selection, unless historical evidence can rule out phylogenetic inertia and we have information at the within-population level that allows us to exclude genetic constraints. (6) Similarly, substantial convergence of reaction norms of different populations can occur within certain ranges of environments, but it is difficult to say to what extent this is due to selection as opposed to being the result of a recent common origin of the same populations or to shared genetic constraints.

For example, Gurevitch (1992) investigated the relative importance of genetic specialization and phenotypic plasticity in explaining the variation of leaf shape between two populations of *Achillea lanulosa.* This plant occurs in the Sierra Nevada range of California, at altitudes from 800 to 3,350 meters. Gurevitch sampled one population from a low altitude and one from a high one, and grew clonal replicates of each in a controlled environment, exposing them to cool and warm temperature regimes. The population living at lower altitudes (i.e., higher temperatures) was characterized by overall larger and more dissected leaves, in accordance with simple predictions based on principles of plant physiology. However, both digitized image analyses of whole leaves and canonical multivariate analyses of leaf traits clearly showed a parallel shift of both populations between a *cold* and a *warm* phenotype, depending on the imposed temperature. Gurevitch concluded that genetic differentiation leading to the formation of altitude- (or temperature-) adapted ecotypes did not imply a trade-off causing a decrease in plasticity. This is what we would expect, especially if the divergence between the two populations has been very recent; the alternative possibility of plasticity being maintained by substantial gene flow contradicts the observation of genetic differentiation for trait means in these populations.

The classic work by Cook and Johnson (1968) (more on this in Chapter 7) represents a clear example of interpopulation differentiation for plasticity with an ecological underpinning. They investigated variation in *heterophylly* (the ability to produce two or more types of leaves) within and among ten populations of *Ranunculus flammula.* Plants living in aquatic environments produced smaller and more linear leaves than the same genotypes raised in a terrestrial habitat. Accordingly, populations exposed to permanent terrestrial conditions displayed the lowest level of heterophylly, whereas all populations living under variable environmental regimes were characterized by high heterophylly. Interestingly, the only population labeled as *constant aquatic* also showed one of the highest degrees of heterophylly. This latter result is unexpected unless the population's typical environmental regime has changed very recently, so that heterophylly is no longer adaptive but the result of the historical path of the population. This is a particularly good illustration of the difficulty of testing adaptive plasticity hypotheses without historical knowledge, given the ubiquitous presence of phylogenetic inertia (Chapter 9).

Schrag et al. (1994) studied the way in which temperature affects phallus production in the snail *Bulinus truncatus.* In this species, both aphallic and euphallic hermaphrodites can self-fertilize, but only the latter can donate sperms to another individual because of the presence of a fully developed phallus. Across a geographical range centered in northern Nigeria, Schrag and colleagues found that euphally ranged from 0 to 81% when estimated in forty-nine populations, with levels increasing during the dry-cool season. Interestingly, these authors also determined a developmental window that may indicate anticipatory plasticity, in that maximum water temperature during the first week of a controlled-conditions study was an excellent predictor of euphally in the next generation of snails. As for variation among populations, these authors carried out detailed studies of three populations, showing that one was essentially insensitive to temperature, whereas the other two had a dramatically altered proportion of euphallic individuals as a response to decreasing temperature (Fig. 4.4). More significantly, the nonplastic population was also the one that remained consistently aphallic during the field studies, demonstrating the repeatability of field results under controlled conditions.

As a final example of among-population differentiation, Blanckenhorn (1991) looked at the differences in life histories in geographically close populations of water striders. He hypothesized that there should not be genetic differentiation between the two populations because the insects would be able to cope with the temperature difference between the cold and the warm stream by phenotypic plasticity. This scenario considers plasticity as "opposed" to genetic variation. It turned out that the null hypothesis had to be rejected—but because of a very interesting combination of results. On the one hand, there was significant genetic differentiation between the two populations in the mean expression of life history traits under *common garden* conditions, and the author found significant heritability for these traits within populations. However, the populations were also *genetically* distinct for their phenotypic plasticities to temperature. Interestingly, these differences in plasticity were significant when the insects were raised in groups, but not singly, highlighting the complex interaction between density of conspecifics and temperature in shaping the local differentiation of reaction norms.

Among-Species Variation

If phenotypic plasticity has any impact on macroevolution (i.e., beyond the fine-tuning of adaptations within species), we have to investigate differences in plastic responses among closely related species. Surprisingly, there is a dearth of studies of this type, particularly when it comes to including the phylogenetic information necessary to disentangle historical accidents

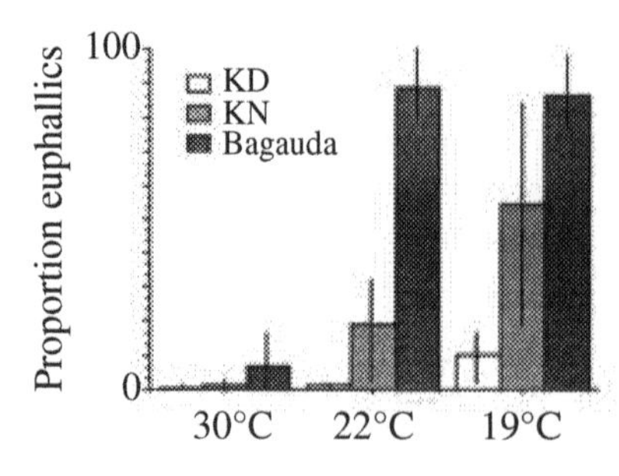

FIG. 4.4. Reaction norms of euphally in response to temperature in three populations of snails of the species *Bulinus truncatus*. The population that is insensitive to the environmental change also shows no variation for euphally under field conditions. (From Schrag et al. 1994.)

from long-term responses to natural selection (see Chapters 9 and 11 for a discussion of this point and an example). The comparative method (phylogenetically explicit or not) has always been a powerful tool in evolutionary biology, so finding that an entire subfield of evolutionary biology has little to contribute to the literature on comparative biology is unexpected. Three kinds of explanations can be entertained for this anomaly. First, since phenotypic plasticity is a property of individual genotypes, one might conclude that it makes little sense when abstracted at the species level. I offered my arguments above for why this is unsound reasoning. Second, it is logistically cumbersome to set up meaningful experiments comparing plasticity across species. It implies sampling different populations and a significant number of genotypes within each population for each species, leading to a project that can quickly become prohibitive in terms of space and personnel. Third, phenotypic plasticity has not yet been explicitly considered as a relevant player in evolution above the species level (with some exceptions [see West-Eberhard 1989 and Chapter 9]), partly because of its perceived extreme lability (after all, it is an *environmental* effect). This, however, is like suggesting that behavior and life history are not important in evolution because they can be dramatically different in closely related species. Quite clearly, we need more studies of the type that I summarize below to attempt to put our ideas on the macroevolutionary consequences of phenotypic plasticity on a firmer footing than some of our Neo-Darwinian assumptions would otherwise allow (Sultan 1992).

Research into differences in plasticity among species has yielded some limited generalizations: (1) Closely related species can differ substantially in their plasticities, thereby demonstrating that plasticity can evolve at a macroevolutionary scale within a very short time frame. (2) On the other hand, closely related species can also show identical or parallel reaction norms. In this case, the outcome of evolution may depend either on similar selective pressures or on some sort of constraint (e.g., lack of the appropriate genetic variation or particular patterns of genetic covariation). In order to decide between the two, we need information on the phylogeny, the quantitative genetics, and the ecology of the taxa in question. (3) There is some indication that plasticity can facilitate the evolution of specialized ecotypes, starting from a generalist strategy. (4) It is possible to identify discrete strategies based on different types of plasticity evolving in closely related species to cope with similar environmental conditions. (5) Closely related species can display very similar reaction norms early during their ontogeny, arriving at divergent phenotypes and plasticities only later in their development. (6) Relevant information on the evolution of adaptive plasticity can be gathered by comparing phylogenetically unrelated taxa with similar and contrasting ecologies. (7) Apparently adaptive plasticity can evolve as the correlated response to different environmental cues.

The first example here is a study by Day et al. (1994) that focused on two closely related species of threespine sticklebacks living in contrasting habitats and aimed at investigating the potential role of phenotypic plasticity in shaping divergence in trophic morphology. One of the two species lives in a benthic environment and feeds on worms; the other inhabits limnetic areas and preys on plankton. When each species was raised on the other's diet, it showed a convergence toward the phenotype of the second species. This could be an indication that the common ancestor of both taxa was a generalist that gave rise to two partial specialists; alternatively, one can hypothesize that the existence of a plastic reaction norm in a benthic-worm feeder taxon could have provided the first step toward the evolution of a limnetic-plankton feeder taxon (or vice versa). More recent evidence suggests that the first hypothesis is correct and that the evolution of benthic and limnetic specialists represents a case of *parallel speciation* (Schluter and Nagel 1995). Interestingly, the species known to live in the most variable habitat also showed the highest degree of plasticity, perhaps an indication that plasticity is being actively maintained by selection in at least one of the two taxa.

My colleagues and I (Pigliucci et al. 1997) also examined the plasticity of two closely related species, this time forming a putative ancestor-descendant pair: *Lobelia siphilitica* and its close derivative, *Lobelia cardinalis.* The reaction norms of these taxa to nutrient availability develop gradually throughout the ten to fourteen weeks that it takes for the main inflorescence to emerge and complete flower production each season (Fig. 4.5). *L. cardinalis* shows more variation for the amplitude of the ontogenetic curves, as well as more plasticity of some of its individual genotypes. Interestingly, genotypes that start with very similar curves end up diverging dramatically toward the end of the growth season; conversely, some genotypes are distinct for most of their ontogeny, but eventually converge toward a similar phenotype, at least in one of the two environments. Both these phenomena are more marked in the derivative *L. cardinalis* than in the ancestral *L. siphilitica.*

Mitchell (1976) conducted a rare study of multiple species, investigating the response to submergence in nine species (and twenty-three populations) of *Polygonum.* He found that different varieties within a species display a distinct pattern of plasticity for plant architecture, which at least in some cases correlates well with the habitat in which the plants are found. Most (but not all) of the tested species survived quite well under submerged conditions, although the ones displaying the highest plasticity for architecture did significantly better. It is noteworthy that Mitchell was able to identify three distinct *strategies* for survival under submerged conditions, adopted by three distinct species and corresponding to three separate plasticity syndromes (Fig. 4.6). In the *amphibious response,* the submerged plant produces enlarged leaves and elongated petioles, as well as increased

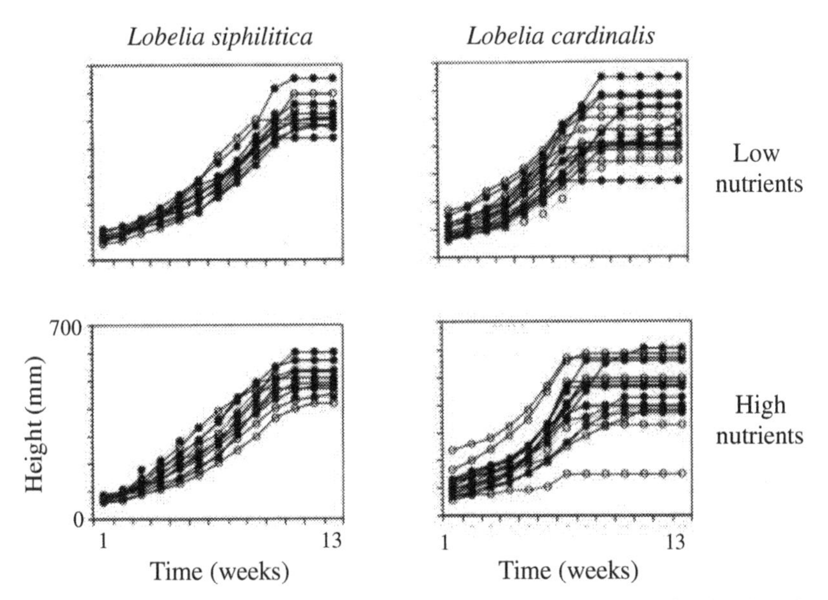

FIG. 4.5. Developmental trajectories of main stem elongation for two closely related species of *Lobelia* raised under high and low nutrients. Solid and open circles distinguish two populations of each species within each graph. *L. siphilitica* on the left, *L. cardinalis* on the right; curves obtained under low nutrients are in the upper panel and those obtained under high nutrients are in the lower one. (From Pigliucci et al. 1997.)

intercalary growth. This allows the upper leaves to float on the water surface (Fig. 4.6, top). In a second strategy, named the *vigorous growth response,* the plant not only produces floating leaves, but the overall growth rate is dramatically increased, so much so that it eventually emerges and starts growing above the surface of the water (Fig. 4.6, middle). The third type of plasticity corresponds to an *emergent response,* in which the plant habit is not altered (no floating leaves are produced), but the growth rate is sufficient to put the upper portions of the plant above water level (Fig. 4.6, bottom). It should be noted that the emergent response is clearly nonspecific, in that it does not entail any environment-induced alteration of the plant architecture and morphology (although it may be the result of plasticity at the physiological or anatomical level). The amphibious response is clearly a specialized adaptation to floating conditions, with dramatic alterations of the morphology of both the leaf and the whole plant. The vigorous growth plastic syndrome is intermediate between the other two, entailing some morphological changes, coupled with an attempt to simply get out of the water. It would be very interesting to explore these syndromes from a phylogenetic viewpoint, to ascertain if they evolved repeatedly in

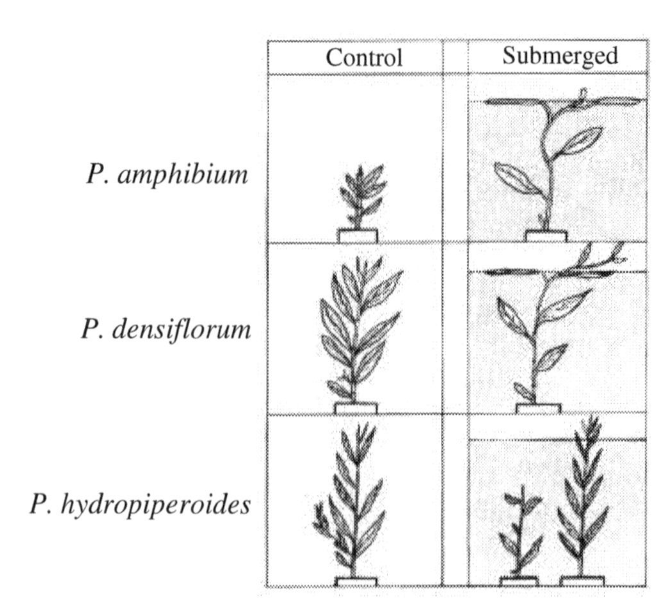

FIG. 4.6. Mitchell's representation of three distinct plasticity syndromes for coping with floods in different species of *Polygonum*. See text for details. (From Mitchell 1976.)

different sections of the genus, or if they represent successful adaptations that were maintained once they appeared.

Blouin (1992) compared the reaction norms of time and size at metamorphosis in three closely related species of hylid frogs from the southeast United States. In nature, the three species utilize distinct larval habitats, ranging from permanent swamps, to ponds that dry infrequently, to very ephemeral bodies of water along roadsides. Despite this divergence of habitats, two of the three species showed parallel reaction norms to both food level and temperature. However, the species living in ephemeral habitats did show a significant departure from the pattern of phenotypic plasticity to temperature of the other two, being much less plastic than its congenerics (Fig. 4.7). At face value, this might be taken as a good case of possible adaptive evolution of reaction norms determined by the lability of the environmental conditions typical of each species. Unfortunately, the reduced number of species and the absence of phylogenetic information make it difficult to decide one way or the other.

An alternative approach to interspecific comparisons of plasticity is to study species that are related ecologically, as opposed to phylogenetically. The advantage, of course, is that whatever convergence of reaction norms one finds (consistent with the ecology of the taxa) are much more likely to be due to independent adaptive evolution than to phylogenetic constraints; although in the absence of phylogenies one cannot distinguish between

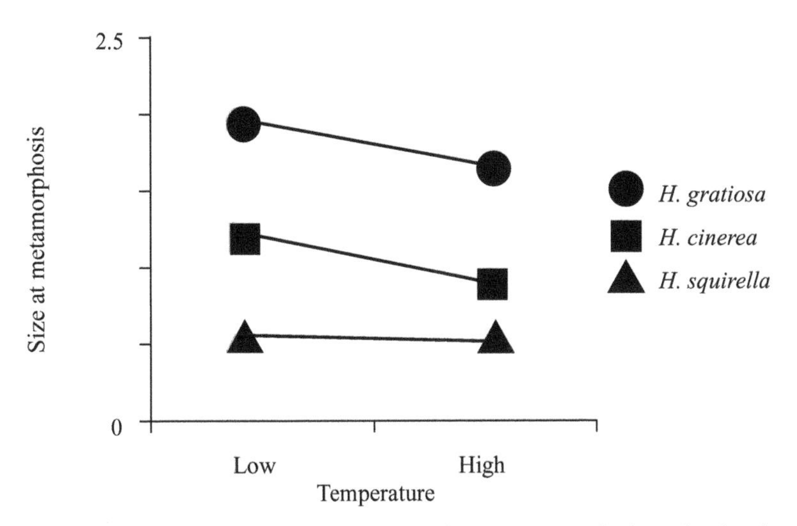

FIG. 4.7. Reaction norms of size at metamorphosis to temperature in three closely related species of *Hyla* frogs. Note the parallelism between the reaction norms of the two species living in permanent habitats (*H. cinerea* and *H. gratiosa*) versus the lack of plasticity of the species living in ephemeral ponds (*H. squirella*): the result of adaptive evolution or of chance? (From data in Blouin 1992.)

adaptation as a process and as simply current usefulness, that is, between adaptation in the proper sense and exaptation (Gould and Vrba 1982). However, an intriguing cautionary tale comes from the work of Reekie and Hiclenton (1994) on the response of flowering phenology to elevated levels of CO_2 (a novel environment for which one cannot imagine plants being adapted in the historical sense of the result of past natural selection). They compared four short-day and four long-day species of plants. Long-day species flower when exposed to photoperiods longer than 12 h, while short-day plants flower when the photoperiod is shorter than 12 h. Reekie and colleagues exposed their plants to the proper inductive photoperiod, while changing the amount of CO_2 they were receiving from 350 (ambient level) to 1000 μL/L. The results were striking. On the one hand, all plants (regardless of the phenology) increased their size when exposed to elevated CO_2; however, the long-day species consistently flowered earlier under high CO_2 than low, whereas the short-day species consistently flowered later under high CO_2 than low. Given the fact that the species within the groups were clearly not phylogenetically related, such parallelism may at first be taken as the result of selection. However, these plants very likely evolved their habit mostly in response to photoperiod, not to changes in CO_2 levels (which in this case represented a novel environment). That being the case, this would represent an interesting instance of parallel (and apparently adaptive in the sense of being useful in the new conditions) reaction norms that actually result from a correlated response to selection under a different environmental stimulus.

Having examined, albeit in a cursory way, variation for plasticity within populations, among populations, and among species, we now turn to a consideration of the constraints that may limit the evolution of plasticity at any of these levels.

Genetic Constraints on and around Phenotypic Plasticity

Genetic constraints can affect virtually any trait that is genetically controlled to some extent (Antonovics 1976; Cheverud 1984), and phenotypic plasticity should obviously not be an exception. I discussed the current theoretical underpinning (as well as the limitations) of the study of constraints on reaction norms in Chapter 1. The idea is that the genetic variance-covariance matrix summarizing the expression of the same trait in multiple environments is a direct measure of the genetic interdependence of different portions of the reaction norm (Falconer 1952; Via 1987).[2] Here I discuss some empirical examples of constraints on reaction norms, while considering two distinct categories: genetic constraints acting on the shape of the reaction norm of a single trait and constraints acting simultaneously on several reaction norms for distinct traits (phenotypic integration [Schlichting 1989a, see Chapter 1). Also of interest are those documented cases in which genetic constraints on character means interact with plasticity to limit the future evolutionary trajectory of a population or species.

Constraints Concerning Simple (Uni- or Bivariate) Reaction Norms

As seen in Chapter 2, a good example of the way to study constraints on reaction norms via genetic correlations across environments is the work of Andersson and Shaw (1994) (see also Roff and Bradford 2000). They showed a range of family-mean correlations between light regimes in *Crepis tectorum* varying from a not significant 0.06 to a highly significant 0.88. Most of the correlations examined in that study were significant, indicating strong constraints on the ability of selection to modify reaction norms. Transplant experiments can also be used to highlight limits to phenotypic plasticity rooted in the genetic differentiation of local populations. Van Tienderen's (1990) study of the relative fitness of hayfield and pasture populations of *Plantago lanceolata* when grown in their own environment and when transplanted into that of the other is discussed in Chapter 6.

A mechanistically oriented approach to the study of constraints on plasticity is exemplified by the work of Dahlhoff and Somero (1993) on the effect of temperature on respiration in five species of abalone belonging to the genus *Haliotis*. These authors found a clear relationship between the reaction norms of respiration rate to temperature and the temperatures of the habitats from which the animals were collected. Specifically, the peak of each species' reaction norm corresponded to the habitat temperature at capture. They then investigated the effect of acclimation under laboratory conditions. They observed a significant change of the temperature inducing thermal inactivation in an adaptive direction following acclimation, where the higher the acclimation temperature, the more the reaction norms shifted their peaks to higher values (Fig. 4.8). However, this was true only for the three species that occur naturally in a wide range of temperatures, whereas the one found only at specific temperatures did not exhibit such a flexible response. Furthermore, even the species that were able to respond to acclimation only did so within a range of temperatures that corresponded to those they commonly encountered in nature. Overall, it seems that the reaction norms of distinct species can be more or less constrained depending on the corresponding species' ecology and, therefore, on past natural selection. This implies a different potential response to selection on plasticity by species with distinct ecological amplitudes.

A fundamental type of constraint is simple lack of genetic variation for plasticity. This is less trivial than it may seem at first glance. Pigliucci and Byrd (1998) studied the plasticity to nutrient availability of several traits in the annual weed *Arabidopsis thaliana*. We compared the response to six specific nutrient stresses (e.g., reduced nitrogen, phosphorus, potassium, and so on) of two inbred lines of *A. thaliana*, Columbia and Landsberg. We

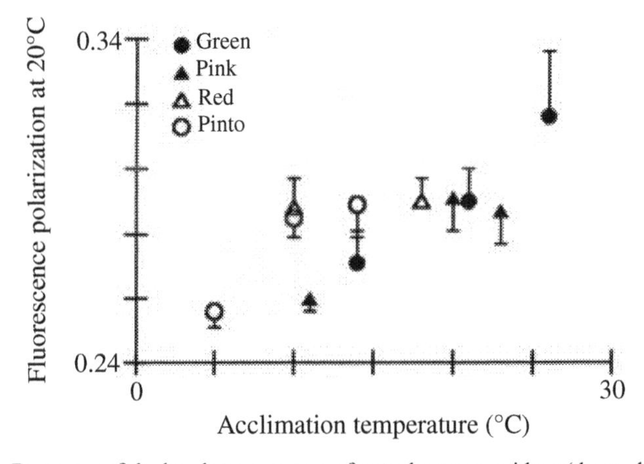

FIG. 4.8. Response of the break temperature of cytochrome c oxidase (the peak of the enzyme's reaction norm) to acclimation temperature in four species of abalone characterized by different ecological amplitudes to temperature. The Pinto abalone is the species with the most restricted natural range. (From Dahlhoff and Somero 1993.)

found little significant genetic variation for plasticity in any of the eight traits in response to any of the six nutrient deficiencies (the two lines were plastic in several cases, but in an identical fashion). This is not surprising if we consider the fact that Columbia and Landsberg were derived from the same original collection from a single population of *A. thaliana* and reflect on the fact that, owing to its high selfing rate, this species does not harbor much within-population genetic variation. However, things are not that simple. We also found extremely significant genetic differences between lines in the *mean* character expression of almost all traits. Since these plants had been raised under stable laboratory conditions for a long time, this result was interpreted as indicating that constraints on genetic variation for plasticity are stronger than constraints on the genetic variation for character means. This may be because many more genes underlie the mean expression than the plasticity of quantitative characters (Jasienski et al. 1997).

Genetic differences in the size of an organism can constrain its fitness reaction norm, thereby offering a link between genetic constraints, plasticity, and fitness (and therefore response to selection). A rather spectacular example of this is offered by a study conducted on the effects of two herbivores on selection for resistance in *Brassica rapa* (Pilson 1996). The author exposed plants to attack by moths and flea beetles, and then assessed relative fitness to estimate the type of selection elicited by exposure to different combinations of the two environments (i.e., density of each type of herbivore). Figure 4.9 shows the fitness landscapes for small (left), average (center), and large plants (right). It is clear that the size of the organism dramatically altered the fitness reaction norms of these plants. For example, a high occurrence of both herbivores is very deleterious for small plants, not too damaging for middle-size individuals, and somehow beneficial for the largest plants examined! On the other hand, higher numbers of moths, especially at low flea beetle damage, were more deleterious for plants of the largest-size class, as can be seen by the augmented curvature and lower level of the reaction norm in the high-moth-number area of the right-hand graph. A different scenario held for augmented damage owing to flea beetles, which was more deleterious for smaller than for larger plants, again as evidenced by the lowering and increased curvature of the reaction norm in the high-damage portion of the relevant axis. The interesting conclusion is that the response to selection exerted by the two herbivores depends very much (qualitatively, not just quantitatively) on the genes (as well as on other environmental factors) influencing plant size.

Genetic constraints can also interplay with plasticity to affect population dynamics and gene flow. Young and Schmitt (1995) investigated several interrelated aspects of variation for pollen release and capture in *Plantago lanceolata.* This plant is wind-pollinated, and wind-tunnel simulations

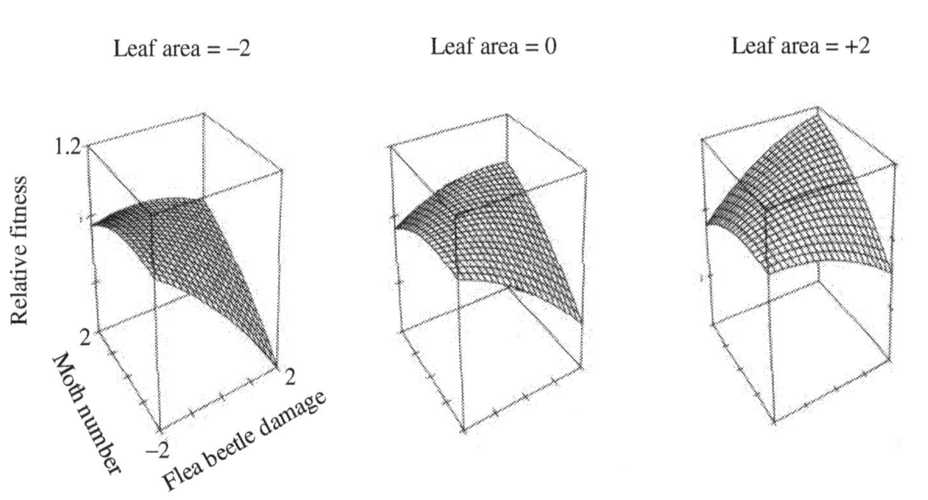

FIG. 4.9. Effect of plant size on fitness reaction norms (and therefore response to selection) to two herbivores in *Brassica rapa.* The diagram on the left shows the fitness landscape of plants of size around two standard deviations below average; the center plot shows plants of average size; and the plot on the right displays the same graph for plants two standard deviations above average. Note that the effect of joint high levels of herbivory is very deleterious for small plants, mildly so for medium plants, and actually favorable for large plants. Insect damage and density are expressed as above average, average, and below average. (From Pilson 1996.)

indicated that there should be selection for increase in the height of the male flowers (which are higher in the inflorescence) because this favors pollen dispersal. Simultaneously, one would expect selection for decrease in the height of the female flowers (usually found lower on the inflorescence) because this maximizes pollen reception. These authors found significant genetic variation for the height of the inflorescence in natural populations of *P. lanceolata,* but there was no heritability for the *distance* between male and female flowers. This would lead to *either* more effective dispersal *or* more efficient reception of pollen, but not both. Plasticity comes into play in this scenario when we account for the fact that *P. lanceolata* genotypes elongate their inflorescences much more under canopy shade than when exposed to full sunlight. Thus, a combination of genetic constraints and plasticity might yield differential gene flow from shaded (better at dispersal) to open (better at reception) natural populations.

It should be clear from this brief discussion that the relationship between constraints and plasticity can be extremely varied and complex. We can study direct genetic constraints on reaction norms and indirect constraints exercised through the sequence and timing of developmental events (Chapter 6), as well as the way in which plasticity interacts with other types of constraints. When considering constraints on the shape of simple reaction norms (i.e., on the expression of the same trait in different environments), three points should be kept in mind. First, if we describe these constraints as interenvironment genetic correlations, we are actually talking about a population-level analysis. Owing to the limitations of quantitative genetics in dealing with mechanistic details of the genotype-phenotype mapping function (Chapter 1), what looks like a constraint at the population level might not translate into much of a limitation at the individual level. For example, new mutations can conceivably rapidly and dramatically alter the shape of given genotypes' reaction norms, while having comparatively little effect on the population interenvironment genetic correlation (because they would appear as *outliers* in the statistical analysis). However, if the mutant reaction norm is favored, it could sweep through the population and dramatically alter the situation as compared to before the selective event.

Second, we *expect* interenvironment correlations to be high. Even though we are using the clever device of assuming that we are dealing with distinct characters connected by a genetic correlation (Chapter 1), we are in fact measuring the expression of a *single* character across many environments. Why should such manifestations of the same biological structure be unrelated? What this implies is a 180° change in perspective. Instead of searching for constraints and being concerned with the degree to which the evolution of reaction norm shape is limited, we should focus on those cases in which we find low genetic correlations across environments, which imply a surprising evolutionary lability of the reaction norm. In other words, given the biology of the system, our null hypothesis should be that the same genes underlie the expression of a character in all environments. When this is *not* the case, then we are presented with interesting questions about which genes change their action and why.

A third point is that it may be difficult to disentangle the effects of genetic constraints on reaction norms from those of stabilizing selection. For example, we may observe the same pattern of plasticity to nutrient availability of characters such as root/shoot allocation: All plants will allocate more to their root systems under poor nutrient conditions, with little variation in the amount of such plasticity. Is this the result of a constraint indicating the absence of genes capable of allocating more resources to shoots under nutrient stress, or is it simply that any such variant would be immediately eliminated by natural selection and therefore not even be observable in studies conducted on samples of natural populations? A partial answer to this question can, of course, come from combined studies of selection on phenotypic plasticity and on the variation in reaction norms inducible by mutation (Pigliucci et al. 1998). Both, however, and especially the latter, are still very rare in the published literature.

Constraints Concerning Groups of Reaction Norms

Schlichting (1986) (see Chapter 1) was the first to point out that another level of constraint on reaction norm evolution may be related to the fact that the plasticities of distinct characters to the same environmental conditions are sometimes correlated, a phenomenon that he called *plasticity integration.* He interpreted his findings about among-species variation in integration to mean that the degree of evolutionary flexibility of the reaction norms of one species is lower than that of another species. Alternatively (and not mutually exclusively), he also hypothesized that selection leads to a higher integration of the phenotype in one species than in the other. Unfortunately, it is difficult to speculate on selective pressures on phenotypic integration, since to my knowledge nobody has attempted a selection analysis at this level of complexity.

Waitt and Levin (1993) repeated previous experiments by Schlichting (1989b) in *Phlox drummondii* to investigate the genetics of plasticity integration. *P. drummondii* has an annual habit and is difficult to clone, so Schlichting's original work was based on phenotypic correlations. Waitt and Levin found a reliable way to clone these plants and investigated the variation between populations of *P. drummondii* in phenotypic integration. Furthermore, they distinguished among intercept, linear, and nonlinear components of the reaction norms, and found that the degree and pattern of integration were different for these three aspects of plasticity. As in Schlichting's original work, they found both positive and negative correlations between plasticities, as well as reversals of sign of the correlation between the same reaction norms when compared across different populations.

Genetic correlations between plasticities were also calculated by Newman (1994) for several morphological and life history traits of larvae of spadefoot toads exposed to different temperatures and food regimes. He found significant correlations between plasticities, mostly but not exclusively positive. Interestingly, the negative correlations were limited to the relationships between the plasticity of size at metamorphosis and the plasticity of all the other traits, regardless of which environmental variable (temperature or food) was considered. Newman interpreted the observed patterns of correlations as suggestive of trade-offs between size and age at metamorphosis, which may be responsible for maintaining genetic variation for plasticity within populations.

Witte et al. (1990) discussed the phenotypic plasticity to diet of anatomical structures in several species of cichlid fishes. This phenomenon was first highlighted by Meyer (1987), who demonstrated that the entire structure of the skull and in particular the jaws of these fishes can be dramatically altered by exposure to alternative diets. Witte and co-workers also explored the implication of such bewildering developmental phenotypic plasticity for the very definition of species in this taxonomically extremely diverse group. However, these authors pointed out that there are limits to the environmentally induced flexibility of the skull, and that these limits are to be found in the fact that distinct anatomical features must be plastic in a coordinated fashion in order to produce a viable phenotype. In other words, the plasticity of a structure can be constrained by the lack of plasticity of the surrounding structures. Witte et al. also noted the occurrence of the reverse phenomenon: The fact that a particular anatomical feature is plastic may force correlated changes in the surrounding features, which are not plastic in response to the environmental stimulus per se but *adjust* to the plastic structure. One can think of such cases as instances of plasticity to the *internal* environment, itself triggered by plasticity to the external environment (for another remarkable example, see West-Eberhard 1989 and Chapter 9).

There is very little we can generalize from the studies of plasticity integration conducted so far, primarily because there are few published studies on this topic (and most of them concern plants). Equally important, as pointed out above, is that it is difficult to gauge the impact of different patterns of plasticity integration on fitness. This, coupled with the virtual absence of studies that account for phylogenetic effects, limits us to the simple conclusion that there is indeed genetic variation for plasticity integration in natural populations. As to what this variation means, how it comes into being, and what consequences it may have for the evolutionary future of a given taxon, we are awaiting further and more detailed research.

I now turn to consider several putative relationships between phenotypic plasticity and other genetic or epigenetic phenomena: heterozygosity, developmental noise, canalization, and homeostasis. While these represent a relatively heterogeneous group, they are connected at a conceptual level and have become integral components of discussions concerning the genetic basis of plasticity.

Plasticity versus Heterozygosity?

Are plasticity and heterozygosity related, and if so, in what fashion? This question has been asked, empirically tested, and theoretically modeled on several occasions, but no clear consensus has emerged. It is reasonable to start by asking why we should expect such a relationship in the first place. Most authors who have been interested in the subject have postulated a negative relationship between phenotypic plasticity and heterozygosity, which follows from the supposition that higher heterozygosity leads to greater environmental *buffering* (Gillespie and Turelli 1989). This idea originated with Lerner's (1954) work on homeostasis, and it applies to plasticity only to the extent that we see plasticity as an unwanted disruption of the phenotype owing to negative environmental influences on an unstable genetic machinery. Of course, *some* kinds of plasticity probably do result from the inability of a genotype to produce an adaptive phenotype across a wide range of environmental conditions. Furthermore, if we are talking about fitness reaction norms, then any plasticity in fitness is, by definition, deleterious, since it means that some environments cause a significant decrease in fitness, a situation certainly not favored by natural selection. Nevertheless, it is difficult to see why such reasoning would apply to plasticity in general.

One of the first empirical approaches to the problem was Pederson's (1968) study on the relationships among stress, heterozygote advantage, and genotype-environment interaction in *Arabidopsis thaliana.* He used ten populations in order to obtain F_2 hybrids and double-cross hybrids, and then compared the three generations (parentals and the two types of hybrids) for their performance under stressful and nonstressful levels of temperature, light intensity, water availability, and nutrient concentration. He found that heterozygotes had a significant advantage under stress in the temperature, light, and moisture gradients, while they were invariant under the nutrient gradient. He then calculated genetic variation for plasticity using a standard analysis of variance, and compared its magnitude to the level of heterozygosity in the sample, where the parentals represented the lowest level of heterozygosity and the hybrids the highest level. The latter comparison was carried out by performing separate F-tests for the interaction term in each generation (P, selfed F_2, and double crosses—F_1 by F_1) and for each environmental factor (Fig. 4.10). The results pointed to an inverse relationship between heterozygosity and the amount of genetic variation for plasticity. It should be noted, however, that this is different from a relationship between heterozygosity and plasticity, in that it implies that if a population is more variable at random loci across the genome, it is also less variable for the plasticity but not necessarily less plastic.

Schlichting and Levin (1984) tested the idea that similarity in phenotypic plasticity among species depends on three factors: the degree of relatedness among species, the similarity in the species' ecology, and the level of heterozygosity. They derived two measures of plasticity for eighteen characters for several species of *Phlox* grown in six distinct treatments, and then tested the three hypotheses by means of Friedman's rank sums tests. They found evidence that a combination of the relatedness and ecology hypotheses is indeed capable of predicting the differences in plasticity among species. However, they did not find evidence of a relationship between plasticity and heterozygosity. The same authors went on to compare the amount and patterns of phenotypic plasticity between cultivated *P. drummondii* with distinct levels of inbreeding (and, hence, of heterozygosity [Schlichting and Levin 1986a]). They considered an outcrossed group and three successive generations of inbreds exposed to a set of six environments (five of

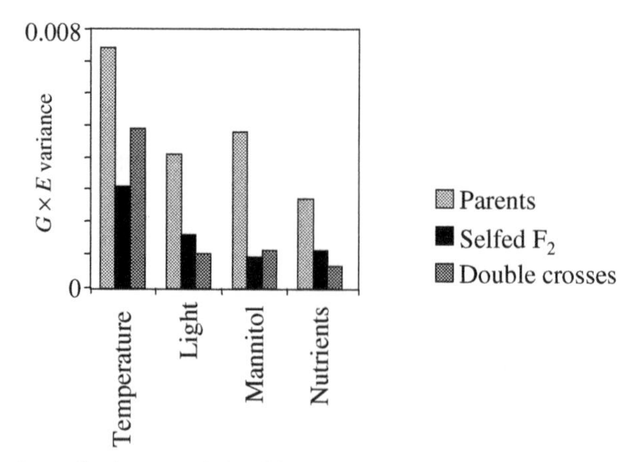

FIG. 4.10. A peculiar inverse relationship between heterozygosity and genetic variation for plasticity in *Arabidopsis thaliana.* Each set of bars represents the amount of genotype-environment variance to a particular environmental gradient across three generations: The parents have the lowest amount of heterozygosity, the selfed F_2 have an intermediate level, and the double F_1 cross have the highest level. The amount of G by E follows almost exactly the opposite trend, with the partial exception of the temperature data. (Data from Pederson 1968.)

which were stressful). The authors measured twelve characters and then focused on those for which they found a significant inbreeding-treatment interaction. They found substantial evidence of inbreeding depression on trait means, but neither the amounts nor the patterns of plasticity were significantly altered in any character. Therefore, neither inbreeding depression nor correlated increase in homozygosity displayed any relationship with phenotypic plasticity.

Yampolsky and Scheiner (1994) found a more ambiguous result in an experiment on *Daphnia magna.* They investigated the relationships among developmental noise, phenotypic plasticity, and allozyme heterozygosity in clones from three natural populations. These authors used only five or three (depending on the experiment) enzyme loci to estimate genome-wide heterozygosity. Nevertheless, they were able to uncover some evidence for a negative relationship between plasticity and heterozygosity, but only for some of the traits studied (especially age at maturity, spine length, and number of offspring). However, one population did not show any correlation between plasticity and heterozygosity for any trait. This ambiguity is reflected in two other studies by Scheiner and colleagues. In a paper on response of plasticity to selection (Scheiner and Lyman 1991) no evidence was found for a correlation between the amount of genetic variation and plasticity in *Drosophila melanogaster.* In a second study addressing the chromosomal location of genes affecting plasticity in *D. melanogaster* it was found instead that plasticity was greatest in the heterozygous stock and lowest in the homozygous isogenic lines (Weber and Scheiner 1991).

A brief survey of the literature by Schlichting and Levin (1984) is also inconclusive. For example, heterozygotes of *Avena fatua* are less plastic than genotypes that are electrophoretically monomorphic, and the same negative relationship between plasticity and heterozygosity has been reported for *Bromus rubens, B. mollis, Limnanthes alba,* and *L. floccosa.* However, the same authors report other cases (in the genera *Collomia* and *Stephanomeria*) for which no relationship was actually found. Furthermore, a reanalysis I conducted of data by Jain (1978) on eight populations of *B. mollis* shows almost no correlation between heterozygosity based on allozyme variation and plasticity of three characters; the only exception yields a *positive* correlation between the plasticity of flowering time and the level of heterozygosity.

At the end of this chapter I discuss what theoretical implications a relationship between plasticity and heterozygosity would have, especially if such a relationship should turn out to be negative, but the jury is still out. Even though most published studies either present evidence for a negative relationship or conclude for the absence of one, positive correlations are occasionally found. While the latter might be expected by chance alone, such findings mean that the published literature spans the whole gamut of possibilities, which should not be surprising given the extreme generality of the question. Heterozygosity as measured in these studies is a property of the whole genome. Plasticity, however, is not, since a given genome can display remarkably distinct patterns of plasticity for the same trait exposed to different environments or for different traits measured within the same set of treatments (while total heterozygosity, of course, remains invariant). Is it therefore surprising that sometimes we find a significant relationship and sometimes we do not? It would be more interesting to ask the question in a more biologically informed fashion. For example, can we uncover any pattern that relates heterozygosity to plasticity for certain environmental regimes but not others, or certain types of traits but not others? Also, is there a relationship between the degree of heterozygosity and the kind of genetic basis of the plasticity being studied? I suspect that we are likely to find such a correlation if we are studying nonspecific plasticities, that is, cases of allelic sensitivity in which the response may be affected by many genes in a nonspecific manner. On the other hand, I would not expect any such relationship to emerge when plasticity is of the regulatory type, since the underlying genetic control may be very specific. Finally, while it is likely that heterozygosity at the genome level might not matter for plasticities controlled by a very reduced number of genes, heterozygosity at the specific loci affecting that plasticity (*plasticity genes*) might make a huge difference.

Plasticity versus Developmental Noise?

Another recurring link, hypothetical or real, in the literature on the genetics of plasticity concerns the possibility that phenotypic plasticity has something to do with developmental noise. The problem with this claim is that it is hard to know exactly what constitutes developmental noise. At least three different phenomena can be lumped under this heading: pure random variation perhaps arising from fluctuations of subcellular components and/or reactions in accordance with the principles of thermodynamics; variation owing to internal environmental differences of a deterministic nature (i.e., biases of the developmental system that lead to pseudorandom fluctuations in the observable phenotype); or microenvironmental fluctuations in the external environment.

While the above distinctions should be kept in mind, in the following I consider noise as being due to either of the first two categories of phenomena and focus on the way it has been measured experimentally. Noise is usually quantified as some kind of asymmetry in structures that should be symmetrical (e.g., left and right wings or bristles, petals in actinomorphic flowers, and so on). The expectation of a relationship between asymmetry and plasticity may have originated from the misconception that plasticity is a "bad" property of a genotype, a somewhat inevitable disturbance of the developmental trajectory that yields different phenotypes depending on the environmental conditions instead of the "optimum" for the species. Similarly, asymmetry of organs caused by developmental noise is very likely to be detrimental, and asymmetry has been the focus of many more or less successful attempts to link it to stress and general degradation of the habitat (Freeman et al. 1993). It must have been natural, then, to look for possible biological links between the two phenomena. However, once again the empirical evidence is rather ambiguous.

Tarasjev (1995) investigated the relationship between plasticity and developmental instability in *Iris pumila* living in open versus shaded habitats. In an elegant experimental design, he measured variation within the same flower, between ramets (clones) of the same genet, between replicas of the same genotype grown within the same macroenvironment, and between replicas of the same genotype grown in spatially and temporally variable macroenvironments. The first type of variability is clearly due to developmental noise; the second and third represent a combination of developmental noise and microenvironmental variation, while the fourth and fifth represent instances of phenotypic plasticity (to light availability). The results that concern us here are that only the asymmetry of one character, anther length, was significantly correlated with just one type of plasticity (spatial), and that the correlation coefficient was not particularly large

($r = 0.38$). Since this was only one out of twenty comparisons, we can fairly confidently ascribe the result to chance.

Scheiner and co-workers published two papers on this subject, one on *Daphnia* and one on *Drosophila.* The paper on *Daphnia* is the same one I discussed in the context of the relationship between plasticity and heterozygosity (Yampolsky and Scheiner 1994). The authors tested the null hypothesis of no relationship between plasticity and developmental noise by calculating the residual variation for each clone averaged across treatments and correlating it with the among-treatment variation. The results concerning developmental noise were as ambiguous as those related to heterozygosity. Plasticity was correlated with noise in only two traits (spine length and age at maturity) in two different experiments, although the correlations were indeed highly significant. In the *D. melanogaster* experiment, Scheiner and his collaborators compared phenotypic plasticity, genetic correlations, and environmental noise for thorax length, wing length, and number of sternopleural bristles (Scheiner et al. 1991). They exposed fourteen lines of flies to 19° and 25°C temperatures. Their table reporting the (full-sib) family and additive correlations indicates four significant coefficients relating noise to plasticity, two of which are negative and two positive, while all the other correlations were not significant. A closer look reveals that the two negative correlations concern developmental asymmetry of bristle number at 19°C and the plasticities of *left* and *right* bristle numbers. Furthermore, the two positive correlations are between the *same* two plasticities and the developmental noise affecting the same trait, bristle number, this time measured at 25°C. Since the plasticities of left and right bristle numbers are significantly correlated, as are the fluctuating asymmetries of the same trait at the two temperatures, the four correlations between plasticity and asymmetry can hardly be considered independent evidence. The authors are therefore perfectly justified in concluding that there is no necessary relationship between plasticity and developmental noise.

A clear-cut case of no relationship whatsoever between developmental noise and plasticity comes from a study by Bagchi and Iyama (1983) in *Arabidopsis thaliana.* They induced mutations in the standard Landsberg laboratory strain of *A. thaliana* by exposure to two levels of gamma radiation. The irradiated seeds were allowed to germinate, and the resulting plants propagated by single-seed descent for several generations. Plants belonging to the M_3 and M_6 generation were randomly sampled, and their progeny (M_4 and M_7) were grown in order to investigate induced genetic variation on quantitative characters. These authors found significant genetic differences in the degree of developmental instability in these plants and selected high- and low-instability lines for plasticity experiments. They ex-

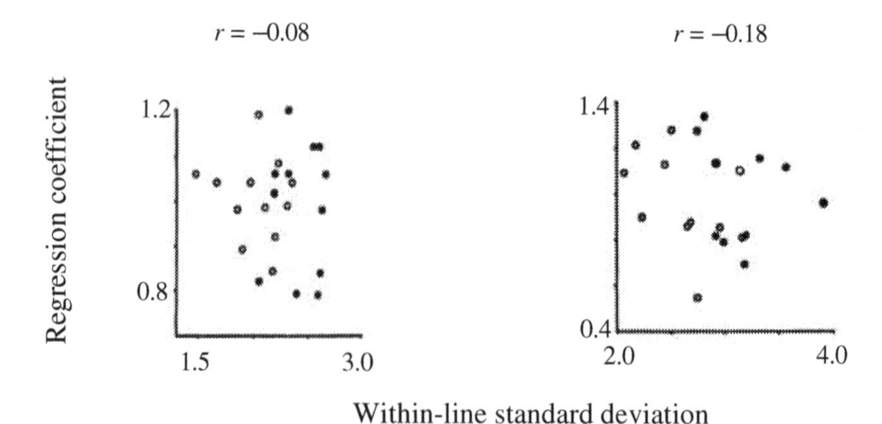

FIG. 4.11. Lack of correlation between developmental noise (on the x-axis) and phenotypic plasticity (on the y-axis: each environment was assigned a value based on the mean performance of all lines in that environment; plasticity of a line was then measured by the regression of that line on the environmental values) in *Arabidopsis thaliana* for two traits: plant height (left) and flowering time (right). Open circles represent lines with high developmental instability and solid circles lines with low developmental instability. (From Bagchi and Iyama 1983.)

posed the lines to different temperatures and calculated the correlation between instability and plasticity. It is clear from Fig. 4.11 that they found no relationship at all.

Overall, I think it is fair to say that there is little evidence for a correlation between developmental noise and plasticity. On biological grounds, we certainly *do not* expect such a relationship. Developmental noise is presumably a manifestation of reduced homeostasis of the developmental program, and as such it represents a phenotype induced by microenvironmental variation. Plasticity, on the other hand, is defined as a response to macroenvironmental variation.[3] While plasticity is of course not necessarily adaptive, we can exclude the possibility of a correlation between adaptive plasticity and noise, in that this would entail a relationship between adaptive and maladaptive environmentally induced phenotypic changes, which would also directly link micro- and macroenvironmental variation. It is conceivable that nonadaptive plasticity, and some kinds of environmental sensitivity in particular, may result from the same source of developmental noise, namely a reduction in the capacity of the phenotype to buffer unwarranted environmental influences. This could explain the few positive results obtained by researchers looking for a link between the two phenomena.

Plasticity versus Canalization versus Homeostasis?

We come now to the final debatable link between plasticity and a host of genetic-developmental phenomena. I am discussing both canalization and homeostasis here, not because they are synonyms but because in my view (and in Waddington's original definition, discussed in Hall 1992b) the latter is the result of the former. As pointed out by Schlichting and Pigliucci (1998), canalization refers to a developmental phenomenon, while homeostasis is the measurable outcome of that phenomenon. When Waddington (1942) originally defined the term *canalization,* he meant the fact that developmental systems tend to limit variation in the final phenotype. He described the fact that it is indeed possible to alter the course of a developmental trajectory, either by an environmental stimulus or by genetic mutation, but that the result is usually an alternative, discrete phenotype and not a wide range of options (Waddington 1952). Some authors later attempted to distinguish between *environmental* and *genetic* canalization (Stearns and Kawecki 1994), but Waddington (1942) used this difference to highlight the fundamental *similarity* between the two: "The constancy of the wild type must be taken as evidence of the buffering of the genotype against minor variations not only in the environment in which the animals developed but also in its genetic make-up. That is to say, the genotype can, as it were, absorb a certain amount of its own variation without exhibiting any alteration in development" (p. 564). He demonstrated that one can trigger a major change in a developmental trajectory by subjecting the organism to particular environmental stimuli (such as a heat shock), but also that *the same* phenotype can be obtained by selection on genetic variants (Waddington 1953). This led him to the concept of genetic assimilation (Waddington 1961), which is based on ideas very similar to those expressed by Goldschmidt (1940) and Schmalhausen (1949), which gave rise to the modern term *phenocopy.*

The potential for confusion about this whole realm of developmental genetics increased even further with the introduction of the expression *homeostasis* (Lerner 1954), also plagued by a variety of environmentally or genetically slanted meanings. Simply put, homeostasis measures the degree of variation of a particular phenotype when it is perturbed either by the environment or by a mutation. It should be clear that this is precisely the outcome of canalization (the process). A more canalized genotype simply yields a more homeostatic phenotype, while a less canalized genotype necessarily yields a less homeostatic phenotype. Therefore, canalization and homeostasis are not synonymous, but neither are they entirely distinct phenomena: One is the outcome of the other.

But the complications do not end here. What do we mean by *environmental* effects on canalization? Waddington (1942) was very clear about the fact that these are not the same kind of environmental effects that one associates with phenotypic plasticity. In fact, he referred to reaction norms as canalized or not canalized, thereby implying that the process of canalization can be seen at work across a series of macroenvironments. In fact, here is what the master said, *verbatim:* "Developmental reactions, as they occur in organisms submitted to natural selection, are in general canalized. That is to say, they are adjusted so as to bring about one definite end-result regardless of minor variations in conditions during the course of the reaction" (p. 563). It should be noted that Waddington was referring to *minor* variations, not to the marked differences between environments that we associate with plasticity. Furthermore, he explicitly used one of the most well-documented cases of plasticity, the environmentally triggered metamorphosis in axolotls, as *an example of canalization.* This is because the axolotl can only produce two distinct phenotypes, metamorphosed or not. In the case of other reaction norms, such as continuous variation along temperature gradients, the distinction between plasticity and homeostasis (and therefore canalization) might become fuzzier. So, for example, van Tienderen (1990) discussed the differences in the plasticity of hayfield and pasture populations of *Plantago lanceolata* in terms of canalization imposing limits on plasticity (Chapter 6). However, the environments he was considering were clearly macroenvironments (variation in density of neighbors and light availability), not the minor variations discussed by Waddington in this context. The limits to plasticity observed in *P. lanceolata* are more likely a matter of costs and trade-offs with genetic specialization (as also pointed out by van Tienderen, see Chapter 7) as well as the result of local selective forces rather than the outcome of canalization.

One of the few cases in which there seems to be a statistical (inverse) correlation between plasticity and canalization is presented in a study of *Drosophila melanogaster* conducted by Garcia-Vazquez and Rubio (1988). They exposed the flies to different temperatures during their development, comparing the plasticity of scutellar extra bristles among distinct lines previously selected for higher or lower degrees of canalization of that trait. In general, these authors did find that the most canalized lines were also the least plastic. This kind of plasticity, however, probably falls into the category of nonadaptive allelic sensitivity, and it is therefore more likely to reveal a link with canalization and homeostasis. This is for the same reasons that I discussed above regarding a possible link between developmental noise and some types of plasticity by allelic sensitivity, namely that the latter may be maladaptive and originate because of reduced buffering of environmental influences, that is, from reduced canalization of development.

Incidentally, it should be noted that developmental noise *is* the opposite of homeostasis, in that it is the result of mutations with small effects or of the reaction to small fluctuations in the internal or external microenvironment. Hence, developmental noise is one result of lack of canalization.

Another important point to consider is that if a reaction norm is canalized, as in the case of the axolotl, it means that *a single genotype* produces a discrete series of phenotypes. Waddington's term does not by any means apply to *a population* of genotypes and therefore to *a series* of reaction norms. Several authors (Lewontin 1974; Levin 1988; Stearns and Kawecki 1994) talk about an ensemble of reaction norms as more or less canalized. For example, Stearns and Kawecki (1994) illustrated different degrees of canalization of a series of reaction norms in *Drosophila melanogaster* (Fig. 4.12). The upper diagram is supposed to show canalization, in that all reaction norms form a tight bundle; the middle diagram depicts a lower degree of canalization; while the lower diagram displays canalization at one extreme of the environmental gradient but lack of canalization at the other extreme.

This approach implies a direct correspondence between the genotypic and the population levels of analysis as far as canalization and homeostasis are concerned. But canalization, by definition, is a property of a genotype and therefore of a reaction norm and not of a population (although, of course, it can be *averaged* across genotypes, which is not done in Stearns and Kawicki's diagrams). The fact that a series of reaction norms converges toward similar phenotypes in some environments may have nothing to do with developmental mechanisms, but may be the result of strong stabilizing selection (on whatever underlying mechanism) in a subset of the environments experienced by the population. Even when convergence of reaction norms is somehow the outcome of a developmental constraint, it represents an epiphenomenon of the constraint and cannot be traced directly back to a property of individual genotypes in any biologically meaningful way. Despite the fact that common usage eventually rules, I would suggest using terms such as *convergence* and *divergence* to indicate the situations depicted at the extremes of the lower panel in Fig. 4.12 and reserve canalization for the restricted meaning originally intended by Waddington.

Phenotypic Plasticity and the Problem of Genetic Variation

Evolutionary biologists have been troubled by the "problem" of genetic variation if not quite since Darwin, shortly thereafter. Dobzhansky set out to investigate the extent of genetic variation in natural populations and how

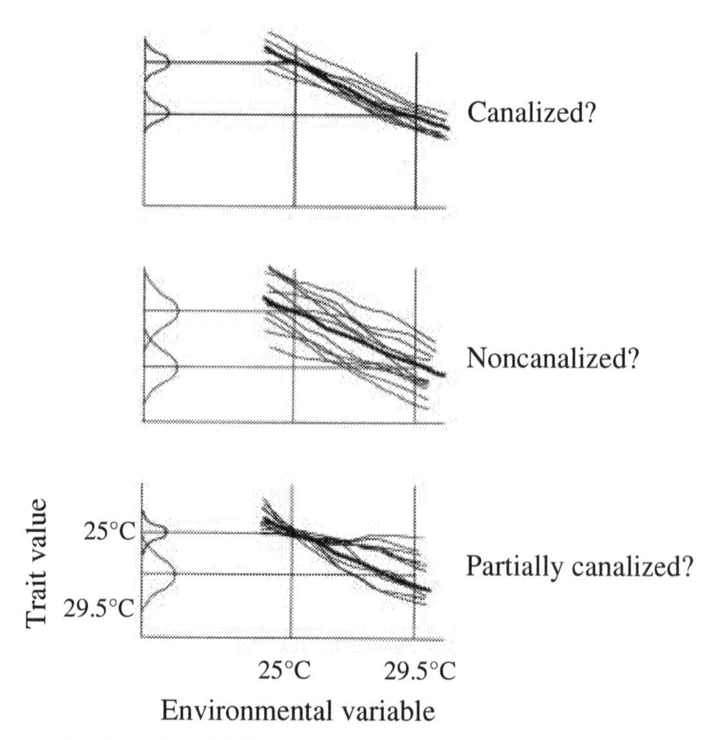

FIG. 4.12. An illustration of different degrees of canalization of reaction norms at the population level (measured by the tightness of the bundle), according to Stearns and Kawecki (1994).

it is maintained and ended up with a multidecade research program that produced the famous GNP series (genetics of natural populations, Provine 1981). He was the first to suggest that—contrary to what was predicted by the then-prevalent Fisherian (1930) paradigm—there is indeed a significant amount of genetic variation to be found in nature.[4] Lewontin, one of Dobzhansky's students, went one step further and by means of the newly invented technique of gel electrophoresis demonstrated that natural populations harbor plenty of variation at key biochemical loci affecting most fundamental pathways in the physiology of an organism (Lewontin and Hubby 1966). However, the question is whether this variation is the result of balancing selective forces or is in fact more or less selectively neutral and maintained by mutation accumulation in relatively large populations. The latter idea had been proposed by Kimura (1954, 1983) well before the experimental results from electrophoretic surveys were published, and the new data renewed the debate and sparked much interest in his theory. It is clear by now that even though many loci can behave as neutral or quasi-neutral under certain conditions, we still do not have a satisfactory explanation of how genetic variation is maintained in natural populations (Hey 1999). The debate is further muddled by the fact that there are different *kinds* of genetic variation, which may require distinct explanations. Are we talking about variation at the DNA sequence level, among variants of metabolic enzymes, among regulatory genes, or among genes underlying quantitative traits? The reasoning and the experimental approach (Lewontin 1984) may be sharply distinct in all of these cases. As phenotypic plasticity has appeared on the horizon of modern evolutionary biology research, it has also become a factor in the genetic variation debate. In the following, I offer a brief discussion of some of the different aspects of the plasticity-genetic variation link, as well as a few examples of how the question has been handled both theoretically and experimentally.

Theoretical Background

The opening salvo linking genetic variation and phenotypic plasticity on the theoretical side was published by Gillespie and Turelli (1989). These authors framed the problem in terms of a negative relationship between plasticity and heterozygosity, which—as we have seen above—has very little empirical support. The idea is that more heterozygous genotypes should be more capable of yielding the same phenotype regardless of environmental *disturbance*. That is to say that there is a relationship between heterozygosity and homeostasis, not plasticity. If we accept Gillespie and Turelli's model for a moment, then it is easy to see why plasticity would imply maintenance of genetic variation. Gillespie and Turelli are really talking about plasticity for fitness, which of course is going to be selected against. Since the model assumes that plastic genotypes are homozygous, it means that the nonplastic heterozygotes should be favored by selection. However, they considered a diploid organism living in a panmictic population, so homozygotes were present in the population as a by-product of the mating system (actually a form of genetic load). Therefore, the existence of plasticity (for fitness), associated with homozygosis, necessarily implies the persistence of heterozygous genotypes and therefore of genetic variation. Simply put, Gillespie and Turelli assumed that there is no single genotype that is most fit in all environments and that the heterozygotes are the fittest on average across environments. The strongest blow to Gillespie and Turelli's paper was Gimelfarb's (1990) criticism showing that the original results hold only under very restrictive symmetry assumptions and are therefore far less general than implied in the 1989 paper. Furthermore, Gimelfarb also demonstrated that, even under the restrictive assumptions, a polymorphism of the kind envisioned by Gillespie and Turelli is not capable of maintaining more than a modest, and in many cases negligible, degree of heritability.

A very different kind of assumption underlies the work of Via and Lande (1987), considered a major theoretical study investigating the plasticity-variability hang-up. These authors concluded that plasticity *cannot* maintain genetic variation because they assume that one particular genotype will be best under all environmental circumstances. However, it is very dangerous to link an unspecified type of plasticity to genetic variation according to a specific mechanistic model. First of all, there are many types of plasticity, at many levels within the whole organism. Second, these usually interact with each other to yield the reaction norm for fitness (through plasticity integration, see Chapter 1). Third, we normally do not know much about the genetic basis of these plasticities, let alone how they relate to genetic variation in the genome at large. The utility of these models, therefore, lies more in spelling out the falsifiable consequences of each particular set of assumptions than in actually solving the underlying problem (see Chapter 10).

Of course, some recent quantitative genetic models tackling the persistence of genetic variation still do not take account of plasticity at all. Lande and Shannon (1996), for example, demonstrated that the assumption of most conservation geneticists that genetic variation is unconditionally "good" for a species can be questioned. In a constant environment, genetic variance for quantitative characters simply creates a genetic load because of stabilizing selection against individuals deviating from the optimum phenotype. In variable environments, on the other hand, genetic variation does permit evolution in response to changing environments, as one would logically expect, but this is by no means always an adaptive response. They link what they call the *evolutionary load* (the deviation of the mean phenotype from the optimum) to the particular pattern of environmental heterogeneity. So, for example, if the environment changes randomly, additive genetic variance simply adds to the load, because a response to selection in any generation will tend to push the population temporarily further away from the optimum defined by the across-generation mean of environmental fluctuations. In other words, the existence of genetic variation under these circumstances allows the population to track the environmental change *too closely*. What happens if the environmental change is somewhat predictable? This can happen if the change is unidirectional (e.g., a warming climate trend), cyclic (seasonal variation), or autocorrelated (so that a change in one direction predicts further changes in the same direction on a short time scale). If the change is unidirectional, Lande and Shannon's model predicts that small amounts of genetic variation will actually reduce the genetic load to a great extent. On the other hand, in the case of cyclical or autocorrelated environmental alterations, the presence of additive genetic variance increases the population mean fitness only when the period of oscillation (or the autocorrelation time) is long; otherwise, we fall back into the problem of tracking the environment too closely. Conspicuously missing from these scenarios is phenotypic plasticity. As pointed out by various authors in different contexts (e.g., Bradshaw 1965; Levins 1968; Lloyd 1984; Gavrilets 1986; Lively 1986b; Sultan 1987; Moran 1992; Scheiner 1993b; Schlichting and Pigliucci 1995; Denver 1997b), adaptive plasticity can evolve precisely when fluctuations in the environment are predictable to some extent. For example, a special case of environmental autocorrelation is when a change in the environment can be used as a cue to predict a seasonal or autocorrelated fluctuation. Relevant cases include the ability to predict seasonal changes by responding to photoperiod in both plants and animals (Cremer et al. 1991; Shimizu and Masaki 1993; Beacham et al. 1994; Temte 1994; Mozley and Thomas 1995), or the reliable prediction of the appearance of predators based on chemical cues available *before* the predator's population density rises significantly (Parejko and Dodson 1991; Bronmark and Miner 1992; Neill 1992; De Meester 1993; Sih and Moore 1993; Sandoval 1994; Tollrian 1995; McCollum and van Buskirk 1996; Trussell 1996).

Quinn raised a second theoretical issue related to the topic of genetic variation (1987): the relationship between plasticity and ecotypic variation. According to Quinn, the very concept of ecotype is outmoded, and we should simply realize that plasticity can explain the existence of what we used to call ecotypes. An ecotype was originally defined (Turesson 1922) as a *genetic response* to a specific habitat. In other words, ecotypes are populations that are *genetically specialized* to particular environmental conditions, such as the case of alpine ecotypes of plants. When plasticity became a more popular subject of investigation, it was seen as the antithesis of ecotypic specialization (see the references in Quinn), thereby once again casting the debate in terms of *either* plastic responses *or* genetic specialization. Quinn, however, correctly points out that a phenotypically plastic genotype could yield what looks like an ecotype under extreme environmental conditions. The only way to find the difference is to grow replicates of the same genotype in a series of environments and see if it maintains the *ecotypic* characteristics regardless of the treatment. Quinn's argument that "we have no further need for comparative studies done only in the field or only under one controlled environment (the so-called 'common garden' experiments), or for studies utilizing only one population of a species or only two or three individuals each from a few populations" is, of course, perfectly valid. On the other hand, I would be uncomfortable drawing the conclusion, as Quinn does, that "it is experimentally impossible to delineate an ecotype as either an ecological or evolutionary unit unless every population of the species is studied." I think a more moderate position in which we view extreme genetic specialization, and therefore ecotypes, as an exceptional degree of reduction of (adaptive) plasticity is more conducive to realistic models of organismal variation. Since ecotypes can show nonadaptive plasticity (or even partially adaptive plasticity to new sets of environmental conditions), the ecotype is just one end of a multidimensional spectrum of possibilities ranging from a rigid, environment-independent expression of traits to a more (but not infinitely) pliable phenotype.

A third aspect of the plasticity-variability debate deals with the often-made observation (e.g., Vela-Cardenas and Frey 1972; Daday et al. 1973; Paulson et al. 1973; Scheiner and Goodnight 1984; Mazer and Schick 1991a,b; Ebert et al. 1993; Simons and Roff 1994; Young et al. 1994; Bennington and McGraw 1996; Merila 1997; Sgrò and Hoffmann 1998; Hoffmann and Merila 1999) that the amount of measurable genetic variation for quantitative traits is itself plastic. In other words, if we measure the heritability of a character in one environment, it is by no means going to be the same in any other environment in which we take the same measurement. Lewontin first discussed the theoretical underpinning of this phenomenon (Lewontin 1974, see Chapter 3), and I come back to its general consequences in the epilogue.

The Empirical Side of the Story

What do we actually *know* about the relationship between plasticity and genetic variation from an experimental viewpoint? As far as the question of a direct link between plasticity and genetic variation, it depends on the level at which the question is asked. The example of the rare species *Vicia pisiformis* is paradigmatic in this respect. Black et al. (1995) studied both genetic variation at the level of DNA and variation for quantitative traits and their plasticity to temperature. They found very little genetic variation for randomly amplified polymorphic DNA (RAPD) markers in this taxon, but a highly significant genetic variance for quantitative traits, as well as plasticity and genetic variation for plasticity. Therefore, the amount of plasticity in this case was not predictive of molecular variation, but it was accompanied by quantitative genetic variation for trait means. Black et al. also argued that conservation biologists should pay attention to the *kind* of genetic variation that is actually going to make a difference in the survival of rare or endangered species. They pointed out that variation for quantitative traits and their plasticities is certainly more germane to a population's dynamics and future survival than variation at neutral or quasi-neutral randomly chosen molecular markers. When the authors used a Mantel test (Manly 1985) to compare differences in plasticity among clones to differences in character means and genetic distances based on the molecular markers, they found no significant association in any case. In this instance, therefore, plasticity has no relationship with broad measures of genetic variation at either the quantitative or the molecular level.

Stewart and Nilsen (1995) conducted another direct empirical test of a possible relationship among molecular variation, genetic variability for quantitative traits, and plasticity, using central and marginal populations (with respect to the species' distribution) of *Vaccinium macrocarpon.* Clones were exposed to variation in nutrient availability and scored for RAPD markers. Once again, no correspondence between data sets was found.

An indirect relation between plasticity and genetic variation measured as heterozygosity can be mediated by the environmental responsiveness of the mating system of some organisms. For example, Holtsford and Ellstrand (1992) investigated the variation in floral traits and outcrossing rate in *Clarkia tembloriensis.* They found that high temperatures (hot summers) shorten protandry, the temporal separation of male and female functions on hermaphroditic flowers. Reduced protandry, in turn, implies a higher degree of self-pollination. To the extent that selfing reduces heterozygosity genome-wide, we have a biologically meaningful link between plasticity and genetic variation at the molecular level.

Parts of the Lande and Shannon (1996) model referred to above had already been tested by Mackay (1981) in an experiment on a large natural population of *Drosophila melanogaster* raised under four environmental regimes: constant environment, spatial heterogeneity (presence or absence of 15% ethanol), and long- and short-term temporal variation. She found that in general more additive genetic variation for quantitative traits was maintained in the variable than in the constant environments and that temporal variation was more effective than spatial heterogeneity in maintaining additive genetic variance. On the other hand, Mackay found no effect of the length of the environmental cycle, contrary to what was later predicted by Lande and Shannon. The role of plasticity in this experiment, however, is difficult to determine. While there was plasticity, the low genotype-environment interaction indicated that reaction norms were mostly parallel to each other. Mackay interpreted this to mean that there was selection on heterozygosity, not on specialized genotypes for the two distinct niches. However, from the context of the article and the description of the experiments, it seems that she equates heterozygosity with plasticity, which means that all variable environmental regimes actually maintained phenotypic plasticity while reducing the genetic variation *for plasticity.* Why plasticity *and* genetic variance within environments were both maintained (i.e., why one specific reaction norm was not the only one to emerge) is not clear.

There is not much in the literature comparing genetic variability and plasticity directly as alternative strategies, and therefore bearing on the question of what represents an ecotype, but some of the papers mentioned above can shed some light on the subject (see also my discussion of Bell's work in Chapter 9, as well as Goho and Bell 2000). For example, Hazel and colleagues (1987) considered genetic variation and plasticity as mutually exclusive pathways toward adaptive evolution of pupal color in the swallowtail butterfly (*Papilio polyxenes*). The results of field experiments showed that dark pupae are more cryptic and have a higher probability of escaping predation. The production of brown pupae can be determined genetically, but can also be obtained by raising the larvae on short photoperiods or by exposing them to dark sites. The results ultimately led to rejection of the hypothesis that genetic variation and plasticity are alternative strategies, confirming that both can occur in a population in response to environmental variation. The taxonomic-ecological work of Trainor and Egan (1991) on the alga *Scenedesmus* clearly shows that not only ecotypes but even supposedly distinct species are in fact the result of different forms of plasticity in response to the density of conspecifics or the presence of bacteria, in accordance with Quinn's suggestion discussed previously. To the extent that different populations of *Arabidopsis thaliana* can be considered ecotypes (usually in relation to their early or late flowering habit), there is now plenty of evidence that they are plastic, although most of the genetic variation for plasticity is between, not within, populations (Pigliucci and Schlichting 1996, 1998), probably because of the highly selfing habit of this species.

An interesting insight into the mechanistic differences between adaptively plastic and ecotypic genotypes has been gained from the work of Emery et al. (1994) on stem elongation in *Stellaria longipes.* The authors compared the response to ethylene (a hormone known to be involved in controlling stem elongation) in alpine ecotypes characterized by low plasticity to wind versus prairie ecotypes with enhanced plasticity. They found that exposure to wind limited growth in both ecotypes but that ethylene has an opposite effect on each of them: It dwarfs the alpine plants when they are exposed to wind (adaptive plasticity), whereas at low levels it stimulates elongation of the prairie plants, independently of wind conditions. It is fascinating to speculate how alpine ecotypes of this species might have originated from the more common prairie plants by altering their production or sensitivity to the same hormone. This is a remarkable example of an understanding of the mechanistic basis of the evolution of plasticity and ecotypic specialization, and it confirms that the distinction between ecotypes and plastic genotypes is not as clear-cut as one might think.

Along similar lines, different ecotypes can show different patterns of plasticity. For example, Neuffer and Meyer-Walf (1996) studied ecotypes of *Capsella bursa-pastoris* adapted to two man-made habitats (hence representing recent evolutionary outcomes): arable fields and trampling areas. The arable field population was very plastic for several traits in response to nutrient supply, which is naturally variable in their habitat. The trampling ecotype, on the other hand, showed no plasticity to the same range of nutrients, which would not normally occur in their habitat.

A link between ecotypic performance or plasticity and the habitat of provenance was also established for two unrelated species (*Festuca arundinacea* and *Dactylis glomerata*) when their temperature sensitivities were examined under controlled conditions (Thomas and Stoddart 1995). Generally speaking, the ecotypes sampled from the coldest habitats showed enhanced sensitivity to low-temperature inhibition.

Finally, what about the empirical evidence on environmentally induced changes in genetic parameters such as genetic variances-covariances and heritabilities? We have already discussed some examples in Chapter 1, but let me summarize a few more results here, starting with a study by Ebert et al. (1993) of *Daphnia magna* exposed to two food levels. Both broad- and narrow-sense heritabilities of adult length, clutch size, and offspring length varied throughout development and because of food level (Fig. 4.13). While

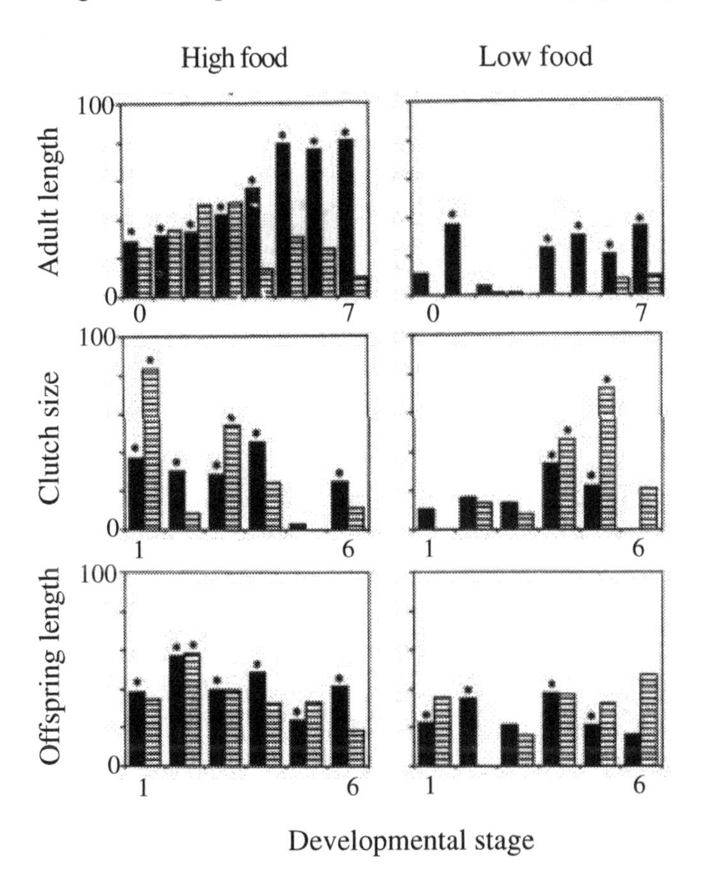

FIG. 4.13. Variation in broad-sense (black bars) and narrow-sense (hatched bars) heritabilities of three traits in *Daphnia magna* through several developmental stages and across two environments. (From Ebert et al. 1993.)

it is clear that the heritability of adult length is markedly lower, and mostly not significant, under low food, the other two traits show a much more complex pattern. For instance, the narrow-sense heritability of clutch size is much higher under low food than under high food at the fourth and fifth instars, but much lower than under high food at the first and third stages.

In general, the data by now overwhelmingly confirm Lewontin's prediction that the observable genetic variance is indeed affected by the environmental conditions. As another example, Merila (1997) found a higher heritability of body size in flycatchers raised under good food conditions than under bad ones. Merila's review of the recent literature established that in most cases stressful environments *reduce* heritability (eight out of nine instances). The same outcome was published by Bennington and McGraw (1996) on field estimates of heritability in *Impatiens pallida* at two sites. The evolutionary consequences of this observation can be multifarious. On the one hand, reduced genetic variation under stress means a lowered ability of the population to respond to selection in a situation in which selection pressure is presumably stronger. On the other hand, since genotypes tend to converge on similar phenotypes under stress, selection will be less likely to distinguish among genotypes and will therefore be less effective at eliminating genetic variation from the population, according to Fisher's fundamental theorem. This, in turn, may help explain the maintenance of quantitative genetic variation for the same traits in other, less stressful (i.e., characterized by reduced selection pressure) environments.

In contrast to the examples described above, cases of *increases* in heritability under stress have also been published. For instance, in a study on *Drosophila melanogaster,* Blows and Sokolowski (1995) crossed laboratory lines that had been maintained in isolation for seventy generations and determined the expression of additive, dominance, and epistatic genetic variance as a function of food availability. Interestingly, while they did not find any relationship between environmental quality and additive genetic variance, both the dominance and the epistatic components were much higher at both extremes of the environmental gradient. However, as we saw in Chapter 1, discussing the results of Mazer and Schick, the situation can be more complex than a simple stress versus nonstress comparison, with different traits showing distinct patterns of plasticity of their heritability.

This barrage of partially contradictory empirical data indicates that even biochemically (rather than statistically) based models such as Ward's (1994) (see Chapter 1) may still not contain enough details to actually reflect the variety of natural conditions they purport to capture. Of course, not everything is variable or unpredictable. Young et al. (1994) showed that the environment does not have any appreciable effect on the heritability of floral traits in *Raphanus sativus.* This is in agreement with the well-known observation that floral traits, as opposed to vegetative or life history ones, tend to show both high homeostasis and very low plasticity.

What happens when one compares variable versus heterogeneous—as opposed to "good" versus "bad"—environments? Simons and Roff (1994) answered that question by exposing a field cricket to controlled laboratory conditions and a field trial. They measured the heritabilities of several traits under both circumstances and found that most (but not all) traits showed a higher heritability in the least variable environment. The usual explanation for this observation (Falconer 1989) is that more variable environments are associated with a higher environmental component of variance, which inflates the denominator of the ratio used to calculate heritability (see Equation 1.1). Simons and Roff, however, surprisingly found that an equally important factor was an actual reduction in the additive genetic variance in the field, that is, the numerator of the heritability term also decreased.

From all of the above, it is very tempting to conclude that the relationship between plasticity and genetic variation is "complex." However, such complexity need not imply that we cannot understand what is going on, since clear patterns do emerge from the literature that we have discussed. For example, fundamental genetic parameters such as heritabilities can change with environmental circumstances, but traits related to sexual reproduction seem to represent an exception that can be readily explained by ecological considerations. Furthermore, additive and nonadditive components of variance react differently to the quality of the environment experienced by the population. So far, most of these generalizations and observations have not been explicitly included in models of phenotypic evolution, chiefly because the empirical evidence has emerged very slowly and only fairly recently. Of course, a better understanding of the relationships between environmental conditions and the parameters describing genetic variation in nature is of paramount importance and an immediate goal for future research.

Conceptual Summary

- Three general kinds of genetic bases for phenotypic plasticity have been proposed at the phenomenological level: overdominance, pleiotropy, and epistasis. The available evidence contradicts the overdominance model, and unfortunately the complexity and variety of the molecular mechanisms underlying plasticity may not allow a distinction between the latter two models.
- There is extensive evidence of genetic variation for plasticity within populations, among populations, and among species.

- The evolution of plasticity, like that of any other genetically based trait, can be affected by the presence of genetic constraints, both at the univariate and multivariate levels. These constraints may manifest themselves, for example, as reduced or absent genetic variation for plasticity, or as tight genetic correlations among plastic traits.
- Contrary to repeated suggestions in the literature, there does not seem to be any consistent link between plasticity and heterozygosity.
- It is clear from empirical results (as well as from theoretical considerations) that plasticity is an entirely distinct phenomenon from developmental noise.
- Plasticity is not the opposite of canalization or homeostasis (in Waddington's sense), and in fact a plastic reaction norm can be highly canalized. Canalization itself is not a population-level property.
- The relationship between plasticity and maintenance of genetic variation in natural populations is complex and depends on the kind of genetic variation one is considering. Simple theoretical models of such a relationship have so far been based on assumptions contradicted by empirical data and have not yielded robust results.
- The relationship between plasticity and heritability is of particular interest for both fundamental evolutionary biology and a variety of applications.

—5—

The Molecular Biology of Phenotypic Plasticity

We vivisect the nightingale
To probe the secret of his note.

—Thomas Bailey Aldrich (T.B.) (1836–1907)

CHAPTER OBJECTIVES

To discuss the available evidence on the molecular basis of phenotypic plasticity, starting with quantitative trait loci studies of genetic variation for plasticity in natural populations. The concept of plasticity genes is introduced and discussed, a series of studies addressing hard-core molecular biology of plastic responses is reviewed, and the molecular biology of epistasis and pleiotropy is addressed. It is suggested that hormones should be the focus of more genotype-environment studies because they represent the interface between the external environment and the genic level of action. The chapter ends with a discussion of the limits of an integrative biology approach, bridging the gap between mechanistic and organismal questions and explanations.

Is There a Molecular Genetics of Plasticity?

The study of phenotypic plasticity has played an increasingly important role in organismal and evolutionary biology, as much of this book clearly documents. However, research on the genetics of plastic responses has long been limited to quantitative genetics and the corresponding use of statistical approaches discussed in Chapters 2 and 4. The very concept of a molecular genetics of plasticity was formally developed only very recently (Smith 1990; Schmitt and Wulff 1993; Schmitt et al. 1995; van Tienderen et al. 1996; Callahan et al. 1997, 1999; Schmitt et al. 1999). However, this state of affairs is a reflection more of how difficult communication between subdisciplines of biology can be than of real lack of effort. The molecular and physiological literature is in fact filled with studies of the mechanistic basis of plastic responses, but they are just not labeled as such and are not framed in the context of evolutionary and ecological questions. Physiologists in particular have always dealt with how hormones mediate environmental signals and effect dramatic phenotypic changes, which have traditionally simply been assumed to be adaptive (Pfennig 1992; Reid 1993; Tollrian 1993; Crews 1994; Gross and Parthier 1994; Voesenek and van der Veen 1994; Koch et al. 1996; Bewley 1997). Molecular biologists have also realized that patterns of gene expression change with environmental conditions, not just with developmental time or tissue specificity (Goransson et al. 1990; Koga-Ban et al. 1991; Chandler and Robertson 1994; Palva 1994; Williams et al. 1994b; Winicov 1994; Fuglevand et al. 1996; Thomas and Villiers 1996; Kreps and Simon 1997; Kuno et al. 2000; McKenzie et al. 2000; Sheldon et al. 2000). Considered together, these studies contain most of the answers that evolutionary biologists have long lamented were missing, dramatically limiting our understanding of phenotypic plasticity. On the other hand, physiologists and molecular biologists are also coming to realize that many of their own old puzzles can potentially be better understood once the proper ecological or evolutionary information is taken into account (e.g., Smith 1995; Ballarè and Scopel 1997). The hope is that physiological ecology will go a long way toward explaining why certain species exhibit some mechanisms of response to environmental heterogeneity and others do not. Furthermore, evolutionary and comparative studies may be able to point out the relationships between genes underlying a variety of responses as well as help explain partial redundancies and nonoptimal performance of the existing systems. This chapter will provide a brief overview of what I consider some key areas of research addressing the mechanics of phenotypic plasticity: QTL studies, *hard-core* molecular genetics of plasticity, the molecular study of pleiotropy and epistasis, and the central role of hormones in mediating genotype-environment interactions. A discussion of the particularly well-known case of phytochrome-mediated shade avoidance responses in plants will be deferred to Chapter 11.

QTL Studies of Genes Affecting Plasticity in Natural Populations

In Chapter 1, I summarized the way in which quantitative trait loci studies are performed, pointed out that this is by no means a novel approach to research on the genetics of natural populations, and discussed some of the limitations of QTL analyses. My cautionary remarks notwithstanding, QTLs are providing a wealth of information on the genetic basis of phenotypic plasticity, and we need to discuss a few examples in detail. A more general set of examples, including the study of bristle number in *Drosophila,* life history trade-offs, heterosis, macroevolution of novel phenotypes in maize, flowering time in *Arabidopsis,* and hybrid breakdown is found in Mitchell-Olds (1995).

One of the first QTL studies to address genotype-environment interactions directly was carried out by Paterson et al. (1991) on tomatoes (Fig. 5.1). They crossed the cultivated variety, *Lycopersicon esculentum,* with a closely related wild species, *L. cheesmanii,* obtaining 350 F_2 individuals to use in the mapping process. While the project was clearly aimed at studying commercially important traits (fruit size, concentration of soluble solids, and pH of the fruit), it provided data relevant to plasticity studies because the F_2 and F_3 progeny were grown in three different "environments": two experimental stations in California (at Davis and Gilroy) and one in Israel. They detected twenty-nine QTLs scattered on eleven of the twelve chromosomes, with each QTL accounting for from 5 to more than 40% of the phenotypic variance of a given trait. Especially interesting from the viewpoint of the genetics of plasticity, four of these QTLs were detected in all three environments, ten were present in two environments, and fifteen were characteristic of only one location. Furthermore, a greater number of QTLs overlapped between the two California sites than between either of these and the Israel site, possibly reflecting some degree of specificity of genetic effects to the local environments. Of course, no insight could be gained on the actual causal connections between gene action and environmental characteristics, since the three environments obviously differed from one another in many respects.

An interesting by-product of the study was a comparison between the relative predictive ability of the genetic and phenotypic information available to the researchers. If the trait had low heritability, the QTLs detected in the F_2 turned out to be much better predictors of the phenotype of the F_3 offspring than the phenotype of the F_2 parent. On the other hand, if the heritability was high, knowledge of the QTLs did not add anything in that respect. For traits with intermediate heritability, the two kinds of information enhanced each other. Another important point about this study is that many of the QTLs detected in that cross turned out to map in positions similar to QTLs previously studied in a cross between *L. esculentum* and another wild relative, *L. chmielewskii.* This has been interpreted as suggesting that the same loci affect the phenotypes of several species of *Lycopersicon* in a similar way. This has obvious implications for our understanding of the macroevolution of phenotypes and of speciation events.

Another study of tomatoes more directly addressed the problem of plasticity and specifically of the ability of the plant to withstand salt stress (Breto et al. 1994; on salt stress see also Liu et al. 2000; Quesada et al. 2000). These

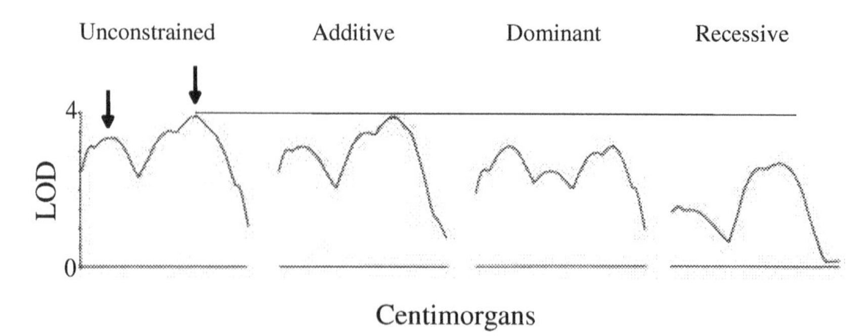

FIG. 5.1. QTL mapping of genes affecting concentration of soluble solids in tomato. The LOD (short for log of the odds ratio) score (on the ordinate) indicates the probability of a QTL existence for each map location along chromosome 6 (on the abscissa). The graph on the left shows the location of two potential QTLs (arrows). The remaining three graphs present the LOD scores of the QTLs when an additive, dominant, or recessive genetic model is compared with an unconstrained model. The data are most consistent with additivity, although dominance cannot be conclusively ruled out. (Modified from Paterson et al. 1991.)

authors crossed *L. esculentum* with the close relative *L. pimpinellifolium* and identified six markers affecting fruit number, fruit weight, and total plant weight. The study uncovered a marker that behaves in a way that is opposite to the expectation from the parental means (it increases rather than decreases fruit weight under saline conditions). This suggests that the cross produced recombinants with transgressive and therefore novel phenotypes, implying that previously unobserved patterns of phenotypic plasticity can evolve quite rapidly. They also pointed out the existence of two interacting QTLs that can break the correlation between fruit number and fruit weight present in the parents, thereby suggesting a role for epistasis in altering genetic constraints under stressful environmental conditions.

A very interesting study of QTLs affecting plasticity was published by Wu (1998) using the *Populus* genetic map constructed by crossing a female *P. trichocarpa* with a male *P. deltoides.* The main objective of the study was to test for the presence and relative role of regulatory versus structural genes affecting plasticity. Wu suggested that the observed genetic variation for plasticity can be accounted for by two components: V, which is due to the heterogeneity of genetic variances between the two environments used, and L, which is related to the genetic correlation (or lack thereof) of the focus trait between the two environments. He then proposed that the loci contributing to V display environment-dependent genetic differentiation and therefore might represent regulatory genes. The loci contributing to L, on the other hand, cause "an inconsistent change in genetic architecture by modifying their expression in varying environments," and they are accordingly considered *structural.* This is equivalent to saying that regulatory loci produce environment-specific genetic effects (some of the loci being regulated are presumably active only in a subset of the environments). While this may be so, it certainly does not cover all the possibilities, and Wu himself acknowledged that a given locus could behave as regulatory while at the same time displaying environmental sensitivity. He further assumed that loci contributing to plasticity do not affect the genetic variance of the trait in a single environment, which is in direct contradiction to his acknowledgment that the same locus can actually play both roles. While Wu's results do identify several QTLs affecting plasticity and his study is useful in characterizing different kinds of effects of these loci on the genetic variances, the gap between a statistical and a mechanistic analysis cannot be bridged so easily. Only direct studies of the function and developmental regulation of the identified loci will tell if they act as regulatory elements and at what level in the complex hierarchy of regulation of these phenotypes they do so.

What Is a Plasticity Gene (and Why Should We Care)?

I have mentioned the idea of plasticity genes several times, and it is now time to briefly address the crucial questions of what they are, if they actually do exist, and why we should bother thinking about them. The idea of plasticity genes, in the sense used here to refer to the molecular basis of plasticity,[1] was proposed by Schlichting and myself (1993) in response to Via's (1993) suggestion that plasticity is not the target but rather the by-product of selection. The controversy that ensued and the implications for our understanding of the evolution of plasticity have been discussed in several places (Via et al. 1995, and see Chapter 1 for a brief synopsis), and they are summarized in Chapter 3 of Schlichting and Pigliucci (1998), so I will not dwell on that aspect of the debate again.

The definition of *plasticity genes* given in Schlichting and Pigliucci (1993) was: "regulatory genes that control phenotypic expression and are independent of trait means." From the context, it is clear that this means genes that control the slope (as opposed to the height) of the reaction norm of a given trait in response to a set of environmental conditions, and in that sense it follows the paper by Scheiner and Lyman (1989). However, further on in our 1993 paper, we provided an example from the molecular literature that implies that plasticity genes are genes whose action is environment-specific, that is, they are expressed in one environment but not in others. In this sense, we made the same mistake that I just attributed to Wu: There is no simple correspondence between statistical properties of genes (action on the slope of a reaction norm or effects on variances) and their mechanistic ones (environmental and developmental regulation, specific function). In

Schlichting and Pigliucci (1995) we proposed a definition that is based more on molecular biology than on quantitative genetics: "loci that exert environmentally dependent control over structural gene expression." Finally, in Pigliucci (1996a) I restricted the concept further to: "regulatory loci that directly respond to a specific environmental stimulus by triggering a specific series of morphogenic changes." In this book, I stay with this latest definition because I still think it is the most appropriate, although *morphogenic change* might turn out to be too restrictive if one wishes to include behavioral and physiological plasticity. However, it should be noted that this does not mean that all regulatory genes are plasticity genes, simply because not all regulatory genes react to changes in environmental conditions. For example, plenty of regulatory genes, especially in animals, initiate predetermined sequences of developmental events regardless of the nature of the environment.

So defined, plasticity genes definitely exist. They are to be identified with any gene coding for receptors of environmental conditions that trigger phenotypic responses, usually (but not necessarily) through the activation of hormone-mediated developmental cascades. Genes coding for photoreceptors in both plants and animals would be primary examples and are among the best characterized in the molecular literature. From this standpoint, many bacterial genes whose products trigger the transcription of operons under specific environmental circumstances would also qualify as plasticity genes. The Salt Overly Sensitive (*SOS*) system in *Escherichia coli,* which causes an increase in the mutation rate in response to environmental stress, is an example that is particularly well understood (McKenzie et al. 2000). Thus, plasticity genes are widespread and very ancient. Furthermore, they have been studied from a mechanistic perspective for a long time, and we know a lot more about them than a casual observer might think. It should be noted that hormones themselves, while key components of plastic responses (see below), do not qualify as products of plasticity genes, unless they also happen to be the receptors of the environmental signal (as opposed to being triggered by the receptor). I do not know of any case of the latter, but it is by no means impossible in principle.

Of course, defining a specific class of genes as plasticity genes does not imply that these are the only genes that affect the expression and evolution of phenotypic plasticity in natural populations. (Similarly, nobody thinks that homeotic genes are the only ones involved in the modification of fundamental structures and body plans in animals and plants.) On the contrary, plasticity genes are likely to represent a minority of such genetic elements, as clearly pointed out in both Pigliucci (1996a) and Schlichting and Pigliucci (1998) as well as by Scheiner and Lyman (1989) in a quantitative genetic framework. Any other gene whose product affects the phenotype in a differential manner in distinct environments is bound to contribute to plasticity, and it will be one of the loci of selection if it is variable in natural populations. In fact, the picture emerging from the molecular literature is that many genes act only in a subset of environmental conditions (see the discussion of QTLs above and examples that follow). This abundance of heterogeneous genetic elements is exactly what makes the link between the molecular and quantitative genetics of plasticity rather tenuous, at least at the moment. The importance of plasticity genes as defined here, however, is that they must be key elements in the evolution of *adaptive* plasticity, for otherwise their very existence does not make any sense (what is the point of an organism perceiving the environment if it does not act on it? I doubt that plants are simply curious about the current photoperiod or that reptiles would simply like to know what the outside temperature is when they deposit their eggs). The connection between any other gene's effects and the environment may be the doing of natural selection, but it could also be coincidental or truly a by-product of other evolutionary processes, or it may be an exaptation (Gould and Vrba 1982). It is this crucial distinction that makes the study of environmental receptors not only fascinating from a mechanistic standpoint (in answer to the "how" question of phenotypic evolution), but also fundamental for ecologists and evolutionary biologists (as an answer to the "why" question).

All this having been said, let me add a word of caution about the whole idea of "genes for" (Sober 1984). The conditions under which we can really talk about genes being "for" something are quite stringent and very rarely met in the evolutionary literature (Kaplan and Pigliucci 2001). First, of course, one has to distinguish between the possibility that the gene in question is carrying out a certain function now as opposed to what it was doing or was selected for in the past. This is the molecular equivalent of the exaptation/adaptation dichotomy, and it can be addressed only by the use of gene phylogenies coupled with detailed information on the phenotypic effects of the genetic element in question. Second, even if we limit ourselves to what the gene is doing now, it may be far from an easy field of inquiry. Strictly speaking, most genes are only "for" the production of a particular protein. One can without difficulty extend the usage to include the alleged function of such a protein (e.g., gas exchange in the case of hemoglobin), but that is as far as one can reasonably go. The more we consider phenotypes that are removed from the primary function of the protein, the more we run into the twin evils of pleiotropy and epistasis and the more difficult it becomes to pinpoint what a gene is actually "for." The hemoglobin β gene, for example, interacts with the α chain to create a heterologous protein. Is that what the gene is for? Clearly, there is much more to the story, because the gas exchange performed by hemoglobin affects many aspects of the phenotype in a more or less direct manner, including the metabolic rate and the behavioral reaction time of an individual. Is this a gene "for" fast reflexes? Even more interestingly, mutated versions of this gene partially cripple the red cells, causing a mild anemia if in heterozygosis. One can hardly suggest that these are genes "for" anemia (or for causing the sickle cell phenotype associated with it), yet they are maintained by selection in certain areas because they also confer resistance to malaria. Are some variants of the β gene "for" malaria resistance in some environments and "for" anemia in others? One can see that this is a real Pandora's box, so the phrase must be used with extreme caution, if at all. A fortiori, this goes for *plasticity genes,* giving us one more reason to restrict their definition within the parameters outlined above.

Molecular Genetics of Genes Affecting Responses to Environmental Heterogeneity

The molecular literature on what evolutionary biologists would term gene-environment interaction is vast, and a thorough review of the field would require an entire book of considerable size. I have chosen just a few representative examples to illustrate the kind of information that molecular biology and physiology can provide for ecological and evolutionary studies, highlighting cases in which the study of the molecular basis of a plastic response actually provides novel insights into evolutionary or ecological questions.

Convincing evidence is accumulating for the existence of at least three types of molecular responses to environmental stresses: (1) specific responses elicited by one particular type of stress; (2) responses induced by a limited number of (sometimes but not necessarily ecologically related) stresses; and (3) generalized responses to a variety of stressful situations. A spectacular example of the first category is provided by the reaction to hypoxia (low oxygen levels) in *Arabidopsis thaliana* (Fig. 5.2). Two enzymes in the alcoholic fermentation pathway, alcohol dehydrogenase (ADH) and pyruvate decarboxylase (PDC), are induced by low-oxygen conditions. This initial observation led Dolferus et al. (1997) to identify two distinct fermentation pathways. One pathway is constitutively expressed in roots and leaves, but the other is induced only by hypoxia and is mainly in the roots, thereby demonstrating environmental as well as tissue specificity. At least seven genes constitute each set. *Arabidopsis* also has two hemoglobin genes, only one of which is induced by low oxygen levels, but the role of such genes in this stress response has not yet been clarified.

A confirmation of the suggestion that differential regulation of gene activity may be a common way of coping with novel environmental circumstances has been provided by a study of the genetic organization of the salt-stress response in the grass *Lophopyrum elongatum* and in its salt-sensitive

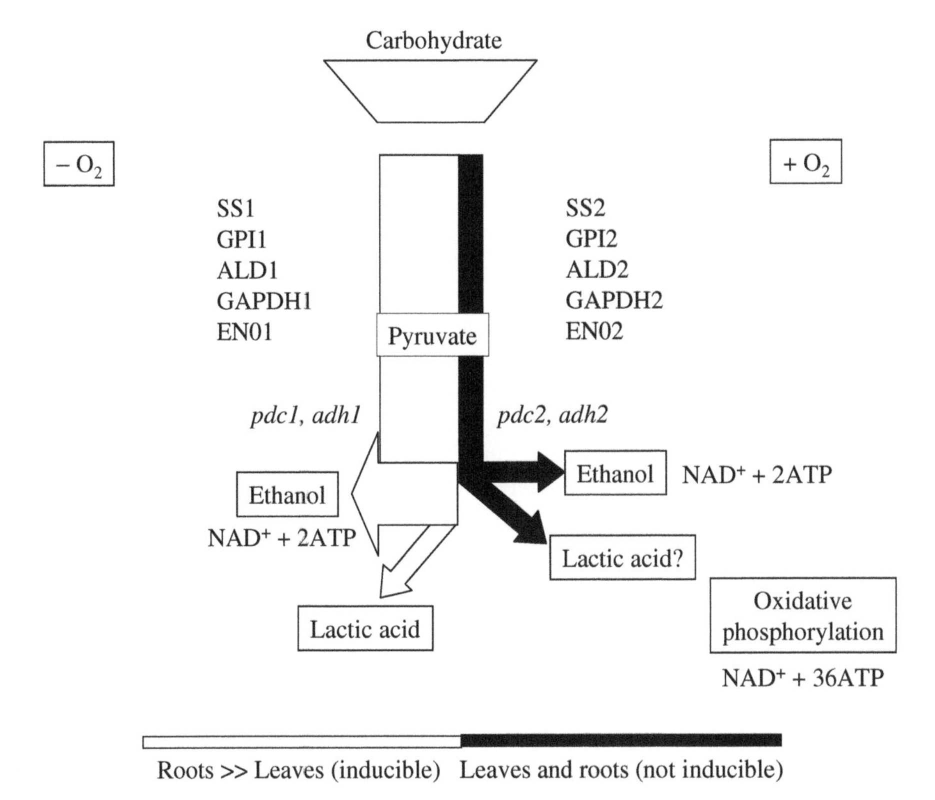

FIG. 5.2. Alternative carbohydrate metabolic pathways in *Arabidopsis thaliana.* The pathway on the right is active by default in both leaves and roots, and the one on the left is activated by exposure to hypoxia and is specific to the root system. At least seven different genes comprise each set. (Redrawn from Dolferus et al. 1997.)

relative, the cultivated wheat (Dubcovsky et al. 1994). The authors were able to identify eleven genes characterized by enhanced mRNA accumulation during the early stages of exposure to salt stress. These genes are present in the salt-sensitive wheat, but they are transcribed at a slower rate. In this case, then, the resistance may not be due to the appearance of novel genes, or even mutated versions of the same genes, but simply to a differential expression of the existing genetic material. Response to salt stress appears to be exceedingly complex even in plants that are not normally exposed to saline conditions, such as *Arabidopsis thaliana.* Liu et al. (2000) demonstrated that the steady level of the *SOS*-2 gene transcript (coding for an active protein kinase) is up-regulated by salt stress in the root, and Quesada et al. (2000) identified seventeen loci across all five *Arabidopsis* chromosomes affecting salt tolerance in this plant, including the *ABI4* gene, which controls sensitivity to the hormone abscisic acid.

A classical example of environment-specific induction of adaptive molecular responses is the heat-shock response, which is common among animals and plants (Rickey and Belknap 1991; Wu et al. 1994; Loeschcke and Krebs 1996; Hong and Vierling 2000). The response encompasses the production of special proteins that help to prevent the denaturation of other proteins when they are exposed to high temperatures. An unusual study by Maresca et al. (1988) detected the presence of the heat-shock-induced HSP70 proteins in three species of Antarctic fishes. The genes are similar to the equivalent *Drosophila* sequence, which was used as a probe. What was provocative about the study is that—unlike the situation in *Drosophila* and in most other species—the Antarctic fishes expressed HSP70 in response to a temperature of only 5°C. While this would not normally be thought of as "hot" from a *Drosophila* standpoint, it clearly qualifies as heat stress for animals acclimated to much lower temperatures. It would be interesting to find out what regulatory elements have made possible such a dramatic shift in the induction threshold of HSP70. In *Arabidopsis,* four genes have been identified as being involved in the acquisition of tolerance to high stress (Hong and Vierling 2000). One of these proteins, codified by *Hot1,* is involved in similar functions in yeast and bacteria, again pointing to a very ancient origin of some of the genes involved in plastic responses.

Perhaps even more intriguing are cases in which a common molecular machinery is being uncovered behind a multitude of responses to environmental conditions, only some of which might be intuitively connected on the basis of a cursory knowledge of the ecology of the species. A very good example of this is provided by a study by Rickey and Belknap (1991), who investigated the pattern of gene expression of four proteins in response to different environmental stresses. They focused their attention on the HSP70, as well as on ubiquitin, phenylalanine ammonia-lyase (PAL), and patatin. The stresses were caused by mechanical injury, cutting injury, heat, and heat shock (Fig. 5.3). All the proteins responded to most of the stresses, with patatin being the only one whose activity decreased under stress instead of increasing. PAL responded to bruises and cuts but not to heat or heat shock, while ubiquitin reacted to all the stresses except heat. Most interestingly, the "heat shock" protein HSP70 responded not only to heat shock as anticipated, but also to bruises and cuts; in fact, the latter reactions were *more* intense than the one to heat shock. Similar patterns of expression of HSP70 were also found by the same authors in two other cultivated plants: tomato and petunia. They demonstrated that the responses could also be evoked by application of exogenous ethylene, thereby linking this hormone to the heat-shock and wounding pathways. Interestingly, the responses to wounding were limited to the specific area being injured, unlike the more general heat-shock reaction. An ecologist would not predict a connection between high temperature and wounding and would not regard

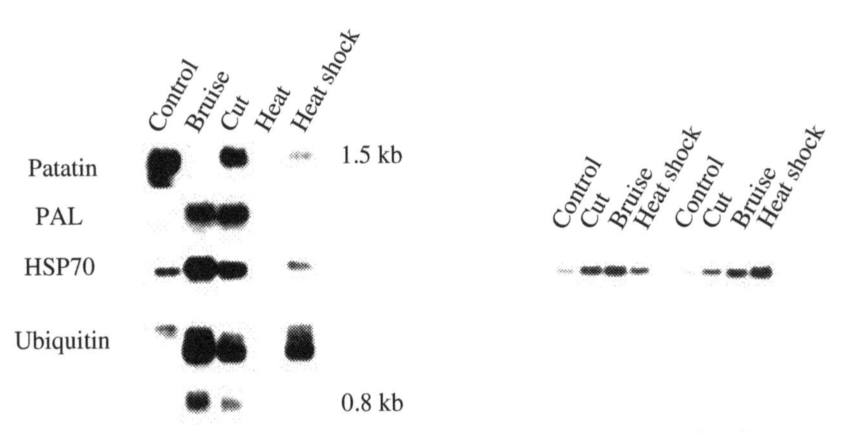

FIG. 5.3. Patterns of gene expression of various proteins in response to a series of environmental stresses. On the left, response of patatin, PAL, HSP70, and ubiquitin to bruise, cut, heat, and heat shock in potatoes. On the right, response of HSP70 to several stresses in tomato and petunia. (From Rickey and Belknap 1991.)

them as similar kinds of stresses. However, the molecular data clearly show that the way organisms respond to stress may be unanticipated on the basis of our understanding of the ecological nature of the stress. It may be that at the cellular or subcellular levels, the effects of a variety of stresses are similar, and they are therefore counteracted with the same battery of molecular tools. This may point to a fundamental discontinuity between mechanistic and organismal levels of analysis and cast some doubt on the extent to which research at one level can shed light on questions at the other.

Another fascinating glimpse into how different things are at the molecular level is offered in a paper by McDowell et al. (1996) on the induction and control of the *ACT7* gene in *Arabidopsis thaliana. ACT7* codes for an actin, a type of protein that is a major component of the cytoskeleton. Actins are necessary during cell division as well as in the process of determination of cell shape. Clearly, we are looking at fundamental cellular functions that are quite remote from the whole-organism level and in particular from the range of phenotypic effects elicited by environmental heterogeneity. Yet, *ACT7* was found to be responsive to light availability and to wounding, two environmental variables that one would not think of as ecologically coupled. Also, the transcription of *ACT7,* which can vary by 128-fold among different tissues, is regulated by a variety of hormones, chief among which is auxin. This regulation is exerted through a number of sequence motifs that apparently function as phytohormone response elements. In this case, of course, the connection between the cellular response and the apparently disconnected ecological situations that elicit it is clear because we know the function of the protein: Actin is required every time there is intense cell division, and as changes in light availability and wounding of the plant require cell division, they both specifically call for enhanced production of actin.

The third level of analysis mentioned at the beginning of this section is provided by inquiries on the responses to a large variety of stresses, either by single genetic elements or genome-wide. The existence of some way to detect generalized stress at the molecular level is suggested by a study of the tobacco Tnt1 retrotransposon in *Arabidopsis* (Mhiri et al. 1997). Retrotransposons are evolutionarily related to vertebrate retroviruses, and it makes sense for such semiparasitic genetic elements to "monitor" the conditions of the host. How this is accomplished is still not clear, but it definitely happens in the case of Tnt1. This element is induced in both tobacco and *Arabidopsis* by pathogen infection, freezing, wounding, oxidative stress, exposure to $CuCl_2$, and exposure to salicylic acid. A clue as to how Tnt1 may perceive the stress incurred by the host is provided by the correlation between Tnt1 induction and the accumulation of camalexin (a phytoalexin) and expression of *ELI3,* a gene induced as a general defense against a variety of stresses. It is conceivable that the retrotransposon is using chemicals generated by the plant under stress to assess the precarious conditions of the host.

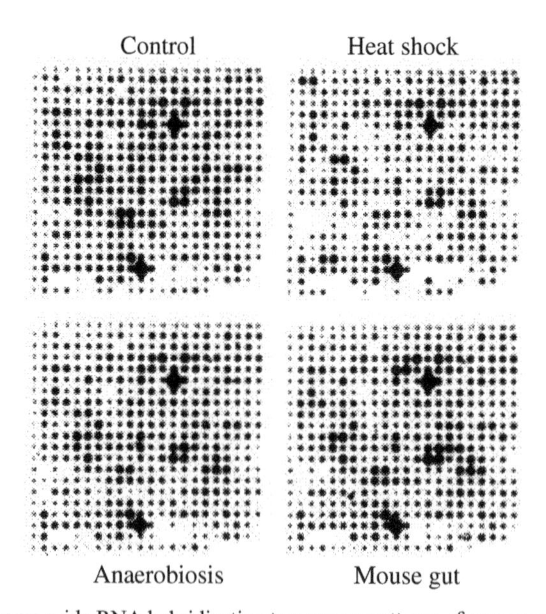

FIG. 5.4. Genome-wide RNA hybridization to uncover patterns of gene regulation induced by environmental stress in *E. coli.* Each dot represents the level of expression of individual RNAs and how they change when the control is compared to heat shock, anaerobiosis, and mouse gut treatments. (From Chuang et al. 1993.)

A final example of how molecular biology makes it possible to study responses to multiple stresses is provided by an investigation into genome-wide control of gene expression in *Escherichia coli* (Chuang et al. 1993). These authors followed the patterns of expression (measured by levels of mRNA) of 400 segments of the genome (Fig. 5.4). The bacteria were subjected to a large variety of environmental conditions: heat shock, osmotic shock, starvation induced by limitation of one of a series of nutrients, growth in mouse gut, and anaerobic growth in a tube, among others. The method was successful in identifying variation in the expression of many known genes, as well as new regions responsive to particular stresses. This general sweep of the relationship between stress and gene regulation may eventually provide a very powerful first screening for genes that affect phenotypic plasticity in other organisms, a possibility now realizable thanks to the more widespread use of microarray techniques. It will be relatively easy to identify batteries of genes responding to specific environmental conditions using this kind of data, as well as to pinpoint more generalized stress-induced elements. We may soon be able to gain decisive insights into the genomic architecture of hierarchical gene regulation under heterogeneous environments.

Epistasis and Pleiotropy under Environmentally Heterogeneous Conditions: The Molecular Side of the Story

The relative importance of *nonlinear* genetic effects such as dominance and epistasis, as well as of pleiotropy, for evolutionary phenomena has been under discussion for quite some time (Wright 1932; Gavrilets and Jong 1993; Pooni and Treharne 1994; Wagner et al. 1994; Cheverud and Routman 1995; Shaw et al. 1997; Rice 1998). In the case of phenotypic plasticity in particular, pleiotropy and epistasis have been called upon as major factors in our understanding of the evolution of reaction norms (see reviews in Scheiner 1993a; Schlichting and Pigliucci 1995). The debate has centered around two complementary questions: First, how much pleiotropy and epistasis is actually demonstrable for plastic responses of evolutionary interest? Second, what difference do such effects make in our modeling of the evolution of phenotypic plasticity? A closer look at the recent molecular literature reveals quite a bit of evidence relevant to the first question. The second question will be addressed in Chapter 10, when we will examine theoretical modeling of plasticity.

Blows and Sokolowski (1995) have investigated the relative importance of additive and nonadditive effects as a function of environmental conditions in *Drosophila*. They crossed isolated strains of flies that had been raised under laboratory conditions for seventy generations and exposed them to different levels of food availability. They found no association between additive genetic effects and environmental quality. On the other hand, their data revealed a dramatic dependence of nonadditive effects on the environment. Specifically, dominance, additive-by-additive epistasis, and dominance-by-dominance epistasis on developmental time increased markedly at both extremes of the environmental gradient. They suggested that the expression of high levels of dominance and epistasis at environmental extremes might be a general expectation for some traits.

How has molecular biology contributed to the mechanistic understanding of these complex genetic effects, once the exclusive province of quantitative genetics? Major advances in the study of pleiotropy have been made possible by the availability of mutants initially characterized because of specific effects of interest to developmental molecular biology research. Ecologists and evolutionary biologists interested in whole-organism-level analyses can use these mutants to pursue organismal questions. An example is provided by the work of van Tienderen et al. (1996) on *Arabidopsis*. They studied the reaction norms to three levels of nutrients in five mutants known to specifically affect flowering time by comparing them to the wild-type line from which the mutants had originally been isolated (Fig. 5.5). The mutants were classified into three categories on the basis of physiological studies: One of them reacts to both daylength and vernalization (exposure to cold), two react to daylength but not to vernalization, and the remaining two are not sensitive to either. The results clearly showed that the mutations can dramatically affect both flowering time and its relationship to the production of rosette leaves, two traits normally tightly coupled in *Arabidopsis* (Mitchell-Olds 1996). When the nutrient levels were altered, fewer nutrients caused a delay in flowering (thereby extending the vegetative phase), without altering leaf production. A further reduction of nutrients had the opposite effect, decreasing leaf production while not altering flowering time. Moreover, the various mutants showed distinct reactions to the three levels of nutrients applied during the experiment. Van Tienderen and colleagues concluded that "flowering time" genes actually have extensive pleiotropic effects on the reaction norms of leaf length, number of both rosette and cauline leaves, and production of axillary shoots on the main inflorescence. Further characterization of the molecular function of these genes will help to elucidate why the mutations examined in this study alter the plant's plasticity in such distinctive ways.

A complex example of environmentally mediated epistasis is provided in a study of the regulation of chalcone synthase gene expression in *Arabidopsis* by Fuglevand and co-workers (1996). Chalcone synthase catalyzes the first step in the biosynthetic pathway that produces flavonoids, an important class of water-soluble plant pigments. The data of Fuglevand et al.

FIG. 5.5. Bidimensional reaction norms of wild-type *Arabidopsis thaliana* and of five mutants affecting flowering time. The three points on each reaction norm refer to 1, 5, and 25% nutrient levels, measured as a percentage of a standard Hoagland solution. Note that not only the mutants' reaction norms (open symbols) shift but also their shape changes when compared to the wild type (solid symbols). (From van Tienderen et al. 1996.)

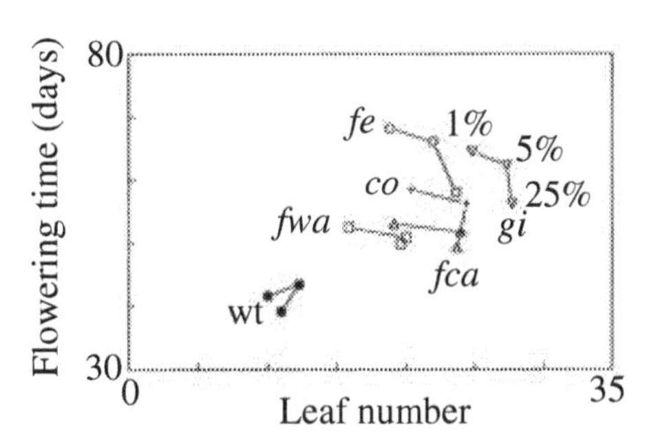

show that separate inductive pathways mediate chalcone synthase response to UV-B and blue light/UV-A, with only the latter involving the CRY1 blue photoreceptor. However, UV-A and blue light interact synergistically with UV-B to stimulate accumulation of chalcone synthase transcript. The synergistic pathways do not involve the CRY1 photoreceptor, since mutants lacking this photoreceptor retain the synergistic effect.

Overall, the number of examples of pleiotropy and epistasis from the molecular literature is in fact very compelling, and every month sees new case studies of complex interactions between biochemical pathways in determining high-level phenotypes and their response to the environment (e.g., see the study by Sheldon et al. [2000] of the role of *Flowering Locus C* in *Arabidopsis* response to vernalization). The general conclusion seems to be that indeed both pleiotropy and epistasis are rampant, as predicted by Wright (1932) and contrary to what was anticipated by Fisher (1930). If gene-gene interactions are dominant, it would seem prudent to incorporate them widely into models of phenotypic plasticity, and indeed of phenotypic evolution in general (Schlichting and Pigliucci 1998), in order to explore their potential consequences for the evolutionary trajectories of populations in more detail. However, a caveat in reference to all of the above needs to be addressed. As Cheverud and Routman (1995) discussed so elegantly, epistasis in a quantitative genetic sense is not directly equivalent to epistasis in a physiological sense (the only one considered in the molecular literature). Indeed, the same is true for pleiotropy and even dominance (Kacser and Burns 1981; Keightley and Kacser 1987; Bourguet and Raymond 1998). Therefore, whereas quantitative and molecular geneticists keep using the same terms as if they referred to the same class of biological phenomena, they actually do not. Ultimately, the meaning that really matters is the mechanistic one, but a better understanding of the relationship between statistical and physiological epistasis and pleiotropy is a key component of any unification of evolutionary and molecular biology.

The Central Role of Hormones

Hormones are the unsung heroes of the nature-nurture field of study. Although a great deal of research is being done on hormones and their effects in both animals and plants, most of it is ignored by evolutionary biologists interested in genotype-environment interactions. Even those who recognize the importance of a mechanistic understanding of phenotypic plasticity tend to focus on the genetic level of analysis, discounting the transduction pathways of which hormonal signaling is a necessary component. I am arguing here that hormones play a central role in the genotype-environment dialectic because of their effect on the following two components of phenotypic development:

1. They shape the organism in response to the genetic programming (G-component).
2. They carry the information from environmental receptors, which triggers the genotypic-specific reactions we call phenotypic plasticity (E-component).

In other words, hormones are literally at the interface between genes and environment. This realization may bring about a paradigmatic shift in the way we think about and work on nature-nurture problems, and perhaps help us to finally distance ourselves from the still dominant gene-centric view without falling into the opposite, nurture-centered, equally misleading position (see the epilogue). I will limit the philosophical speculation and focus instead on some selected examples of how a better understanding of the functions of hormones can further our study of phenotypic plasticity and its evolution. I do not pretend that the following is even a representative survey of studies on hormones—the field is too well developed to attempt that in less than a full book-size treatment (by now this may look like a regular excuse in my writings, but it is also the truth). However, I hope a few cases from the animal and plant literature will make the point sufficiently clear.

Pfennig (1992) studied a diet-induced polyphenism in southern spadefoot toad tadpoles. These organisms can develop as either omnivores or carnivores, with the latter favored in ephemeral ponds abundant in macroscopic prey such as shrimp. The carnivores are more efficient at catching the macroscopic prey and develop more rapidly than the omnivores. Pfennig conducted allometric studies that showed that the difference between the two morphs is mostly reducible to acceleration or retardation in the growth of a few anatomical features. These changes in the localized rate of development mimic the effects of the thyroid hormone. When he exposed tadpoles to a diet that would normally induce omnivores, but added thyroxin, he obtained the larger carnivorous morphs instead. He concluded that an endocrine signal might be the key that allows individual tadpoles to assess and properly respond to changing environmental conditions. In other words, the thyroid hormone may be responsible for carrying the information from the external environment and triggering the adaptive response based on the genetic makeup of the individual. In this case thyroxin is likely to be the coupling element between genes and the environment.

Denver (1997b) offered a more general discussion of the role of hormones in controlling the reaction to environmental changes in amphibians and other vertebrates. It is clear from his review that plasticity in response to a variety of environmental conditions, including water availability, is not just a passive response to stress, but is actively managed by the animal's endocrine system. This system, comprised of the thyroid and the interrenal, is the same one that controls the process of metamorphosis, thereby tightly linking plasticity with genetically triggered epigenetic changes. According to Denver, a single neurohormonal stimulus, the corticotropin-releasing hormone (CRH), is responsible for activating both endocrine systems. CRH is a vertebrate stress neurohormone, and its role is highly phylogenetically conserved among vertebrates, ranging from amphibians to mammals. This provides yet another direct link between phenotypic plasticity and macroevolution (Chapter 9).

In Chapter 4, I mentioned another interesting polyphenism, the variation in eyespot and wing coloration of the butterfly *Bicyclus anynana* (Monteiro et al. 1994; Holloway and Brakefield 1995; Roskam and Brakefield 1996; Brakefield 1997). Koch et al. (1996) selected under constant temperature for butterflies always producing the dry-season form, for insects always appearing as wet-season form, and for fast development regardless of the seasonal form. They obtained dramatic responses to the selection protocol, potentially indicating the presence of genes with major effects underlying this type of seasonal plasticity. Interestingly, they found important differences among the selected lines in the levels of ecdysteroids in the hemolymph. The line selected to display the wet-season form showed an earlier increase in the hormones, which reached twice the level typical of the line selected for the dry-season form. They then injected the insects with the hormones at different concentrations. The results suggested that nonlethal levels of ecdysteroids significantly shorten the pupal stage in all lines. When administered to the dry-season line, ecdysteroids caused a shift in the wing color pattern toward the one characteristic of the wet-season form, with a more conspicuous transverse band on the wing and an increase in the size of the eyespot (Fig. 5.6). Koch et al. concluded that "the results demonstrate that ecdysteroids appearing early in the young pupa produce the wet-season form of the wings. The same hormonal system mediates both developmental time and wing pattern determination." The role of these hormones in actually relating the information about temperature and photoperiod, the environmental determinants of the two forms, remains to be

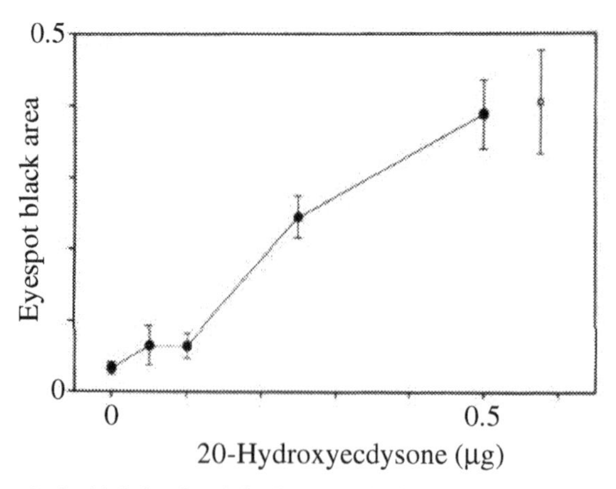

FIG. 5.6. Increasingly high levels of the hormone 20-hydroxyecdysone shift the eyespot phenotype of *Bicyclus anynana* butterflies selected for a low eyespot black area toward the genetically fixed phenotype typical of a line selected for a high eyespot black area (open symbol). (From Koch et al. 1996.)

ascertained. I would not be surprised, however, if such a role were demonstrated soon.

In Chapter 6, I discuss the hormonally mediated sex determination characteristic of several reptiles, and the reader is referred to that chapter for more information on what may turn out to be a major role of plasticity in the macroevolution of mating systems in vertebrates. Plants, of course, are another group of organisms in which hormones have been extensively studied, and they provide excellent material for thinking about the mechanistic aspects of phenotypic plasticity (Reid 1993; Gross and Parthier 1994; Kieber 1997). A nice example of how complex the interaction between plant hormones and environments can be is provided in a study by Williams et al. (1994a) in *Arabidopsis.* They focused on changes in gene expression at the A1494 locus in response to wilting and low temperature. Wilting increases the transcription of A1494 mRNA and simultaneously increases the presence of the hormone abscisic acid (ABA). This tight correlation, however, is only partly causal. By studying ABA-deficient and ABA-insensitive mutants, they were able to tease apart the relative contribution of several factors. A1494 mRNA was not present as a background in ABA-deficient mutants, but it accumulated in response *both* to wilting and to exogenous ABA, thereby indicating additive effects of the environment and the hormone. However, in the wild type, exogenous ABA causes an increase of A1494 mRNA with a kinetic dynamics that closely resembles the one induced by wilting alone. The two results suggest a model in which A1494 can be induced by both ABA and wilting per se, obviously through some other signaling system. The ABA-A1494 interaction is, therefore, *both* additive and epistatic. The situation is even more complex when we add the response to low temperature to the picture. In this case, Williams et al. found that the response of A1494 is mediated exclusively by ABA, because of the lack of the proper mRNA in cold-treated ABA-deficient mutants. Interestingly, heat shock does not significantly affect the accumulation of A1494, thereby decoupling low- and high-temperature stresses, while partially coupling (via ABA) responses to low temperature and drought.

The roles of blue light and abscisic acid in producing different leaf types in the heterophyllous plant *Marsilea quadrifolia* have been teased apart by Lin and Yang (1999). Heterophylly is the production of leaves with distinct shapes in response to changes in water availability, and it is a common strategy in amphibious plants (Wells and Pigliucci 2000). Lin and Yang found that either exposure to blue light or application of abscisic acid induce the development of aerial leaves in *M. quadrifolia.* However, when the effects of the hormone were blocked by an inhibitor, the plant was still perfectly capable of going through the developmental switch, suggesting that blue light and abscisic acid work by way of independent pathways (Fig. 5.7). The whole picture, however, is more knotty: When Lin and Yang transferred the plants to a nutrient-enriched medium (thereby changing a second environmental parameter), the level of ABA increased gradually, similar to what happens during the transition between aquatic and aerial leaves. Therefore, blue light and nutrient availability (the latter through the mediation of abscisic acid) can independently cause the same sequence of morphological changes. What happens if the two environmental signals oppose each other, and under what ecologically relevant circumstances might they do so? These questions were not addressed by Lin and Yang and await further studies.

A fascinating insight into the intricacies of environmental, developmental, and hormonal regulation of gene transcription is provided in a study of accumulation of specific transcripts in tomato by Parsons and Mattoo (1991). They characterized the gene expression of three cDNA clones, pT52, pT53, and pT58. The transcription of pT52 is induced by wounding in the early-red and red stages of fruit ripening, as well as by the hormone ethylene (a gas). On the other hand, pT53 is *repressed* by wounding at the same stages of development, but is induced by ethylene. Finally, pT58 is induced by both wounding and ethylene, but only in the early-red stage of fruit development. To add to this complexity, pT52 is also induced by exposure to light, unlike the other two genes. A hormonal role has been demonstrated in this case, but we still do not understand how the hormone is relevant to the response to specific environmental signals and especially how these signals are perceived by the plant and transduced to the point that they can be further carried to the target tissues by the hormone.

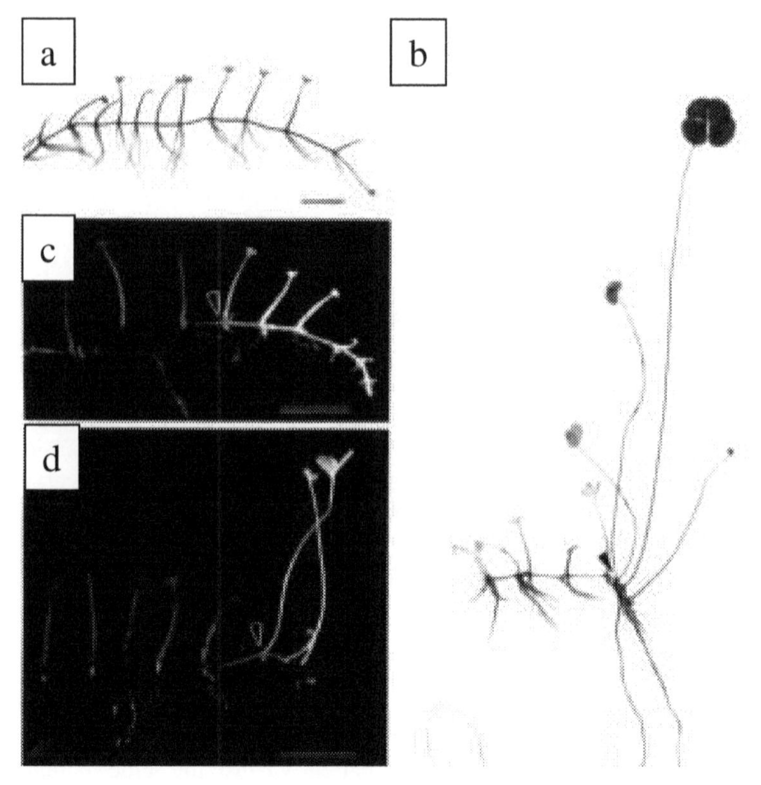

FIG. 5.7. Induction of heterophylly in *Marsilea quadrifolia.* (a) A plant kept under conditions that maintain the aquatic leaf morph. (b) The switch to aerial leaves induced by blue light. (c) A plant grown under aquatic-morph conditions and then subjected to an inhibitor of ABA activity—note the reduction in size and coloration of the organs. (d) As in (c) except that blue light was added to trigger the heterophyllous switch. Arrows indicate the time of application of the treatment. (From Lin and Yang 1999.)

Overall, hormones seem to play a fundamental interface role between the environment and appropriate genetic and epigenetic responses. This role is evident in both plants and animals, despite the considerable differences in the hormonal systems of the two kinds of organisms. Of course, many questions yet remain unanswered or barely touched upon. In most cases we do not even know exactly how the environmental signal is perceived or how it triggers the hormonal system. More is known about what happens from the hormone downstream, including the structure of some of the hormonal receptors and their tissue specificity. At the genetic level, a general pattern seems to emerge: Researchers have repeatedly identified specific short upstream sequences that regulate the response of genes to a variety of environmental signals and limit gene transcription to specific organs or tissues.

I believe these partial findings justify my expectation that hormones will soon be reconsidered as pivotal factors in explaining genotype-environment interactions.

From Phenotypes to Molecules and Vice Versa?

Can we really go from phenotypes to their molecular basis and vice versa (Pigliucci 1996)? Moreover, what are the advantages, if any, of doing so? The answer to the first question is a definite no, at least at this point in time. However, the real question is: Can we hope to bridge the gap in the foreseeable future? The answer to this more qualified question depends on whom one asks. There has been quite a bit of justified excitement about the recent progress in molecular developmental genetics. There is a real feeling that we are finally peeking through the enigmatic "black box" of development. Furthermore, at least some of this research is clearly interdisciplinary, with reductionist biologists partially interested in an ecological/evolutionary perspective, and organismal biologists eager to incorporate molecular data in their papers (Callahan et al. 1997; Purugganan 1998). However, there are theoretical reasons for caution on this matter (Schlichting and Pigliucci 1998). Since gene action is only local, epigenetic effects are the true link between genes and phenotypes. Epigenesis, however, is an astoundingly complex set of events that cannot at the moment (if ever) be reduced directly to the gene level, the focus of most current molecular genetics. Furthermore, the effects of the environment are not limited to direct alterations of gene transcription or translation, but echo throughout the epigenetic process. Instead of prying the black box open, it seems to me that a better metaphor would be poking a few holes in its surface.

As usual, looking at specific examples helps to fix our otherwise wandering thoughts, pulling them away from speculation and anchoring them to science. The best case studies I can think of as paradigms of what an integration of molecular and evolutionary biology could do for the study of plasticity are the research on phytochrome-mediated reaction to light in plants and the studies on the developmental and environmental control of eyespot in butterflies. The first is discussed at length in Chapter 11, while the second is summarized in Chapter 4. The amount of work done in the second case by a single group of researchers (Brakefield's laboratory in the Netherlands) is astounding and provides an excellent model for what can be done with a novel system in the span of a few years (Monteiro et al. 1994; Holloway and Brakefield 1995; Holloway et al. 1995; Brakefield and Breuker 1996; Brakefield et al. 1996; Koch et al. 1996; Roskam and Brakefield 1996; Brakefield 1997; Brakefield and Kesbeke 1997; Monteiro et al. 1997a,b,c).

However, these same examples also elucidate the limitations of the "evo-devo" approach. For one thing, it is clear that mutant analysis can bring us only so far in terms of understanding the whole-organism phenotype. Several crucial genes may never be identified because mutations at those loci are likely to be lethal or sublethal owing to major disruptions in development. Also, many of these mutations are characterized by such widespread pleiotropic effects that it often becomes difficult to even be sure that what one is studying is in fact a gene "for" the phenotype one was originally interested in. It is true that epistatic interactions can be addressed by the use of double and sometimes triple mutants. However, it is hard to build genotypes with more than two or three mutations, simply because the combined effects of the mutant genes become highly deleterious. Even the existing double and triple mutants are difficult to analyze because the gene-gene interactions, the level of pleiotropy, and the effects mediated by environmental conditions are so complex (Callahan et al. 1999). Further, we must take into account the fact that most of these mutants are high-level regulatory elements controlling vast arrays of pathways, and we do not even know how to begin to untangle their detailed structures. Finally, we simply lack conceptual tools for studying epigenesis and properties emergent from gene action. Our entire way of thinking about development is still gene-centered, and the current widespread acknowledgment of environmental and epigenetic components—while certainly an improvement upon the past—is still based on the idea that the genes reign supreme.[2] They do not, and we will not make any major leap forward until we internalize (and develop tools to deal with) the fact that genetics, epigenetics, and environment are *equally important* components in determining the phenotype.

The second general question asked at the beginning of this section was: Should we care? The answer is yes and no. I think that an argument can be made that the more we can learn about the different levels of complexity of living organisms, the more complete our knowledge of those systems will be. For example, it is generally appealing and intellectually satisfying to be able to consult books that will describe anything one might want to know about *Drosophila, Arabidopsis,* or *Caenorhabditis* from the molecular to the evolutionary level (see Chapter 8 in Schlichting and Pigliucci 1998), so multilevel approaches should be pursued independently of other considerations. Also, one can advance the reasoning that at least under some circumstances the solution, or a new insight, to a given problem can be provided by results obtained at a different level while studying the same organism. Let us consider the case of the evolution of shade avoidance plasticity: As we shall see in Chapter 11, the question of how this complex adaptive system came into existence (an evolutionary question) can begin to be addressed only by looking at the gene phylogeny of the phytochromes and other photoreceptors, an eminently molecular kind of analysis. However, the argument has been made (e.g., by G. Bell, pers. comm.) that ecology and evolution do not really need much input from cell and molecular biology, and vice versa. Most of the questions of interest to ecologists and evolutionists are of the type, *why* has a given trait evolved? That is, they address ultimate causes. The answer does not really depend on which particular cellular or molecular events or components made such evolution possible. The components are simply assumed to be there, as implicit in the fact that phenotypes are (in part) the result of molecular mechanisms interacting with the environment. Most of the progress in quantitative genetics and optimization theory throughout the twentieth century has been made possible by such decoupling. Analogously, most questions asked by cell and molecular biologists fall into the category of *how,* that is, of proximate causes. In this case, one can safely assume that such structures evolved for a function (especially if a current adaptive meaning can be attributed to them) and that, regardless of how that function changed through time, the researcher can find out how the pieces are put together. Again, the spectacular progress of molecular biology in the second half of the twentieth century does not owe much directly to ecology and evolutionary biology.

I will not go as far as saying that the two types of questions are completely decoupled. However, I think in the long run that the most reasonable model for the interaction between "how" and "why" questions in biology is that they *inform* each other. Therefore, while it is not necessary to know why something evolved in order to understand how it works, it is good to have independent evidence that what we are studying actually did evolve for adaptive reasons. Similarly, while we might not be too concerned about how a certain genetic-epigenetic system is put together in order to ask meaningful questions about what it is doing, knowledge of the mechanisms may help our research to some extent, at least by cutting down on the number of alternative hypotheses that we are considering.

Conceptual Summary

- While most molecular biologists do not frame their work with reference to genotype-environment interactions and their evolution, a substantial amount of molecular and physiological work during the past decades has directly addressed the molecular basis of phenotypic plasticity.
- Quantitative trait loci studies have been conducted on genotype-environment interactions, and have started to unravel the complexity of the genetic response to environmental heterogeneity in natural populations.

- Plasticity genes are defined as regulatory loci that respond directly to a specific environmental stimulus by triggering a specific series of morphogenic changes. Not all regulatory genes are plasticity genes, and the genetic basis of any plastic response will necessarily include significantly more genes than those directly sensing the environment. However, the latter category is conceptually important because its very existence cannot be explained without invoking the action of natural selection.
- Molecular studies of genotype-environment interactions have highlighted the existence of specific responses elicited by one particular type of stress, of responses induced by a limited number of stresses, and of generalized responses to a variety of stressful situations.
- Epistasis and pleiotropy at the molecular level have been extensively documented and make the interpretation of patterns of plastic reactions very difficult if one does not have access to molecular information. As a cautionary note, it should be kept in mind that molecular epistasis and pleiotropy do not necessarily correspond to the same concepts as understood in quantitative genetics.
- Hormones are the true interface between the genetic level of action and the external environment in that they perform two important roles: They shape the organism in response to the genetic programming (G-component), and they carry the information from environmental receptors, which triggers the genotypic-specific reactions we call phenotypic plasticity (E-component).
- As much as molecular biology can inform questions in evolutionary biology (and, to a lesser extent, the other way around), the two disciplines are characterized by distinct types of question (how versus why), which can be pursued to a large extent independently of each other.

6

The Developmental Biology of Phenotypic Plasticity

Organic life, we are told, has developed gradually from the protozoon to the philosopher, and this development, we are assured, is indubitably an advance. Unfortunately it is the philosopher, not the protozoon, who gives us this assurance.

—Bertrand Russell (1872–1970)

CHAPTER OBJECTIVES

To discuss the relationship between phenotypic plasticity and a variety of developmental phenomena. A series of examples are introduced to highlight the way in which a developmental perspective helps to understand plasticity. The mechanisms underlying developmental plasticity above the genic level are discussed, including heterochrony and hormonal actions. Evidence for the adaptive significance of developmental plasticity is introduced by example. The concept of ontogenetic contingency is discussed, together with its relevance for discussions of limits of adaptive plasticity. The idea that the manifestation of plasticity depends on developmental *windows* is considered, and a few actual instances are examined.

Which Plasticity Is *Not* Developmental?

It can be argued that *all* morphological plasticity is developmental in nature, and therefore a chapter devoted to developmental phenotypic plasticity is at least partially redundant. After all, any trait that we measure in the "adult" or reproductively mature form of an organism had to develop from somewhere, and that development occurs during a more or less long period of time, not instantaneously. Indeed, I have argued elsewhere that a key to understanding adult differences in reaction norms is precisely to follow how these reaction norms gradually diverge throughout the ontogeny (Pigliucci et al. 1996). Nevertheless, pushing this line of reasoning too far would essentially reduce *all of biology* to developmental biology, a prospect that has not yet attracted many supporters. I think the limit lies in the diminishing returns that are associated with reducing our inquiry from adult phenotypes all the way back to the zygote. Ultimately, differences in adult phenotypes originate somewhere along the time line connecting zygotes to formed individuals. However, some traits develop much later than others and are so close to the mature morphology (or behavior) that in studying them we are far from the realm of developmental biology sensu stricto. Furthermore, the adult types are often quite distinct organisms from their embryonic or larval counterparts. In many cases, different stages of the life cycle occupy distinct ecological niches. This may imply distinct selective pressures, so that phenotypic plasticity at the larval stage can be much more than just the precursor of plasticity of adult forms. This is essentially why it is important to determine when and in what form plasticity occurs during the ontogeny of an organism. Furthermore, obviously most physiological plasticity and some behavioral plasticity barely include a developmental component—additional evidence that plasticity and developmental plasticity are not synonymous.

The developmental plasticity question could be asked from an entirely different perspective: Instead of considering development as an explanation of observable patterns of plasticity, one could legitimately ask why we should care about environmental effects on developmental systems. The answer may depend on which conception of organismal biology one embraces. Van der Weele (1993) pointed out that there are three schools of thought regarding the importance of environment-development interactions. The classic Neo-Darwinian approach is built around the so-called *gene-centric* view: The genetic program is central to our understanding of development, and environmental effects are background noise to be minimized. This argument has retarded the study of plasticity in general (Chapter 3), but it has lost much of its lure within the evolutionary community. The second attitude is the so-called *internalist* perspective. According to this view, development is determined more by the environment than by genes, but this is the *internal* environment of the cell and cell-cell interactions, not the external, biotic or abiotic, environment typically addressed by plasticity studies. Finally, the *constructionist* approach sees the external environment as the dominant player in a continuous interaction with the genotype. I agree with van der Weele that the three perspectives are not mutually exclusive, but in fact represent the three players constantly interacting to shape ontogenetic trajectories: genes, epigenetic effects mediated by the internal environment, and external environmental conditions (what Schlichting and I [1998] termed the developmental reaction norm, or DRN).

Furthermore, each aspect may in fact prevail in a particular organism or affect a particular trait or segment of the ontogeny. Clearly, the external environment does not contribute much to the determination of the number of nostrils in vertebrates, and even the internal conditions simply follow a strict genetic determinism. On the other hand, several species of vertebrates (e.g., many fishes) alter the shape of their skulls throughout development as a direct response to dietary conditions, a case in which factoring the external environment is essential to an understanding of the overall equation. Throughout this book, I focus on the classical definition of plasticity as a response to external environmental conditions. However, see Schlichting and Pigliucci (1998) for a broader perspective that treats the internal and external environments as two sides of the same coin.

In the following, I consider four aspects of plant and animal developmental plasticity. First, I review a few examples to delineate more clearly what I mean by developmental plasticity. This section amounts to an *empirical* definition of the subject area. Second, I discuss some of the proximate mechanisms yielding developmental plasticity. Such a discussion stops short of the molecular level of analysis, as Chapter 5 was devoted entirely to that topic from a more general perspective. Third, I examine the case for the adaptive meaning of developmental plasticity. A broader discussion of the ecology of plasticity and its adaptive role follows in Chapter 7. Last, I address the suggestion that developmental mechanisms can pose a limit to plasticity in the form of various constraints and costs. The increasingly important topic of costs to plasticity in general is discussed in more detail in Chapter 7.

Case Studies: Development *Is* Plastic

Examples of developmental plasticity are now accumulating rapidly, in part as the result of a renewed interest in developmental biology in general, spurred by recent advances in molecular genetics, and in part by the lingering frustration of evolutionary biologists after the failure of the Neo-Darwinian synthesis to explicitly incorporate development and embryology (Mayr and Provine 1980). The following is by no means meant as an exhaustive review, but simply as a brief survey of the variety and extent of ontogenetic phenotypic plasticity studies. Each example has been chosen to illustrate one particular aspect of developmental plasticity and its relationship to broader questions of phenotypic evolution.

First, it is clear that the amount of phenotypic plasticity of a given character can change throughout ontogeny, as was demonstrated in a study by Cheplick (1995a) on the corm-forming grass *Amphibromus scabrivalvis*. He compared the number of ramets (individual branches of a clone) produced by clones grown with and without fertilizer. Increased nutrient availability enhanced production of ramets as expected. However, the *intensity* of the response also increased throughout ontogeny (Fig. 6.1), showing that the amount of both plasticity and genotype-environment interaction depends on exactly when these quantities are measured.

Developmental plasticity can be understood in mechanistic terms by studying the interactions between environmental stimuli and their transducers, especially hormones (Chapter 5). Perhaps one of the best examples of this is the phenomenon of temperature-dependent sex determination in reptiles (Crews et al. 1994; Girondot et al. 1994; Janzen 1995; Crews 1996). In some reptiles the sex of the hatchling is not determined genetically, but by the temperature experienced by the eggs (see Fig. 9.5). There usually is a very steep threshold temperature, on each side of which the developing animals become either male or female. For example, in the common snapping turtle, temperatures between 23° and 27°C yield only males, females are obtained from eggs raised at or above 29.5°C, and mixed sex ratios are produced in between, with an increasing proportion of females at higher temperatures. Rhen and Lang (1995) investigated the relationship between temperature, sex, and growth rate in the snapping turtle, to test the hypothesis that the proximate (mechanistic) cause of sex dimorphism is a simple alteration in the developmental trajectory and, specifically, a hormonally mediated differential growth rate between males and females. In order to test their hypothesis, however, they had to somehow decouple temperature and sex, which are normally completely confounded from a statistical point of view, since only individuals of one sex occur within a particular range of temperatures. To achieve the decoupling, they produced males and females

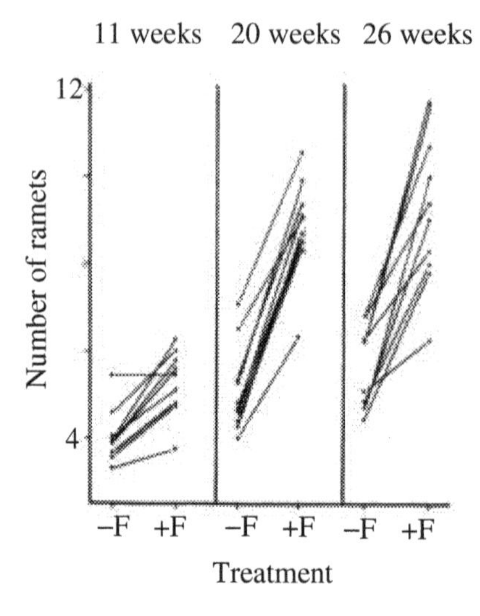

FIG. 6.1. The plasticity to nutrient availability (F = fertilizer) of number of ramets produced by the grass *Amphibromus scabrivalvis* depends on the time during ontogeny at which it is measured. (From Cheplick 1995a.)

at will at the "wrong" temperature by applying appropriate hormonal manipulation. When the eggs were treated with estradiol, they produced females, regardless of the temperature; when treated with an aromatase inhibitor, the eggs developed mostly into males at temperatures that would normally induce females, but kept producing males at male-inducing temperatures. This cunning scheme allowed Rhen and Lang to clearly demonstrate that hormonal action catalyzed by differences in temperature results in differential growth rates that indeed produce sexual dimorphism. Developmental plasticity in this case is mediated by relatively simple temperature-hormone interactions.

Developmental plasticity can also play a major role in explaining morphological differentiation of geographically separated organisms. Martin-Mora and James (1995) investigated plasticity in shell morphology in the queen conch, *Strombus gigas.* They started by characterizing the multivariate phenotypic space occupied by six natural populations of this species in the Bahamas, and found differences in the length of the spines, the width of the shell, the length of the aperture, and the length of the spire. They then transplanted individuals from one site to the other five and followed them throughout their development. Interestingly, the final morphology of the transplanted populations resembled the morphology of each of the resident populations, although the shift in phenotype did not completely bridge the morphological gap between transplants and residents. This is strong evidence that both phenotypic plasticity and genetic differentiation have to be considered in explaining geographical patterns of morphological diversity.

Another intriguing paper on developmental plasticity addressed the possibility of divergent patterns of plasticity between organs that are functionally equivalent but developmentally distinct. Dong and Pierdominici (1995) compared the plasticity to light availability in two types of stems produced by three species of grass. *Agrostis stolonifera* only produces stolons, a type of stem that grows horizontally—typical, for example, of strawberries. A second species, *Holcus mollis,* grows only by rhizomes, horizontally creeping underground stems that bear both roots and leaves. The third species considered by these authors was *Cynodon dactylon,* a grass that forms both stolons and rhizomes. Dong and Pierdominici exposed the three species to different levels of light intensity and studied the developmental morphology of the two types of stems. Both types responded plastically by increasing branching under high light levels, but only stolons elongated more than the control under low light levels, *regardless of the species.* The authors then analyzed the patterns of resource allocation in these species, and found that partitioning to rhizomes decreased under low light, whereas stolons received the same amount of resources, again independently of the taxon. The interpretation of these results was that rhizomes serve mostly as storage organs, whereas stolons function as foraging organs for light (hence their enhanced plasticity). If confirmed by studies on additional species, this would be an example of the way in which developmental phenomena such as organ identity and allocation patterns result in different patterns of plasticity, possibly because of differential selective pressures.

Developmental plasticity can also interact with genetic differentiation to yield a complex pattern of within- and between-species morphological variation, as demonstrated in a study by Pigliucci et al. (1997) in two species of *Lobelia* exposed to different nutrient availability. We characterized the ontogenetic trajectories of several traits in two populations each of the closely related species *L. cardinalis* (the cardinal flower) and *L. siphilitica* (the blue lobelia). We found that aspects of the ontogenies affecting the timing of flowering, the production of rosettes, and the rate of growth of the flowering stem were clearly distinct between the two species, but largely invariant to environmental influences. On the other hand, components of the ontogeny affecting traits such as rate of leaf production, branching, and fruit production were highly responsive to the nutrient treatments, but in a similar fashion in both species. This complex scenario was summarized in a multivariate analysis of the data, which showed that the two species differed more dramatically in their development at high nutrient levels, but almost converged when raised under stress (Fig. 6.2). This kind of study raises questions concerning the relationship between environmentally dependent constraints and selection: Was the convergence of ontogenies under stress due to a shared selective history under similar conditions between the two species, or was it due to some kind of fundamental developmental limitation intrinsic in the body plan of *Lobelia* sensu lato? Given that the

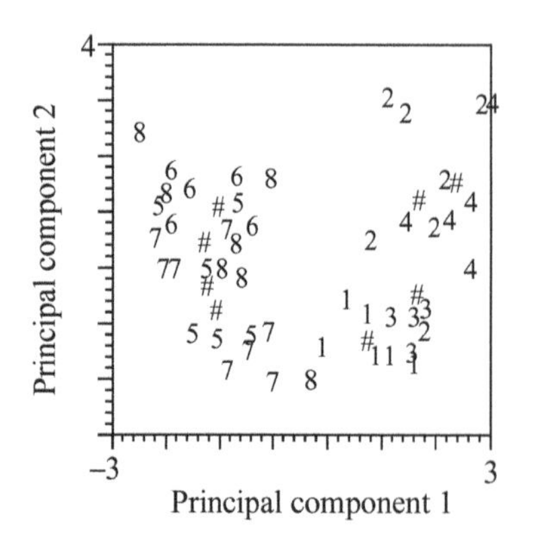

FIG. 6.2. Principal components analysis of developmental phenotypic variation in *Lobelia cardinalis* and *L. siphilitica.* The first principal component explains variation in flowering time, rosette production, and rate of elongation of the flowering stem. This component distinguishes between species (*L. cardinalis* on the left), but not between treatments. The second principal component summarizes variation in the rate of leaf production, branching, and fruit production. It does not distinguish between species, but separates response to high nutrients (top) from low nutrients (bottom). 1, 3 = *L. siphilitica,* low nutrients; 2, 4 = *L. siphilitica,* high nutrients; 5, 7 = *L. cardinalis,* low nutrients; 6, 8 = *L. cardinalis,* high nutrients; # = centroids. (Redrawn from Pigliucci et al. 1997.)

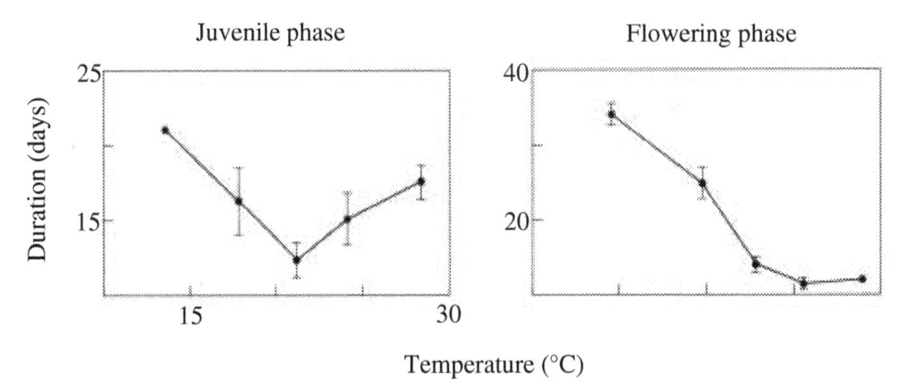

FIG. 6.3. Effect of temperature on the duration of two phases of development (juvenile, left and flower, right) in *Petunia* × *Hybrida.* Note how the pattern of plasticity to temperature is dramatically different depending on when during the ontogeny it is considered. The plant is not sensitive to photoperiod in either phase, but responds to it in the time between these two phases (not shown). (From Adams et al. 1999.)

ecology of these taxa is distinct, we tended toward the latter explanation, but studies of other species in the group will be necessary to confirm the developmental constraint hypothesis.

As a final example, let us consider a study by Adams et al. (1999) on the effect of temperature, total light integral, and photoperiod in *Petunia* × *Hybrida.* The authors carried out reciprocal transfer experiments between short (8-h) and long (16-h) photoperiods, while simultaneously manipulating the temperature as well as the total amount of light that the plants were receiving. This taxon shows three phases of growth that were found to vary in their sensitivity and response to the environmental conditions (Fig. 6.3). During the juvenile phase, the plants were insensitive to photoperiod but responded to both light integral and temperature; after this initial phase, time to flowering was markedly influenced by photoperiod as well as by the other two factors; during the final phase of flower development, however, plants were again photoperiod-insensitive and responded chiefly to temperature. Taken together, these data highlight the fact that one has to be careful when studying phenotypic plasticity because not only can the same trait (flowering time in this case) be influenced in a different way by distinct environmental factors, but its response to such factors may dramatically depend on *when* the environment changes.

It is clear from these and similar examples that ontogenetic trajectories can indeed be plastic, and that the internal conditions of the organism do to some extent interact in a continuous process with the external environment to yield morphological differentiation among adult individuals. But which mechanisms are responsible for such developmental plasticity?

Mechanisms of Developmental Plasticity above the Genic Level

When we ask questions about causality in biology we can refer to any of a number of levels. Aristotle's distinction between proximate and ultimate causes is certainly applicable here, with the caveat that biological phenomena have a hierarchical series of proximate causes. In evolution, the equivalent of an ultimate cause (when there is any) is the increase in fitness brought about by natural selection.[1] On the other hand, as far as proximate causes go, we have a panoply of choices. In some sense, one can think of the field of developmental biology itself as being based on this search for hierarchically nested causes. For example, what is the "cause" of the vertebrate eye? One set of explanations may refer to the internal anatomy of the structure and another to the tissue-tissue interactions that occur during development in order to generate that structure; or we could think of the patterns of tissue and cell differentiation or of the hormonal signals and receptors that cause such patterns; or we could investigate the function and structure of the genes underlying the whole process. In this section I discuss a few investigations of the proximate causes of developmental plasticity above the genic level of action. Chapter 5 is devoted entirely to the last, the most basic—and most remote from the whole-organism—level of proximate causality.

The first example of a study on the mechanistic causes of developmental plasticity concerns temperature-dependent determination in reptiles. As we have seen, in the common snapping turtle low temperatures produce males, whereas higher temperatures produce females, and this is due to hormonal action. Crews (1996) describes the specific mechanism in terms of the differential kinetic response to temperature of two enzymes that compete for the same hormonal substrate. The hormone testosterone can be the substrate of both 5α-aromatase (in which case it is transformed in estradiol) or of 5α-reductase (in which instance it turns into dihydrotestosterone). It turns out that this is a main switch along the developmental pathway that leads to gender differentiation. At higher temperatures, 5α-aromatase is a better competitor for the substrate, and the resulting product, estradiol, is directly causally linked to the differentiation of female gonads. But at lower temperatures, 5α-reductase competes efficiently for testosterone; the resultant dihydro form of the hormone leads to two forms of androstanediol, which lies on the developmental pathway to male gonads. It is (by and large) as simple as that, and this explains the results obtained by Rhen and Lang with hormonal manipulation discussed above.

Another example that helps toward an understanding of the proximate causes of developmental plasticity involves the classical and singularly understudied class of phenomena known as heterophylly in plants (Cook 1968; Cook and Johnson 1968; Mitchell 1976; Kane and Albert 1982; Deschamp and Cooke 1985; Winn 1996a,b; Wells and Pigliucci 2000). This is the situation in which a particular leaf can acquire one of two (or more) possible morphologies depending on the external environmental conditions. Leaf shape can respond to different environmental stimuli, and to various degrees, but one of the most common effectors is water availability. Goliber and Feldman (1990) studied the proximate causes of heterophylly in response to water in the aquatic plant *Hippuris vulgaris* (Fig. 6.4, lower panels). They found that the leaf primordia are fully competent to develop into either type of leaf up until the tenth plastochron[2] (i.e., when they are about 300 μm long). After that, the ontogenetic trajectories deviate sharply from one another (Fig. 6.4, upper panel). Between the tenth and the twenty-first plastochrons the meristems are apparently partially competent, since hybrid leaves are formed, with the basipetal portion of the tissue of one type and the apical portion of the other type. This indicates that the developmental decision is local, not made at the whole-leaf level. Since the effects of various hormones (internal environment) on heterophylly are well known, this type of plasticity allows the researcher to connect three different levels of biological phenomena with the external stimulus: morphological, anatomical, and hormonal. Essentially, all the pieces of the puzzle are in place except for the genes coding for the appropriate environmental receptors (which are hitherto unknown in any heterophyllous species). In *Proserpinaca palustris* (Fig. 6.5), for example, exogenous applications of gibberellic acid mimic the patterns of heterophylly elicited by changes in photoperiod, whereas exogenously applied abscisic acid serves as a surrogate for fluctuating water levels. Joint applications of these hormones suggest that their relative concentration may mediate the expression of heterophylly in *Proserpinaca* species. A genetic analysis in this group would be particularly feasible because the genus is composed of three species, including a heterophyllous one, a nonplastic one, and a hybrid showing intermediate levels of plasticity.

A good example of proximate causality of developmental plasticity at the level of allocation to different growth fields within the organism is provided by a study by Thompson (1999) on diet-induced plasticity in the grasshopper *Melanoplus femurrubrum.* Thompson investigated the possibility that plasticity in size and shape of the body and head may be mediated by a combination of changes in overall body size and of alterations in the localized growth rate of individual morphological structures. The author exposed individuals belonging to thirty-seven full-sib families to a diet of either soft (high nutrients) or hard (low nutrients) plants. The results demonstrated that the plasticity of head shape in response to diet is due to environmentally dependent changes in allocation to tissue growth. The animals were able to maintain growth of head size on low nutrients by simultaneously reducing overall growth of body size. Thompson also found

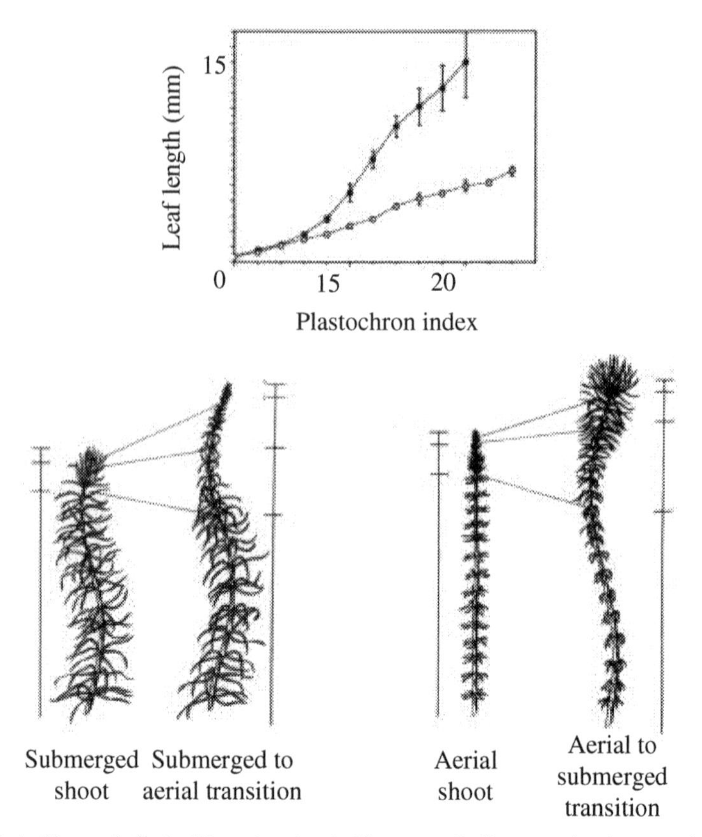

FIG. 6.4. Heterophylly in *Hippuris vulgaris.* Upper panel: divergent developmental trajectories of aerial and submerged leaves, measured as leaf length versus plastochron index (a measure of developmental stage). Lower left: fate of leaves of different age when the plant is shifted from submerged to aerial conditions. Lower right: fate of leaves of different age when the plant is shifted from aerial to submerged conditions. (From Goliber and Feldman 1990.)

significant variation among families for the plasticity of the shape of their ontogenetic trajectories, with different genotypes using distinct combinations of body size and head growth, analogous to the case of *Lobelia* discussed above. The data also showed a marked dependence of the degree of genetic variation for plasticity on the ontogenetic stage, with some periods of development associated with significant genotype-environment interactions and others that lacked variation in this parameter. The feeling is inevitable that, when one considers development and environment simultaneously, things get much more complicated, but that one also gains insights that would not have been possible by considering the two factors in isolation.

Perhaps the most frequently called upon mechanism for developmental plasticity is *heterochrony.* The term indicates a variety of alterations in the rate or timing of developmental events, resulting in an array of peramorphic (overdeveloped) or paedomorphic (juvenilized) forms (Haldane 1932; Gould 1977; Alberch et al. 1979; Raff and Wray 1989; McKinney and McNamara 1991; Conway and Poethig 1993; McKinney and Gittleman 1995; Raff 1996; McNamara 1997; Friedman and Carmichael 1998; Cubo et al. 2000; Porras and Munoz 2000), and it represents a hierarchically higher level of causality compared to the examples discussed so far. There are basically six types of heterochrony (Alberch et al. 1979): The onset of a developmental event can be moved up or postponed; the offset of the same process can occur earlier or later than in the reference taxon; and the rate at which said event occurs can be slowed down or accelerated. Furthermore, combinations of different kinds of heterochronic phenomena can also occur, thereby considerably complicating the analysis of developmental trajectories and increasing the array of possible morphological outcomes (e.g., Guerrant 1982; Jones 1992). Not surprisingly, heterochrony has also been invoked in the study of developmental plasticity. For example, Strathmann

FIG. 6.5. Heterophylly in *Proserpinaca palustris.* The sequence shows the dramatic changes in leaf morphology from below water (left) to above water (right), passing through a rapid intermediate stage (middle). (Photo: Carolyn Wells.)

et al. (1992) studied the effect of food supply on the development of sea urchin larvae. They exposed larvae in the laboratory and in the field to different amounts of food and observed heterochronic shifts owing to alterations in the timing of development of larval and juvenile structures. Most interestingly, they noted that the direction of these plastic shifts paralleled morphological and ecological changes across closely related species in evolutionary time.

Another example of the role of heterochrony in developmental plasticity is offered by the development and facultative metamorphosis of some amphibians, particularly several species of salamanders. Reilly (1994) discusses changes in head shape in salamanders that occur during the rather rapid transition between larval and adult forms (Fig. 6.6, upper left panel). Some salamanders display a diet-induced plasticity regulating the timing of their metamorphosis. For example, individuals of *Ambystoma talpoideum* are capable of delaying the onset of metamorphosis and can mature sexually when still morphologically at a larval stage (Fig. 6.6, lower left panel).

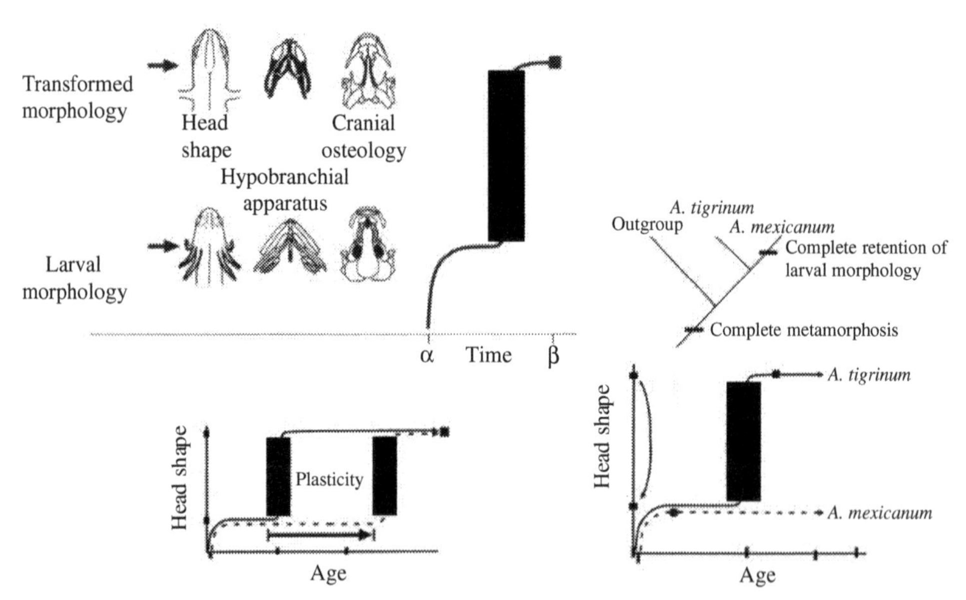

FIG. 6.6. Possible heterochronic changes affecting head shape in salamanders. Upper left panel: larval versus adult morphologies of head shape, hypobranchial apparatus, and cranial osteology. The curve on the right shows the ontogenetic trajectory that leads from one morphology to the other (between times α and β). Lower left panel: heterochrony induced by phenotypic plasticity to diet in *Ambystoma talpoideum.* Some individuals simply delay the onset of metamorphosis, thereby retaining a larval morphology for a longer period of time. Right panel: an evolutionary transition associated with loss of plasticity in the derived *Ambystoma mexicanum* when compared to its sister taxon *A. tigrinum.* (From Reilly 1994.)

What is interesting from an evolutionary point of view is that similar gains or losses of plasticity mimic the direction of evolutionary change in some lineages of salamanders. Thus, whereas *Ambystoma mexicanum* has lost the capacity to achieve metamorphosis, thereby permanently retaining its larval morphology, its sister taxon *A. tigrinum* still displays the typical S-shaped ontogenetic trajectory (Fig. 6.6, right panel).

Overall, heterochrony may indeed be a powerful causal link between developmental phenotypic plasticity and more mechanistic explanations of plasticity such as those discussed above or in Chapter 5, in terms of hormonal action or molecular genetics, respectively. However, one must exercise caution in adopting an indiscriminate use of heterochronic explanations to understand the mechanics of plastic shifts in the phenotype. There are three limitations to the heuristic use of the concept of heterochrony. First, often one does not see "pure" heterochronic changes, but mixtures of different heterochronic phenomena, which result in possible ambiguities of interpretation of the developmental processes underlying the observed patterns (e.g., Guerrant 1982; Jones 1992). Second, we know (Nijhout et al. 1986) that heterochrony at the developmental level may not correspond to heterochrony at the genetic level (i.e., to shifts in the timing or rate of the action of genes), and vice versa (for a recent example, see Cubo et al. 2000). Third, other, less studied and more difficult to detect, developmental phenomena such as heterotopy (Wray and McClay 1989; Fang et al. 1991) or the differential location of gene expression (Guerrant 1982) may have the same or even wider implications than heterochrony.

Finally, I refer the reader to my previous book (Schlichting and Pigliucci 1998) in which the concepts of plasticity, development, and character correlations are integrated in the idea of developmental reaction norms (DRNs). What I have included here is necessarily a subset of examples and considerations focused on plasticity of what is a more extensive treatment of phenotypic evolution in its multifarious and complex aspects, which is found in *Phenotypic Evolution: A Reaction Norm Perspective.*

Is Developmental Plasticity Adaptive?

The general framework for studying plasticity as an adaptation is laid out in Chapter 7. However, we have to discuss a few examples here of developmental plasticity that are seemingly adaptive because the intricacies of some developmentally plastic responses to environmental heterogeneity strongly suggest an adaptive explanation for their historical origin. This makes developmental plasticity research one of the best approaches to an understanding of the role and limits of natural selection in shaping the response of organisms to changes in the environment.

As we saw above, one of the clearest examples of developmental plasticity is facultative paedomorphosis in salamanders (Semlitsch 1987; Semlitsch and Wilbur 1989; Whiteman et al. 1996; Goodwin et al. 1997). These animals may follow two distinct life history strategies: metamorphosis, with the consequent transition from a larval-aquatic to an adult-terrestrial, sexually mature stage; or paedomorphosis, that is, the retention of larval characteristics (and habitat occupancy) while reaching sexual maturity. Whiteman (1994) discussed the conditions under which facultative paedomorphosis is expected to evolve extensively, and the following is a brief summary of his arguments. He suggests that three scenarios would lead to developmental plasticity in salamanders. First, what he terms the *paedomorph advantage hypothesis,* according to which paedomorphosis is maintained by a combination of biotic and abiotic factors. If the animals are living in a fairly stable aquatic environment and are surrounded by harsh terrestrial conditions, paedomorphosis is obviously an advantage. However, the occasional year in which the pond's conditions are critical (such as a particularly dry season) would be enough to maintain the facultative option of metamorphosis. The second scenario he calls *best of a bad lot hypothesis.* This is a situation in which metamorphosis is again the favored strategy, but only large larvae can achieve it. The second-best option would then be represented by larvae that are sufficiently large to reach sexual maturity, but not large enough for metamorphosis to take place, an example of developmental constraint. These would be the paedomorphs. Even smaller larvae would have to forgo both metamorphosis and paedomorphosis, and would incur the greatest disadvantage. Why would metamorphic individuals in the population at all sizes not simply replace the paedomorphs? The answer is that in unfavorable habitats the advantage of reaching sexual maturity at an early age (paedomorphs), and therefore of having ensured reproduction, may balance the advantage afforded by metamorphosis at a larger size and later age (since the animal may never get there). The last situation leading to the maintenance of facultative paedomorphs is the *dimorphic paedomorph hypothesis.* This applies under conditions intermediate between those described by the previous models, so that two types of paedomorphs may exist in the population. Here the largest individuals become paedomorphic because they have an advantage over all the others and can stay in the most stable environment; medium-size larvae would go through metamorphosis in order to escape such intense competition; finally, smaller larvae would not be able to achieve metamorphosis and would be forced to stay paedomorphic (however, they would retain the advantage of not having to delay sexual maturity). This discussion illustrates that an apparently simple series of developmental choices can be the result of a complex set of circumstances and be maintained by more than one environmental regime in a given population. The significance of Whiteman's scenarios is that they lead to predictable hypotheses relating the distribution of body size in these animals to the number of individuals choosing one strategy over another, as well as to the prevalent environmental conditions.

Predator-induced alterations in morphology of a prey are perhaps the most intuitive instances of adaptive developmental plasticity. Indeed, the very first case of phenotypic plasticity ever described was the alteration in body shape in the crustacean *Daphnia,* which is related to the presence of a predator (Woltereck 1909, Chapter 1). A more recent example was discussed by Bronmark and Miner (1992), who studied changes in the morphology of the crucian carp when in the presence of its main predator, the piscivorous pike. Both field and laboratory experiments confirmed that the carp increases its body depth, an effective predator-avoidance strategy since the pike is a gape-limited predator. However, the carp incurs costs in becoming deeper-bodied, since it lives in resource-limited environments. The balance between costs of expressing the wrong phenotype and advantages of the predator-limiting strategy explains the evolution of phenotypic plasticity (see Chapter 7).

Variable patterns of predation can also instigate the evolution of a complex series of developmentally fixed and plastic behavioral patterns. A good example is the population variation in vertical migration patterns of copepods. Neill (1992) reckoned that the optimal pattern of migration in a lake would depend on the type and frequency of predation experienced by the prey at different stages of the ontogeny. By comparing the behaviors of different populations inhabiting two different lakes, he found evidence of predator- and population-specific patterns of migration that were distinct by the life history stage. Furthermore, he uncovered both developmentally fixed and predator-induced patterns of migration, again in a fashion dependent on the particular life history stage. For example, copepods from a given location changed their responsiveness to predators during their ontogeny: Small (young) individuals migrated through the water column when a predator was introduced, but large (older) copepods were obligatory nocturnal migrators regardless of the presence of predators. The ontogenies of different populations were related to the kind of predators experienced by these populations. Overall, these animals must have had a complex evolutionary history of genes increasing or decreasing plasticity throughout their ontogenies.

A splendid example of developmental plasticity that can hardly be interpreted as nonadaptive is the case of the diet-induced developmental polymorphism in the caterpillar of *Nemoria arizonaria.* Greene (1989) described how caterpillars that develop in the spring on oak catkins assume a phenotype that mimics the plant's inflorescences. Other caterpillars of the same species develop during the summer, when the favored host has lost the catkins. Accordingly, they mimic the tree's twigs. Furthermore, the cue is offered to the caterpillar by the chemical composition of its diet. Larvae raised on catkins apparently detect the low levels of tannins in these structures and develop as catkin-mimics. On the other hand, larvae raised on twigs perceive high levels of tannins and therefore mimic twigs. Conclusive evidence that tannins are the actual cue was found when Greene raised the caterpillars on catkins artificially augmented with tannins; predictably, the larvae developed into twig-mimics (Fig. 6.7).

Phenotypic plasticity mediated by maternal effects has also been invoked as an example of adaptive plasticity, for instance in the case of the seed beetle, *Stator limbatus* (Fox et al. 1999). The argument is that if the female's environment is a reliable indicator of the environment likely to be encountered by the offspring, then female-mediated manipulation of offspring phenotypes will increase the latter's fitness (Agrawal et al. 1999). Fox and co-workers studied the female's manipulation of egg-size plasticity across host plants as a way of influencing the phenotype of the next generation. They found that even though plasticity was partially constrained by a high genetic correlation of egg size across hosts, there was evidence for host- and population-dependent variation in plasticity and that the genetic constraint was not such as to preclude further evolution of this trait. The same argument of adaptive plasticity through female manipulation of offspring resources (and, again, in particular egg size) has also been advanced to solve

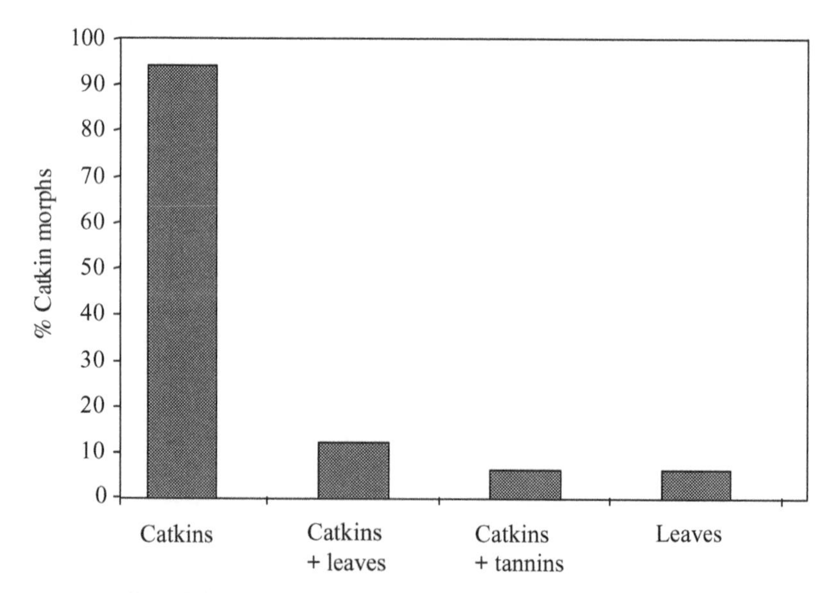

FIG. 6.7. Effect of diet on mimicry in caterpillars of *Nemoria arizonaria.* The chemical cue is provided by tannins, naturally present in twigs and leaves. (From data in Greene 1989.)

the problem of why some reptiles show the marked temperature-dependent determination of sex ratio that we have already examined (Shine 1999).

It seems clear from these and similar examples that it is difficult to argue that some instances of developmental plasticity not only are adaptive in the sense of being currently useful, but indeed are the result of a historical process of adaptation (Gould and Vrba 1982). Far from me to advocate a Panglossian-adaptationist research program (Gould and Lewontin 1979; Pigliucci and Kaplan 2000), but it is hard to conceive how else the intricate proximate causal mechanisms of these types of plasticity could have evolved if not in response to specific challenges posed by predictable environmental variation. Of course, it is always possible, indeed likely, that the *elements* now constituting the plastic response (e.g., chemical sensors of predator presence or of tannins) did evolve for different reasons and at one time had a different function (Ganfornina and Sanchez 1999). However, they must have been co-opted long ago, in order to ensure the sophisticated trigger-response mechanisms that we have discussed. Of course, further confirmation of the historical nature of these adaptations requires accurate phylogenetic tests, along the lines indicated earlier for the case of the evolution of metamorphosis in *Ambystoma* salamanders.

Ontogenetic Contingencies and Limits to Developmental Plasticity

Several categories of limits to plasticity have already been identified. In this book, I discuss genetic constraints (mostly in Chapter 4), various categories of costs (mostly in Chapter 7), and developmental limitations (mostly here). These types of limitations to which plastic responses are subjected actually slide into each other. It is easy to imagine how a developmental constraint can be mechanistically reduced to a genetic one (see Schlichting and Pigliucci 1998, Chapter 6). In addition, costs related to resource allocation can themselves be interpreted as limitations of the developmental or genetic machinery. Having said this, I will now discuss the subject from as much of a developmental perspective as will be useful here.

One of the first papers to explicitly point out developmental limitations to adaptive phenotypic plasticity when an organism is exposed to a novel environment was van Tienderen's discussion of morphological variation in *Plantago lanceolata* (van Tienderen 1990). *P. lanceolata* in The Netherlands can occupy distinct habitats, two of which are characterized as *hayfield* and *pasture*. Plants growing in hayfields experience intense competition from a high density of neighbors, and their season can be cut short by early harvests. On the other hand, plants from pastures experience little competition from other vegetation, but have to withstand heavy grazing pressure by herbivores. Van Tienderen demonstrated that the two types of populations have evolved genetically specialized strategies to cope with their specific environments (i.e., they have evolved toward the status of ecotypes [Turesson 1922]). Accordingly, hayfield populations display an erect growth habit and produce few rosettes (vegetative growth), long leaves, and few but large spikes (reproductive structures); these features allow these plants to compete effectively for light and be reproductively active early in the season, before harvest time. The morphological syndrome of pasture individuals of *P. lanceolata* includes prostrate plants, short leaves, many rosettes, and many but small spikes, almost exactly the opposite of the hayfield plants. This second suite of traits allows the pasture populations to respond effectively to grazing and trampling. Van Tienderen then performed reciprocal transplant experiments in which hayfield populations were grown in pastures, and vice versa. He found (Fig. 6.8) that pasture plants displayed a *hayfield phenotype* when grown under the high vegetation typical of hayfields; similarly, a hayfield population clearly showed a *pasture phenotype* when grown under the low vegetation characteristic of pastures. This obviously indicates some degree of adaptive plasticity even in these specialized populations. However, the plasticity was not sufficient to entirely close the gap between the two genetic types, so van Tienderen interpreted this as a limit to already existing plasticity: While *P. lanceolata* displays adaptive plasticity, this is not sufficient to overcome the local adaptation of each ecotype to its most

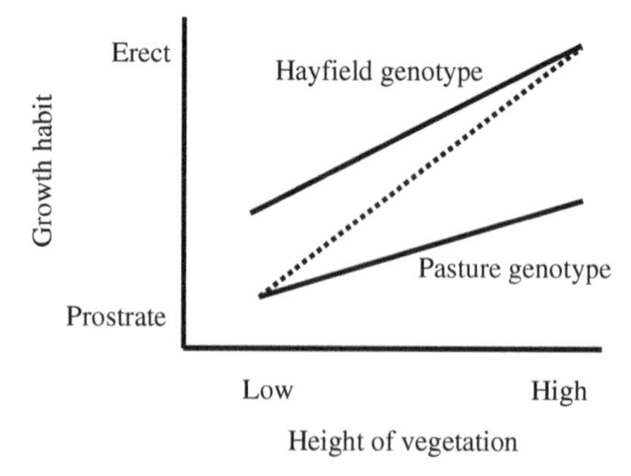

FIG. 6.8. Limits to developmental plasticity to novel environments in two populations of *Plantago lanceolata.* The pasture plants shift toward a hayfield phenotype when grown under hayfield conditions; similarly, the hayfield population shifts toward a pasture phenotype when exposed to pasture conditions. However, the plasticity of neither population is sufficient to entirely bridge the morphological gap between the two, which thus requires some degree of genetic specialization or further selection on the reaction norm. (Redrawn after van Tienderen 1990.)

commonly encountered environment. Of course, one can argue that if these populations were in fact to experience the novel environment for a sustained period of time, their reaction norms might be modified by selection in order to overcome the initial limit. This is a very likely scenario indeed, but it does not invalidate the conclusion that currently existing plasticity implies a limited adaptability to novel environmental conditions. Furthermore, most constraints, be they developmental or genetic in nature, are probably *local,* in the sense that they may be overcome by further mutation-selection (Camara and Pigliucci 1999; Camara et al. 2000)—unless they are actually defined by fundamental physical limits imposed on the biology of the organism.[3]

One of the most important conceptual contributions to the study of developmental plasticity and its limitations on the production of adaptive phenotypes is the result of a series of papers by Diggle on sex determination in *Solanum hirtum* (Solanaceae: Diggle 1993, 1994). *S. hirtum* is an andromonoecious plant, producing two types of flowers: staminate (male) and hermaphroditic. Diggle (1993) explored the alternative possibilities that sex determination (i.e., the proportion between male and hermaphrodite flowers) in *S. hirtum* is due to genetically fixed patterns of allocation to one or the other type of flower or to plasticity in response to environmental heterogeneity. She concluded that sex determination is, indeed, plastic, and that this plasticity is at least in part a response to nutrient availability within the plant. She also found genetic variation for developmental plasticity among individuals, thereby affording *S. hirtum* the potential to respond to selection. Diggle (1994) then went on to further investigate the relationship between development and plasticity in this system. She established that sex determination depends not only on plasticity to resources, but also on the developmental architecture of the plant. Floral sex is undetermined until the floral buds reach 9–10 mm in length; that is, the floral primordia retain the potential to respond plastically until fairly late in the development of the flower. However, she found that only some primordia are plastic, in particular those produced distally (i.e., away from the main stem of the plant). If the buds are developing in a basal position, they turn into hermaphrodite flowers *regardless of the environmental conditions.* Diggle termed this dependency of plasticity on the previous developmental history of the organism *ontogenetic contingency.* Interestingly, the developmental cause of the intergenotypic differences in plasticity in this case is reducible to variation in the architecture of these plants: Some plants initiate the *distal* (and therefore plastic) portion of the inflorescence earlier or later than others, which translates into the macroscopically observable genetic variation for plasticity of sex determination (Fig. 6.9). Development, plasticity, and apparent genetic variation for a morphological trait are therefore connected in a complex, yet understandable, causal web.

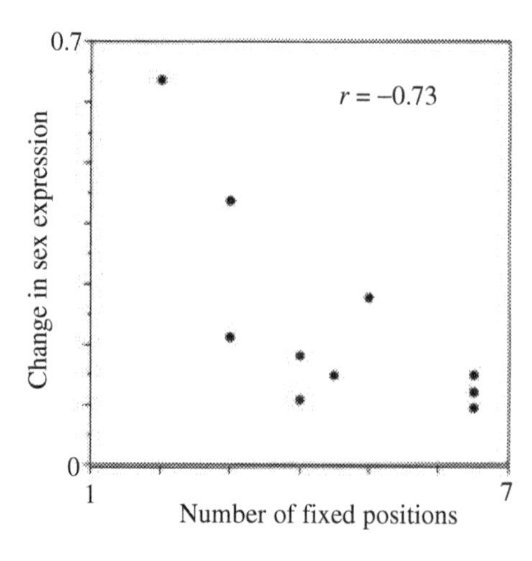

FIG. 6.9. The relationship between fixed developmental programs and phenotypic plasticity termed *ontogenetic contingency* by Diggle (1994). The number of fixed positions (abscissa) refers to the number of flowers characterized by invariant developmental patterns in genotypes of *Solanum hirtum:* the higher this number, the lower the plasticity expressed by a given genotype (ordinate). Each point represents a different genotype.

Ontogenetic contingency is not limited to plants, as demonstrated by a study of *Drosophila melanogaster* by David and colleagues (1990), which actually preceded Diggle's conceptualization of the phenomenon. David et al. addressed the decrease in pigmentation of *Drosophila*'s body segments with increase in temperature. They found that only the last three abdominal segments actually react in the population they studied. Furthermore, this response was not uniform, with the shape of the reaction norms being statistically significantly different among the three segments. They concluded that while pigmentation in response to temperature is probably generally adaptive, the segment-specific differences represent developmental constraints reflecting distinct abilities of different developmental units to react to the external environment.

A very elegant study of the relationship between developmental constraints and adaptive plasticity was conducted by Gedroc et al. (1996) on two annual plants, *Abutilon theophrasti* and *Chenopodium album.* These authors set out to test the classical optimization theory that the ratio between roots and shoots should be higher when the plant is nutrient-limited (because nutrients are acquired through the roots). They therefore grew the two species under low or high nutrients, and found that in both cases plants growing under low nutrients not only predictably do so at a slower pace, but indeed allocate more resources to roots than to shoots. This result is consistent with the optimization hypothesis. They also found that the allocation to roots occurs very early in ontogeny and that both species exhibit what Gedroc et al. termed *ontogenetic drift:* The root/shoot ratio decreases steadily throughout ontogeny, regardless of the initial allocation. The most interesting results, however, were obtained when the authors switched plants from high to low or from low to high nutrients midway through the experiment. In this case, plants that were switched to high nutrients grew more vigorously, *but they did not change their allocation to roots versus shoots*. Furthermore, plants that were shifted to low nutrients not only did not alter their allocation pattern, but also kept growing at the same pace. These results indicate that there is a window of opportunity early in development for the plant to be plastic, after which the architecture of the individual is sufficiently well established to preclude adaptive phenotypic plasticity from manifesting itself—a phenomenon similar to Diggle's ontogenetic contingency. Consequently, in this case the observable differences among adult plants are the outcome of very early developmental decisions, and the fitness of the individuals is influenced much more strongly by the environment they experience as young plants than by the conditions encountered later in their ontogeny.

The existence of *windows for plasticity* is of course very important because it has consequences for the timing of genetic variation for plasticity and thus on the response of plastic traits to selection. Developmental windows can be studied by switch experiments of the kind just discussed, which are unfortunately rare in plasticity research. Another example of this approach is the analysis of developmental plasticity for flowering time in response to photoperiod in *Arabidopsis thaliana.* Mozley and Thomas (1995) detected a window for this plasticity by focusing on the effect of mutations at phytochrome and blue receptor loci in *A. thaliana.* Their results led to the formulation of a complex model of interactions between light signals and photoreceptors in determining flowering time in this plant (Fig. 6.10). In *A. thaliana,* germination is controlled antagonistically by phytochrome A, phytochrome B, and at least one blue receptor. During the first few days of growth, from germination to de-etiolation (the opening of the cotyledons), the plant is insensitive to changes in daylength. However, between circa 4 and 45 days into the life cycle (for the ecotype Landsberg *erecta*) the plant's sensitivity to photoperiod increases, with phytochrome A and the blue receptor favoring early flowering and phytochrome B favoring more vegetative growth. The balance in the antagonism among photoreceptors is determined by the length of the light portion of the day, with days shorter than 12 h of light maintaining vegetative growth and longer days inducing the switch to reproduction. Eventually, however, the plasticity window "closes," and the plant flowers regardless of the photoperiod after about 45 days. This is consistent with *A. thaliana* being considered a facultative long-day plant and provides a fascinating insight into the sophistication of the interactions between external environment and developmental program.

A final example of plasticity windows during development is provided by the study of Novoplansky et al. (1994) of the Mediterranean annual *Onobrychis squarrosa* in response to shade. They found that the effect of shade depends on the season, with 15–20 (but not 5 or 10) days of exposure to the

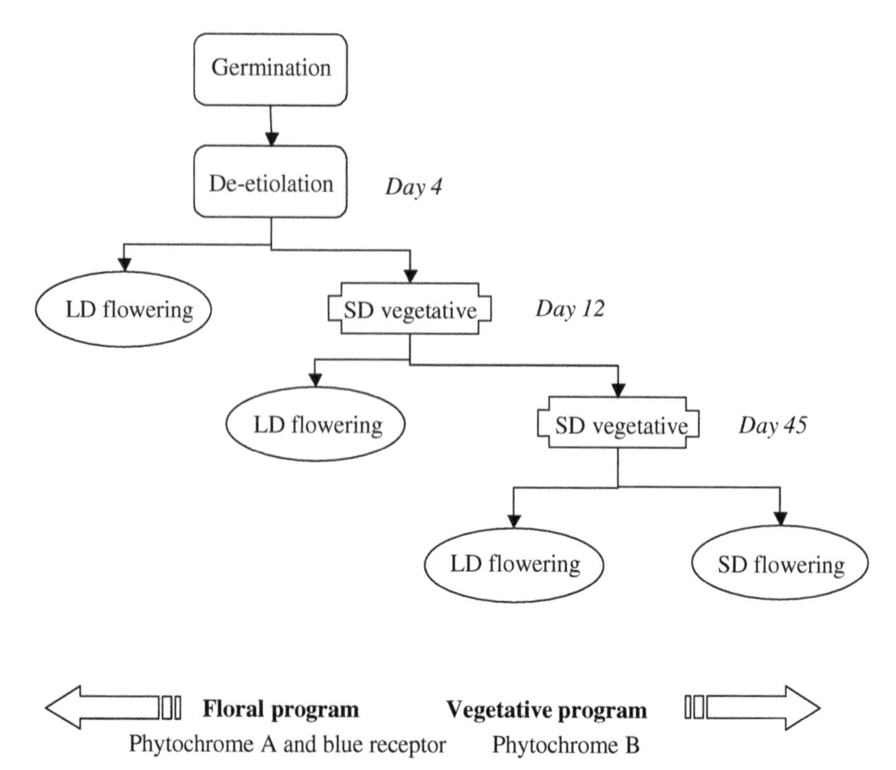

FIG. 6.10. Summary of the results of experiments on the window for plasticity to daylength in *Arabidopsis thaliana.* The plant is insensitive to photoperiod between germination and day 4. Afterward, long days trigger early flowering mediated by phytochrome A and a blue receptor, whereas short days maintain vegetative growth mediated by phytochrome B. After 45 days, the window closes and the plant will flower regardless of photoperiod. (Adapted from Mozley and Thomas 1995.)

treatment being sufficient to cause an increase in apical dominance during the fall. Later in the season, however, even longer periods of shade failed to elicit the classical shade-avoidance phenotype, indicating that the window of sensitivity to that particular environmental signal had closed and the developmental trajectory of the plant was no longer responsive.

Overall, these examples show that the study of developmental plasticity is a fundamental key to our understanding of the limitations that affect the evolution of plasticity itself. First, analyses of variation in developmental trajectories may highlight the existence and define the kind of genetic constraints that affect the developmental system, thus delineating the limitations on the short- and mid-term evolution of plastic responses. Second, developmental constraints may themselves be the foundations of observable trade-offs, and thus affect the ecological performance of plastic genotypes. Therefore, one can think of the developmental level of analysis as a powerful interface between the ecology and the genetics/molecular biology of phenotypic plasticity. Without a careful understanding of developmental phenomena, neither aspect will shed light on the other, and ecological genetics will not achieve the synthesis that seems now to be within reach.

Conceptual Summary

- Not all phenotypic plasticity is necessarily developmental (e.g., physiological and some aspects of behavioral plasticity are not). However, a developmental perspective is fundamental to a full understanding of genotype-environment interactions.
- It is important to understand the mechanisms underlying developmental plasticity above the genic level, such as hormonal action and heterochrony. These provide an interface between the most basic level of phenomenology (investigated by genetics and molecular biology) and whole-organism patterns of plasticity.
- There are clear examples of adaptive developmental plasticity, especially in cases of diet- or predator-induced morphological adjustments. While the elements involved in these cases of plasticity may have originally been co-opted from different ancestral functions, there can be little doubt that natural selection has shaped such systems in their current fashion.
- Diggle proposed the term *ontogenetic contingency* to describe situations in which phenotypic plasticity is manifested only conditionally to the occurrence of specific developmental events. This phenomenon may be widespread in both animals and plants and may provide a major explanatory framework for the observable limits of adaptive plastic responses.
- Experimental demonstrations of developmental *windows of plasticity* also show conclusively that the plasticity of an organism depends not only on the traits and environments considered, but also on the time frame during ontogeny when such environments are experienced.

—7—

The Ecology of Phenotypic Plasticity

Ecology is rather like sex—every new generation likes to think they were the first to discover it.

—Michael Allaby (b. 1933)

CHAPTER OBJECTIVES

To discuss the ecology of phenotypic plasticity, in terms of environmental conditions that may lead to the evolution of plasticity, and the relationship between plasticity and fitness. The relationships among plasticity, ecotypes, generalists, and specialists are probed in some detail. Three ways of approaching the problem of determining the adaptive significance of a plastic response are evaluated with some empirical examples. The Bradshaw-Sultan effect is introduced. The chapter ends with a discussion of limits and costs of plasticity and of the relationship between plasticity and trade-offs.

Why Ecology?

It may strike the reader as odd to talk about the ecology of phenotypic plasticity. After all, plasticity is—by definition—the result of environmental influences on the genotype. Ergo, in the same way in which all morphological plasticity can be considered developmental (Chapter 6), ecology (in the broad sense of interactions between organisms and their environment) is really embedded within the definition of plasticity itself. However, if we consider plasticity as a trait in its own right, then we have to address the question of the impact of this trait on the fitness of the organism. Such an impact is going to depend on the particular range of environmental conditions experienced by the organism and its immediate descendants, that is, on its realized niche. Furthermore, the Hutchinsonian niche (the total *potential* of multidimensional environmental space that could be occupied if there were no competition) is in fact a major player—together with the genetic basis of plasticity (Chapter 4)—in determining the evolutionary trajectories of reaction norms (Chapter 9). At this point the reader might begin to appreciate why it would be advantageous to be able to write this book using hyperlinks.

The general ecological significance of phenotypic plasticity is discussed by Grime et al. (1986). Although their focus is on plants, similar arguments can easily be extrapolated to other forms of life. They suggest three distinct roles of plasticity in contrasting ecological scenarios, one for each of the three basic ecological strategies that plants can adopt (Fig. 7.1). If the plant is a *stress-tolerant,* its plasticity should be manifested through reversible mechanisms such as acclimation, which allow the individual to simply survive the period of stress and wait for better conditions in order to further its reproduction. This type of plasticity is mostly physiological in nature. In the case of a *competitive* strategy, morphological and developmental plasticity would function as the plant equivalent of a foraging mechanism, maximizing competitive ability by dynamically redirecting growth onto favorable patches and away from unfavorable ones. Finally, a plant adopting a *ruderal* strategy should be characterized by plasticity in life history and phenology, which would trigger premature reproduction to escape stress and allow at least some reproductive fitness. Grime and colleagues provide excellent examples of each category (although see Stanton et al. 2000 for a different perspective).

A more general argument for the relevance of plasticity in ecology was advanced by Sultan (1987). In discussing the nature of reaction norms, she pointed out that the very fact that the environment *interacts* with genotypes to produce phenotypes adds a whole new component to both ecology and genetics (and, therefore, to evolutionary theory). Organisms that adjust to their environment through phenotypic plasticity alter their own relationship with the environment in a play and counterplay fashion. As Sultan put it, paraphrasing Lewontin (1978; Levins and Lewontin 1985), "by virtue of this capacity for response (i.e., plasticity), the relation of organism to environment is not confrontational, but interactive; each is both cause and effect of the other." Therefore, the environment is no longer to be considered a "problem" that the organism has to "solve." On the contrary, the organism defines, is affected by, and affects the environment in a continuous feedback (dare I say dialectical?) process. This view is perfectly in line with modern concepts of niche in ecology. The niche is no longer seen as an entity independent of the organism, which can be "filled" or be left "vacant." Instead, the niche is defined by the very presence of that organism. Phenotypic plasticity provides a strong argument in favor of the latter view.[1]

A topic often related to discussions of ecology and plasticity is the question of *ecotypes.* As we saw in Chapter 2, ecotypes were originally defined by Turesson (1922) as the "ecological unit to cover the product

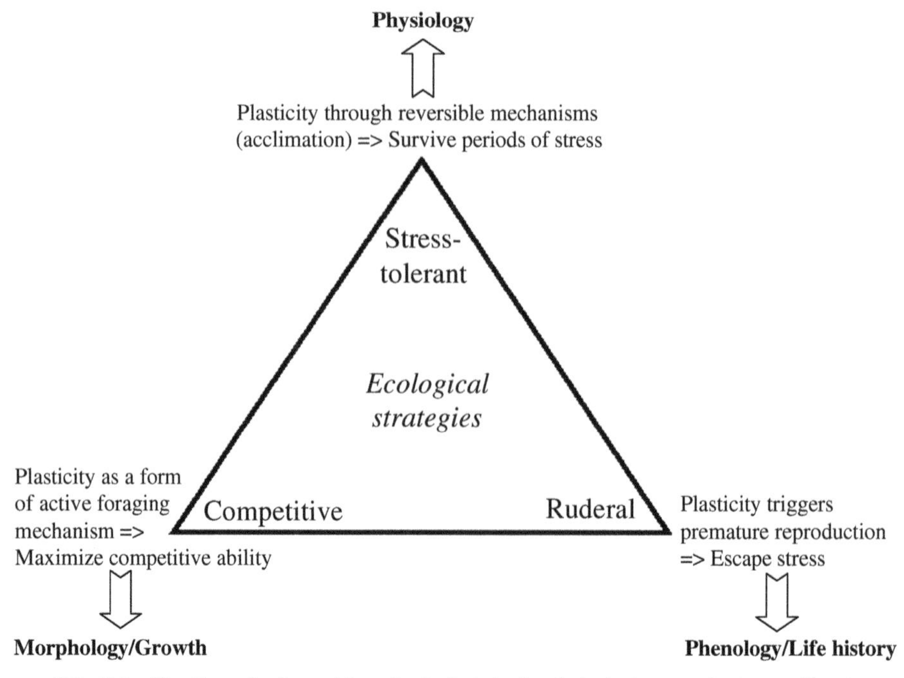

FIG. 7.1. The three fundamental ecological strategies that plants can adopt according to Grime et al. (1986). These strategies are associated with different types of plasticity, which yield to distinct ecological scenarios and act at different levels of the phenotype. Very similar scenarios can be drawn for animals or other life forms.

arising as a result of the genotypical response of an ecospecies to a particular habitat." By *ecospecies* he meant "the Linnaean species or genotype compounds as they are realized in nature." Modern understanding of the ecotype concept refers to a *genetically specialized* local population that is adapted to a particular set of environmental circumstances (Galen et al. 1991; Diggle 1997). As we saw in Chapter 4, Quinn (1987) suggested abandoning the term *ecotype* entirely in favor of recognizing the existence and role of phenotypic plasticity. I disagree. As much as plasticity is an important component of the ecology of organisms (especially, but not limited to, sedentary ones), an ecotype is at the opposite end of the spectrum of an adaptively plastic genotype. (An ecotype can show maladaptive or neutral plasticity, and in fact the evolution of specialists [ecotypes] and generalists [plasticity] may be linked by way of intermediate populations with previously nonadaptive reaction norms.) As such, the idea of ecotypes provides a useful conceptual null hypothesis for an understanding of the ecological relevance of phenotypic plasticity. What I am suggesting here amounts to a fourfold classification of reaction norms to blend plasticity and ecotypic specialization:

1. A genotype can have a mostly flat (nonplastic) reaction norm for a focal trait (say, plant architecture) and be particularly adapted to a special set of environmental circumstances (e.g., alpine conditions). This would be an ecotype, with no plasticity for the focal trait, a high fitness in the special environment, and a low fitness elsewhere. Ecologically, this organism would be a specialist.
2. On the other hand, a genotype can be highly plastic for a focal trait, and this is what confers on that genotype the ability to colonize multiple environmental settings or to withstand major changes in the environments it experiences. For example, consider a heterophyllous species able to live below and above water. Plasticity would then be adaptive, and the organism would be considered a generalist.
3. A third possibility is a genotype that is plastic for a given trait, but is in fact adapted only to a particular set of environmental circumstances. The plasticity, therefore, would be maladaptive or neutral, and the organism should still be considered a specialist. This situation is probably a more common one in ecotypes than case 1.
4. Finally, we can envision a genotype that is not plastic, but lives in a variety of environments. The organism would adopt an "average" phenotype that is suboptimal in all environments, but not too deleterious in any. This would be a case of a nonplastic generalist. Such a genotype—as counterintuitive for researchers accustomed to thinking in an adaptationist mode as it may be (Gould and Lewontin 1979)—is actually found in many populations in which plasticity studies are conducted, though it may not be the prevalent genotype. This category reflects the idea that evolution is not about the survival of the fittest, but the survival of whatever works well enough (i.e., selection is not an omnipotent force; its efficacy depends on historical accidents and a host of other phenomena such as drift, migration, and mutation [den Boer 1999]).

These categories reflect a population genetic view of the problem, in that genotypes classifiable as any of the above are indeed repeatedly found in surveys of natural variation for plasticity. If and under what conditions any of these patterns becomes prevalent within a population is a different question. Be that as it may, we can put the above classes together as shown in Table 7.1. Again, these considerations are valid provided we are focusing on an ecologically relevant trait, in that there is no such thing as a plastic or nonplastic genotype across characters. This scheme also allows us to hypothesize some of the possible evolutionary pathways that may lead from one type to another (Chapter 9). For example, a generalist plastic genotype could evolve into a specialist ecotype by losing plasticity owing to the fact that it experiences only a subset of the original environments. (Loss of plasticity may be due to costs [see below] or to the cumulative effect of neutral

TABLE 7.1

Classification of the Relationships among Generalists, Specialists, Plasticity, and Ecotypes

	Adapted to a restricted set of environments	Adapted to a larger set of environments
No plasticity (for the focal trait)	Ecotype, specialist	Jack-of-all-trades–master-of-none, generalist
Plasticity (for the focal trait)	Ecotype, specialist, maladaptive or neutral plasticity	Adaptive plasticity, generalist

mutations and random drift.) Alternatively, a specialist ecotype could yield a specialist with maladaptive or neutral plasticity if new mutations alter the reaction norm but are not exposed to purging natural selection because the relevant environments are not experienced frequently enough. This latter type, however, could evolve adaptive plasticity if it starts colonizing environments for which it is partially *preadapted* through previously unutilized portions of its reaction norm. Finally, a Jack-of-all-trades–master-of-none generalist may evolve from a nonplastic ecotype if the new environmental conditions are not too different from the original ones. It can then, in turn, be selected for adaptive plasticity to better match the new environmental conditions and evolve into a plastic generalist.

Phenotypic Plasticity as an Adaptive Strategy

If plasticity has any ecological (and, indeed, evolutionary) consequence, it has to impact organismal fitness. This truism aside, the study of the relationship between plasticity and fitness is a complex and controversial subject (see Chapter 6 for an introductory discussion confined to cases of developmental anticipatory plasticity). For one thing, the very definition and especially the measurement of fitness constitute one of the recurring nightmares of organismal biologists (Lande and Arnold 1983; McNeill Alexander 1990; Ollason 1991; Rausher 1992; de Jong 1994; Schluter and Nychka 1994; Burt 1995; Brodie III and Janzen 1996; Bell 1997a,b). Furthermore, the concept of adaptation and the process of adaptogenesis are among those least understood in modern evolutionary biology (Lewontin 1978; Gould and Lewontin 1979; Gould and Vrba 1982; Levins and Lewontin 1985; Huberman and Hogg 1986; Gittleman and Kot 1990; McNeill Alexander 1990; Harvey and Purvis 1991; Travisano et al. 1995a,b; Wagner 1995; Huynen et al. 1996; Reznick and Travis 1996; Rose and Lauder 1996a,b; Rose et al.

1996; Seger and Stubblefield 1996; Gavrilets 1997; Gotthard 1998; Orr 1998). When we combine these two problems together with the relatively new field of phenotypic plasticity, we have the recipe for a long and controversial scientific debate.

Let us start by recalling some major points presented by Sultan in a particularly lucid discussion of the relationship between phenotypic plasticity and adaptation (Sultan 1995). First, she makes a distinction between two modes of adaptive evolution: allelic substitution *a la neo-Darwinian synthesis* and (genetically based) phenotypic plasticity. Then, she further distinguishes between *adaptive* and *inevitable* phenotypic plasticity. (In Chapter 2 we discussed similar distinctions made on the basis of genetic or developmental considerations and the relationships among these views of plasticity.) Sultan then correctly points out that many studies on plasticity completely ignore the fitness question. Accordingly, plasticity is simply treated as "a neutral metric of phenotypic differences in various environments, of unknown and possibly little functional significance."

If one wishes to avoid the trap outlined by Sultan and address the relationship between plasticity and fitness, one can follow three possible approaches to the study of adaptive phenotypic plasticity: historical, fitness-correlate, and optimality. The *historical* approach relies on the comparative analysis of plasticity in different species whose phylogenetic relationships are known. I discuss this in Chapters 9 and 11. The *fitness-correlate* approach is the one usually employed to investigate the adaptiveness of any trait within a given environmental setting (i.e., using some sort of regression of fitness on traits), following the principles originally laid out by Lande and Arnold (1983) and modified by several other authors (Crespi and Bookstein 1989; Rausher 1992; Brodie and Janzen 1996; Willis 1996). Sultan dismisses the fitness-correlate approach based on the fact that across-environment comparisons of phenotypes and fitnesses will invariably result in negative correlations between plasticity and fitness, even when plasticity is advantageous. This is sometimes referred to as the *silver spoon effect* (Grafen 1988). Let us suppose that plasticity for leaf size is advantageous as a response to light availability. Plants grown under low light will produce *relatively* larger leaves (i.e., they will allocate a larger portion of the total biomass to leaf production). Simultaneously, however, plants experiencing high light levels will be *absolutely* larger, because of the silver spoon effect, even though they will allocate a smaller portion of total biomass to leaf production. Therefore, despite the fact that the allocation shift actually minimizes the damage caused by the stressful conditions (adaptive plasticity), the overall appearance will be that plasticity by differential allocation to leaves is accompanied by a reduction in fitness as an inevitable consequence of stress. The point is very well taken, but I suggest that the dismissal of this approach may be a bit premature. All that is needed to avoid the silver spoon problem is a comparison between the degree of plasticity and the *within-environment* fitness. The question then becomes, given the inevitable loss of fitness caused by a stressful environment, do plastic genotypes lose less than nonplastic (or less plastic) ones? For examples of applications of the fitness-correlate approach to plasticity, see the works of Weis and Gorman (1990) and Dudley and Schmitt (1996, Chapter 2), also discussed later in this chapter.

The *optimality* approach (Fig. 7.2), according to Sultan, depends on "engineering principles and ecophysiological interpretation." In other words, either by mathematical modeling (Chapter 10) or by carefully constructed verbal arguments, we can "predict" which phenotype *should* be favored by natural selection in each environment. All we have to do then is compare the observed phenotypes to the expected optima to judge the degree of adaptation of a particular plastic response. As Sultan realized, this is easier said than done. For example, there may be constraints or costs affecting plasticity acting simultaneously with natural selection, making the resultant phenotype both inevitable and—to some degree—adaptive. I discuss costs to plasticity later in this chapter.

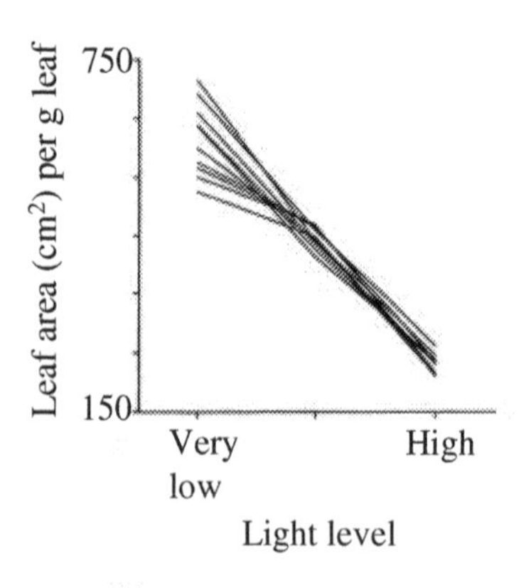

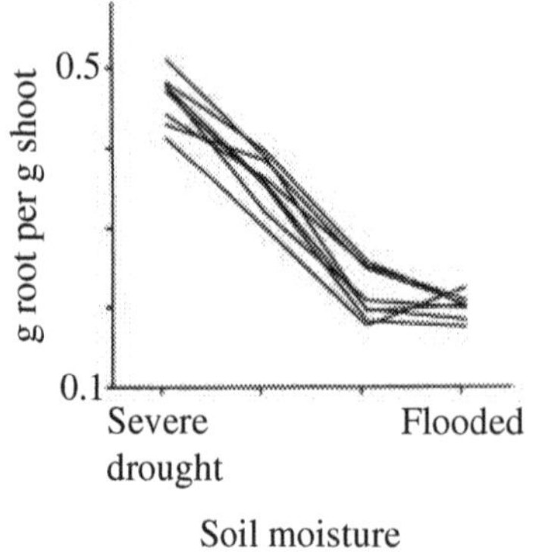

FIG. 7.2. An illustration of Sultan's "engineering" approach to characterizing adaptive plasticity. Logical and biological considerations would lead us to expect a plant under stress induced by low light to increase allocation of resources to leaf area (upper diagram) and one under drought stress to increase allocation to roots (lower diagram). This is exactly what is observed when reaction norms of genotypes of *Polygonum persicaria* are studied in the appropriate environments. (From Sultan 1995.)

However we wish to approach the study of plasticity in an ecological context (by optimality, fitness-regression, or historically), one of our most frustrating experiences as students of the ecology of plasticity is that we often find plenty of genetic variation for almost any kind of plasticity in response to any type of environment. Thus it becomes very difficult to make sense of the results. Sultan suggests several reasons for this regrettable state of affairs:

1. We need to concentrate on traits that best summarize the ecological characteristics of the organism under study. Too often we pick traits that are simply the most convenient to measure (see Wagner 2001 for a discussion).
2. An equal degree of care must be taken in setting up the experimental treatments. The environments need to be relevant to the natural conditions encountered by the organism; otherwise the observed responses will be to a "novel" environment (Service and Rose 1985; Holloway et al. 1990; Joshi and Thompson 1996; Hawthorne 1997), and therefore difficult to interpret ecologically or evolutionarily. Some characterization (and, better, quantification) of the natural environments prior to setting up an experiment is always preferable over arbitrary treatments. We cannot imitate nature closely, but we should avoid Pindaric flights of imagination.
3. In nature, the intensity of selection on different portions of the reaction norm may vary as a function of the frequency of occurrence of different environments. While the latter is usually (but not always [Weis and Gorman 1990; Bell and Lechowicz 1991; Lechowicz and Bell 1991]) difficult to estimate, we will do well to remember that experimental environments imply a flat, equal frequency distribution, which may dramatically affect our results. (For a theoretical treatment of this point see Via and Lande 1985.)
4. The strength of correlations with fitness will depend on the quality of the environment. It is more likely that an adaptive phenotype will be significantly correlated with fitness under conditions that actually tax the organism and—presumably—led to the evolution of that adaptation to begin with. Many common garden experiments are conducted under semi-ideal conditions in order to maximize survival, but if the environmental quality is high, it may not matter which particular phenotype one happens to express.
5. A common observation in plasticity studies is the occurrence of inconsistent results, in which the correlation of a trait with fitness can change sign across environments or go from zero to significant. Apart from the reason given in point 4, it is very possible that different developmental or physiological mechanisms intervene in different conditions, so that

the observed plastic response is actually underlined by distinct biological processes (see Stearns et al. 1991).

6. I will add one more category to Sultan's discussion. Many studies on plasticity tend to consider the reaction to a particular environmental variable (e.g., temperature) as a coherent ecological, genetic, and therefore evolutionary unit. But it is known from physiological and molecular studies that the response to one extreme of an environmental gradient may be completely different in kind from the response at the other extreme (Chapter 5). For example, while they are both reactions to temperature extremes, cold stress (Welin et al. 1994; Mizoguchi et al. 1996) and heat stress (DeRocher et al. 1991; Rickey and Belknap 1991; Starck and Witek-Czuprynska 1993; Gombos et al. 1994; Krebs and Loeschcke 1994b; Loeschcke and Krebs 1994; Bubli et al. 1998; Downs et al. 1998; Rutherford and Lindquist 1998) rely on entirely distinct molecular machineries. Why then should we treat them as the ends of a continuum in plasticity experiments?[2]

Given all of the foregoing, it is no wonder that there are few empirical studies demonstrating adaptive plasticity, and that they do not as yet present a sharp picture of the ecology of phenotypic plasticity. Nevertheless, let me attempt to discuss some selected examples below, to fix our ideas on different aspects of this complex problem.

Empirical Studies of Adaptive Plasticity

Even though the literature on adaptive phenotypic plasticity is relatively limited, it is hard to convey the breadth of ecological and evolutionary situations that have been uncovered so far. Therefore, the following is by no means an exhaustive list of these cases.

One of the classic examples of adaptive plasticity is provided by Cook and Johnson's (1968) research on heterophylly, the production of distinct types of leaves above and below water, in *Ranunculus flammula.* While widely cited, the breadth and impact of this paper can hardly be overstated. As I noted in Chapter 4, these authors studied heterophylly within and among several populations of the amphibious *R. flammula.* Their approach utilized a rare combination of controlled experiments (under terrestrial, aquatic, and mixed conditions), transplant experiments, and crosses to determine the heritability and investigate the genetic control of heterophylly. They found a recurrent association of heterophyllous individuals with unpredictable environments or environments that were *immature* from an ecological standpoint (i.e., early in a succession). Cook and Johnson also suggested that disruptive selection produces genotypes especially adapted to either completely aquatic or completely terrestrial conditions. The reciprocal transplants and the controlled experiments were designed to test the degree of specialization or generalist behavior of different genotypes. The results strongly suggested that heterophyllous plants are characterized by the largest ecological valence, putting them into the category of plastic generalists (with respect to leaf morphology) referred to above. Cook and Johnson's classification of F_1 hybrids as heterophyllous or homophyllous genotypes was consistent with the action of dominant genes inhibiting lateral cell expansion; interestingly, the environmental factor controlling such action was temperature (specifically, low temperature), not water availability. These authors' general findings established heterophylly on the list of the strategies available to a plant to invade new, variable environments, early in a succession. Later on in ecological time, specialist genotypes arise and supplant the heterophyllous ones. The natural hybrids between specialists and generalists would then survive only in ecotonal situations linking a completely terrestrial to a completely aquatic environment.

Heterophylly sensu lato (i.e., in the general sense of variation in leaf shape), however, need not be adaptive, as elegantly demonstrated by Winn (1999), who investigated seasonal changes in leaf shape and anatomy in the annual plant *Dicerandra linearifolia.* The observed phenotypic variation is a response to temperature, and it is consistent with an adaptive plasticity hypothesis formulated according to Sultan's engineering criterion encountered earlier. The problem is that in this particular case the relationship among temperature, seasonality, and leaf phenotype appears to be entirely accidental. Winn conducted phenotypic selection analyses to test the hypothesis that seasonal phenotypic variation is adaptive (therefore using the fitness-correlate approach discussed above) and found no evidence in favor of the adaptive hypothesis since selection favored thick, large leaves regardless of the season. She concluded that seasonal variation in leaf form may persist in nature solely due to the absence of selection against individuals in which leaf traits are variable. This is a cautionary tale that should be kept in mind whenever one draws conclusions on adaptive plasticity based solely on one type of evidence. Another situation in which plasticity is actively selected against was described in arthropods inhabiting chronically metal-stressed soils (Kohler et al. 2000), while Decker and Pilson (2000) showed that a case of hypothesized adaptive plasticity (environmental sex determination in some euphorbs) did not stand up to scrutiny.

A by now classic example of the relationship between plasticity and fitness is the study conducted by Weis and Gorman (1990) on the insect-plant system *Eurosta-Solidago.* While the results in this case point toward selection *against* plasticity, the methodology adopted certainly stands out as one of the best examples available to date. The biological system consists of a gall-forming insect, the dipteran *Eurosta solidaginis,* and its host plant, the composite *Solidago altissima* (common goldenrod). Weis and Gorman used an attribute of the plant—the lag-time between the insect's attack and the plant's response—as a quantitative measure of the environment experienced by the insect. That way, they were able to put together a remarkable data set comprising not only the reaction norms of several full-sib families of insects, but also an estimate of the fitness function relating gall diameter to survival, as well as the frequency distribution of the different environments (i.e., the lag times). They then applied the Lande-Arnold (1983) method of regression of fitness against trait values to estimate selection coefficients on phenotypic traits. The only difference from the standard procedure was that the phenotype was characterized by two attributes: the elevation (mean) and slope (plasticity) of the full-sib reaction norms. Weis and Gorman found that there is indeed directional selection to decrease phenotypic plasticity of gall size, but that this is only one-quarter as strong as directional selection on mean gall size. Their conclusion was that the actual intensity of selection on plasticity in the wild depends on the environmental gradient experienced by local populations in species characterized by a broad distribution.

There are few other examples of studies of plasticity in which convincing estimates of fitness are taken into account, either by doing actual field work or by measuring fitness under controlled but realistic conditions. One is the work of Hazel and colleagues (1998) on the environmentally dependent color of the pupae of *Papilio polyxenes,* which has been hypothesized to be adaptive because of preferences for pupation sites and selection for crypsis at these sites. These authors tested the adaptive hypothesis by measuring field survival rates of pupae of different colors placed on different sites near or above the ground. They found that green pupae above ground (where they would normally be found) had higher survival rates than any other color/site combination. They also detected a strong dependency of predation rate on the time of day: Whereas above-ground pupae were most subjected to predation during the day, ground-level pupae were in danger chiefly at night, which suggests an influence of predation patterns on site selection. The conclusion of Hazel et al. (1998) regarding the plasticity was that environmentally cued coloration should indeed be favored as long as diurnal predators locate the pupae visually. Another convincing example of a test of the adaptive plasticity hypothesis, which includes naturalistic estimates of fitness, is van Buskirk and Relyea's study (1998) of tadpoles of *Rana sylvatica.* They raised tadpoles in the presence or absence of dragonfly larvae, which act as predators. The plastic response of the tadpoles included significant shifts in size and shape, with the development of smaller and shorter bodies and deep tail fins (Fig. 7.3). These authors clearly demonstrated that the dragonflies were particularly efficient at killing tadpoles that did not show the altered body morphology, with traits showing

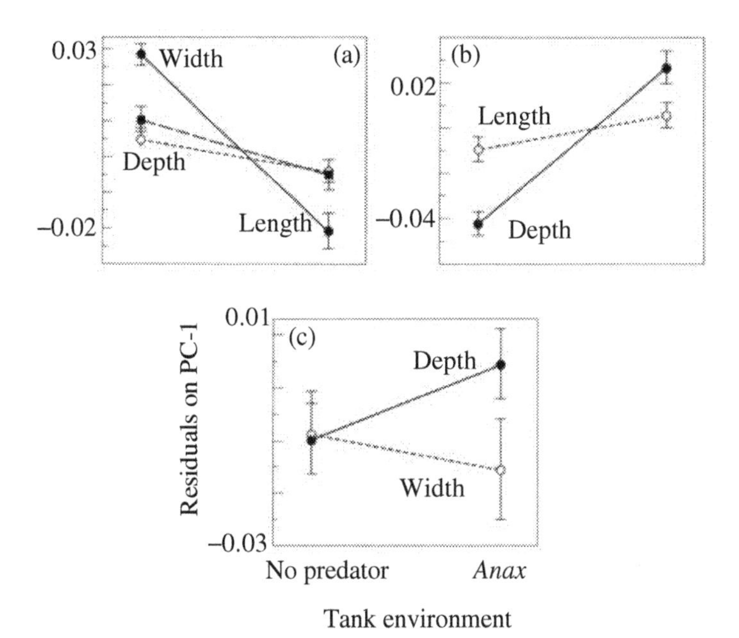

FIG. 7.3. Morphological plasticity in response to the presence of predators in *Rana sylvatica.* Each line represents the response of characters describing: (a) body shape; (b) tail fin shape; (c) tail muscle shape. Note that all measurements are corrected for general body size by first regressing against the first principal component and retaining the residuals. (From van Buskirk and Relyea 1998.)

the strongest degree of plasticity also being under the most intense selection when predators were present. However, they also demonstrated a cost to the induced phenotype (see below), since the predator-triggered morphologies fared poorly if the environment did not actually include dragonflies. This is an excellent example of how studies of adaptive plasticity are particularly efficacious when the cue can be decoupled from the selective agent.

As I mentioned in Chapter 3, one of the very first examples of phenotypic plasticity ever studied as such is the phenomenon known as *cyclomorphosis,* a seasonal shift in morphology in response to predator presence in the crustacean *Daphnia* (Woltereck 1909). This parthenogenetic zooplankter is still at the center of contemporary plasticity studies (Burks et al. 2000). One of the most interesting modern papers on *Daphnia* concerns the relationship between *D. pulex* and the dipterian predator *Chaoborus americanus,* the phantom midge (Parejko and Dodson 1991). In the presence of *C. americanus, D. pulex* undergoes a developmental switch that leads to a protuberance with spines on the dorsal carapace, termed *neckteeth.* When Parejko and Dodson isolated clones of *D. pulex* from different ponds, they found that plasticity for the production of neckteeth was stronger in ponds infested with the predators, in accordance with the adaptive plasticity hypothesis. Also in agreement with the adaptive scenario, *D. pulex* is able to detect the presence of predators before the predator density becomes high (anticipatory plasticity), since the plastic response is catalyzed by water in which the predator had been grown but is no longer present (Fig. 7.4). However, the relationship between prey and predator is not the only factor in the observed plasticity. Under laboratory conditions, the plastic response for *D. pulex* is induced more often at lower food levels than when food is abundant (provided, again, that the predator cue is present). This unexpected and counterintuitive result is actually accounted for by a closer look at the biology of *D. pulex*. This crustacean is particularly susceptible to predation when food availability is low because these conditions lead to a dramatic limit in population growth—or possibly they cause a reduction in the size of the larvae, making them more susceptible to the predator (S. Scheiner, pers. comm.) Furthermore, another species of *Daphnia, D. cucullata,* has been shown to be able to influence the phenotype of its own offspring as a function of the environment encountered by the parental generation, a case of transgenerational adaptive plasticity (Agrawal et al. 1999). The overall scenario, therefore, suggests a complex interaction among predator presence, food availability, population dynamics, and maternal effects to shape the selection patterns on reaction norms for predator avoidance in this crustacean.

Another clear-cut example of adaptive phenotypic plasticity is the one offered by Sandoval's (1994) study of web design in the spider *Parawixia bistriata.* He observed that this spider produces two distinct types of web, a small one characterized by fine mesh and a larger one with wide mesh.

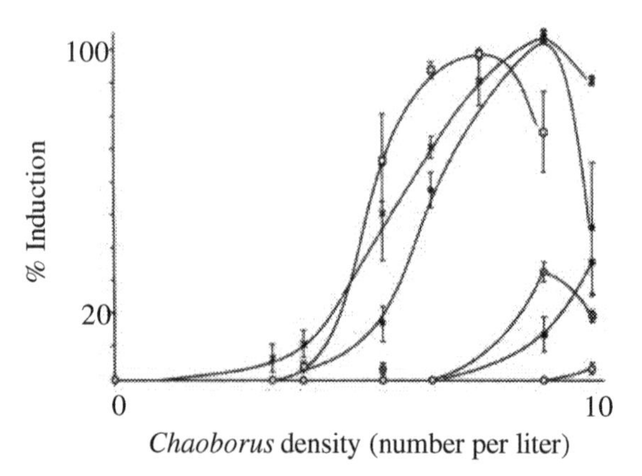

FIG. 7.4. *In vivo* reaction norms of *Daphnia pulex* to predator density. The three higher curves are for clones originating in predator-infested ponds, and the lower ones are from predator-free environments. (From Parejko and Dodson 1991.)

Further investigation revealed that the two types of webs are produced at different times, the small ones daily at sunset and the large ones only during diurnal termite swarms. In fact the two types are better suited to capturing different kinds of prey: termites for the large webs and mostly small dipterans for the small ones. Sandoval also found that the timing of web building is synchronized with the peak activity of each type of insect, in accordance with the adaptive plasticity hypothesis. From a conceptual standpoint, I see Sandoval's examples of dimorphic webs as equivalent to heterophylly and many other cases of so-called *within-individual* plasticity (except, of course, that one change is behavioral [Chapter 8] and the other is morphological [Winn 1996a]). In these cases the same organism experiences both environments, one of the clearest conditions leading to the evolution of phenotypic plasticity (Schlichting and Pigliucci 1995; see also Scheiner 1993a). In both web dimorphism and heterophylly the morphological changes are rather abrupt, and in both instances the adaptive interpretation is quite clear; although we know what cues trigger heterophylly (Kane and Albert 1982), we do not know what signal causes the spiders to switch to the termite-catching web (perhaps the daily light cycle?).

Many of the examples considered so far nicely illustrate the contrast between two of the approaches Sultan described for studying adaptive plasticity: the fitness-regression method (e.g., Weis and Gorman's or Winn's papers) and the optimality-engineering approach (web production and *Daphnia*'s response to predation). The following example demonstrates the power of a combined approach utilizing both quantification of the fitness landscape and experiments inspired by optimality criteria. The system studied by Dudley and Schmitt (1996) was shade avoidance plasticity in *Impatiens capensis* (briefly discussed in Chapter 2). Shade avoidance, which represents one of the best-understood examples of adaptive plasticity (Givnish 1982; Casal and Smith 1989; Schmitt 1997; Smith and Whitelam 1997), will be discussed again in a different context in Chapter 11. If shade avoidance is adaptive, one should observe selection for decreased plant height under low densities of conspecifics and for increased height under high densities. This is indeed what is shown in Fig. 7.5. The fitness isoclines are increasing from the lower left to the upper right when neighbor density is high, which means that taller plants are favored under these environmental circumstances. However, when plant density is low, the isocline pattern is exactly the opposite, favoring shorter individuals, even though longer leaves are favored under both circumstances. Therefore, the fitness-regression approach confirms the adaptive plasticity hypothesis. Dudley and Schmitt also used a phenotypic manipulation experiment to tackle the same problem. They induced or suppressed elongation in the plants and then exposed them to both high-density (i.e., favoring elongation) and low-density

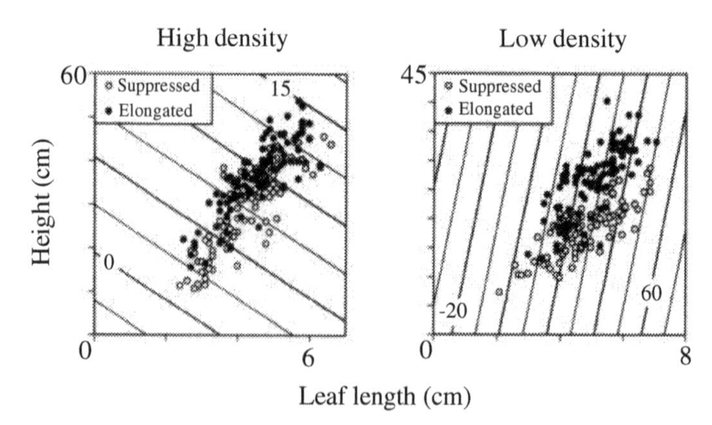

FIG. 7.5. Selection gradients for a combination of leaf length and plant height in *Impatiens capensis* under two environmental regimes. Note that while longer leaves are favored in both environments, high density of conspecifics favors taller plants, whereas low density enhances the fitness of shorter plants. (From Dudley and Schmitt 1996.)

(i.e., favoring suppression) environments. That way, they forced plants to express the "wrong" phenotype in either environment. As expected, suppressed plants did better than elongated ones at low density, but not as well at high density (Fig. 7.6). This further supports the idea that the plasticity in this case is maintained by a complex system of reciprocal advantage of

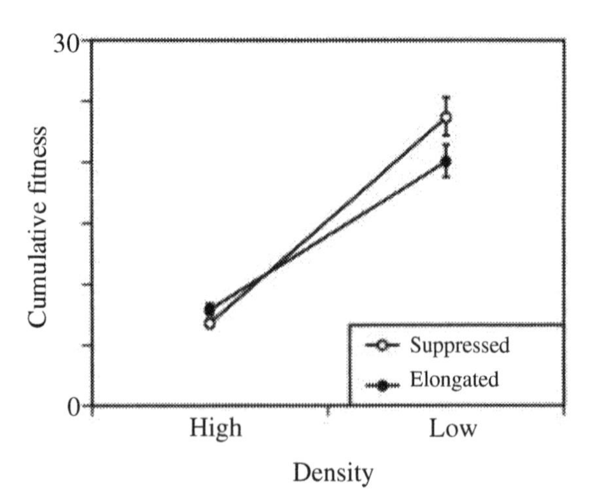

FIG. 7.6. Dependency of cumulative fitness in *Impatiens capensis* on neighbor's density. Suppressed plants do better than elongated ones at low density, where they can exploit the lack of competition while being structurally more sturdy, but not as well at high density, where they cannot outgrow their competitors in the search for light. (From Dudley and Schmitt 1996.)

alternative phenotypes in two environments, exactly as in the cases of heterophylly, antipredator morphology, and web design.

Let me conclude this section with a classic example of adaptive plasticity from two papers published by Lively (1986b,c) on the developmental morphology of the acorn barnacle (*Chthamalus anisopoma*) in response to its predator, the carnivorous gastropod *Acanthina angelica* (Fig. 7.7). The bent form of the barnacle is demonstrably efficacious in defending the animal from the predator (Lively 1986c), but Lively then inquired into why such an advantageous morph was not fixed regardless of the environmental conditions. The answer is that the bent morph is also characterized by slower growth rate and reduced fecundity, which puts it at a disadvantage when no predator is around (Lively 1986b). The spatial patchiness of the predator itself therefore maintains the advantage of a plastic form over either specialist. In fact, in simulations aimed at further exploring the dynamics of this system, Lively (1999) demonstrated that the induced defense is an evolutionarily stable strategy, even if the environmental cue is relatively unreliable. Furthermore, conditions leading to a stable genetic polymorphism for the two morphs (specialists) are rather narrow, although mixtures of inducible and specialist genotypes can coexist. Lively also showed

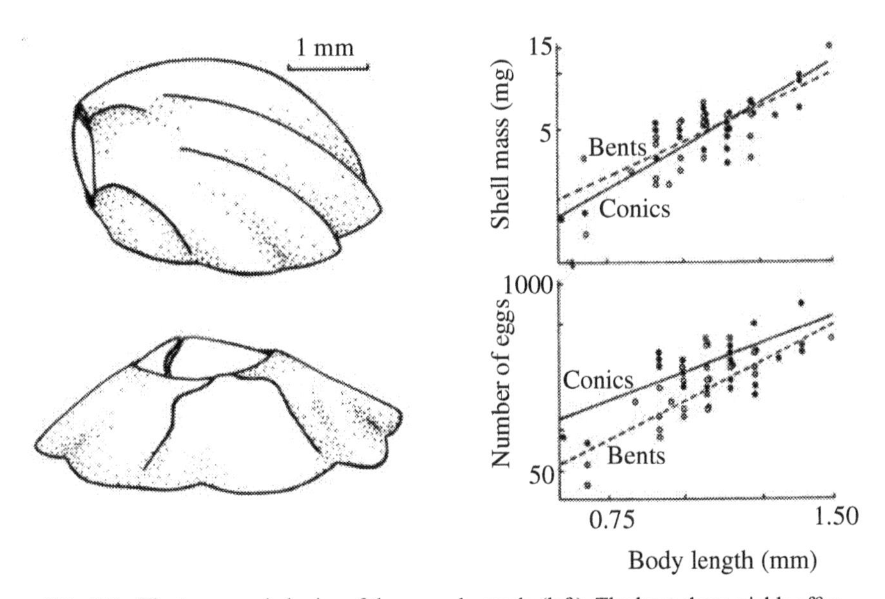

FIG. 7.7. The two morphologies of the acorn barnacle (left). The bent shape yields efficacious protection against predation from a carnivorous gastropod. However, it also comes at a cost manifested in lower growth rate (upper right) and fecundity (lower right), which explains why the bent morph is produced via a plastic response rather than being fixed regardless of the environmental conditions. (From Lively 1986b,c.)

that the conditions for the evolutionary stability of the plastic genotype depend on the costs associated with the plasticity itself (see below).

So far we have been concerned mostly with relatively simple cases of adaptive plasticity, in which the alteration of one focal trait directly affects an organism's fitness. More generally, however, living organisms display a tendency not only for different traits, but also for different plasticities, to be correlated. How does this *plasticity integration* (Chapter 4) bear on ecological questions concerning responses to environmental heterogeneity?

Plasticity Integration and the Bradshaw-Sultan Effect

Plasticity integration (Schlichting 1986, 1989), as we discussed in Chapter 1, is the idea that the reaction norms of different traits may be correlated (by definition, *across* environments). As with any correlation, this may result from many possible causes. The two major deterministic phenomena leading to character (and therefore plasticity) correlations are genetic constraints and selection. (Of course, a genetic correlation may also be due to linkage and may originate from nondeterministic phenomena such as random drift [Chapters 1 and 4 and references therein].) In fact, even if the current pattern of character correlations in a population is the result of selection, such correlations do constitute a constraint toward further evolution of the traits themselves, as I explained in the preface.

What concerns us here is the ecology and natural selection of plasticity integration, topics on which very few authors have commented and on which little experimental work has actually been done. This dearth of studies notwithstanding, I submit that a powerful explanatory principle for at least some patterns of plasticity integration is what I termed the *Bradshaw-Sultan effect,* since it was first discussed by Bradshaw in his seminal 1965 review and empirically demonstrated more recently by Sultan (1995). This is a situation in which across-environment stability of a given phenotypic trait (e.g., fitness) is achieved because of marked plasticity in another trait (e.g., ability to produce different leaves under different conditions). The best way to understand the effect is by examining in detail the actual example provided by Sultan (Fig. 7.8). Sultan exposed plants of *Polygonum persicaria* to two light treatments, one of which represented an extreme stress for this species (8% of full sun). She then looked at the plasticity of traits that characterized the seedlings in order to study intergenerational phenotypic plasticity. Surprisingly, she found that the reaction norms for seedling biomass were almost flat, that is, there was no plasticity despite a marked decrease in light availability for the parent plants, which should have negatively affected the seedlings. The reaction norms for achene (the particular type of fruit produced by this plant) mass did not help elucidate

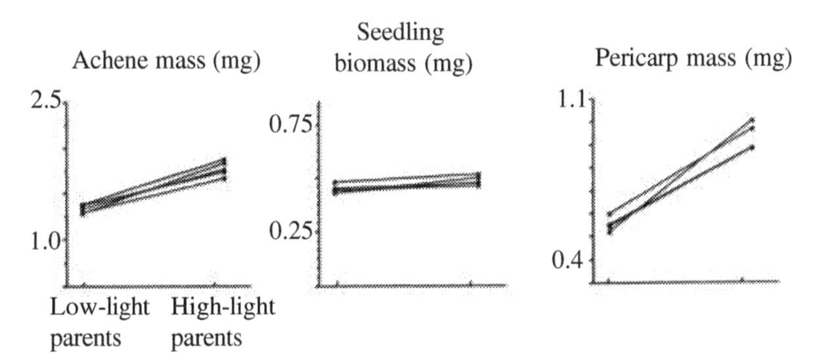

FIG. 7.8. The Sultan effect exemplified by the relationships among the plasticities of three traits in *Polygonum persicaria* genotypes raised under two different levels of (parental) light. The across-environment phenotypic stability shown by seedling biomass (center) is due to the fact that the achene's mass has been only slightly altered under low light (left). This is possible because the mother plants drastically diminished pericarp mass (right) while maintaining seedling provisioning. (From Sultan 1995.)

the situation either, since even though there was a slight decrease in achene mass, the reaction norms were still almost flat. Part of the key to the puzzle was found by plotting pericarp mass against light availability. The pericarp is the thick shell enclosing the embryo and its nutritive tissue, the endosperm. As expected, low light availability had forced the plant to "cut" fruit biomass. However, instead of decreasing the more crucial seed provisioning, the mother plant had reduced the less important outer shell. While this still affects the seed's germination behavior, and possibly the seed's longevity in the soil, it does not significantly affect offspring survival and establishment, definitely a smart strategy for a brainless organism. What we are observing here is a situation in which plasticity for one trait leads to homeostasis for a fitness-related character. To put it as Sultan did: "This response by parent plants evidently constitutes functional homeostasis: adaptive constancy of phenotypic traits central to fitness, achieved by plasticity in developmentally related traits." It is important to realize that the Bradshaw-Sultan effect has the potential to explain the common occurrence in the literature of counterintuitive reaction norms for some traits and to provide a key to a better understanding of the complexity of plasticity integration. However, detecting the effect may not be trivial, because one has to have good working hypotheses about which characters may be more relevant to the organism in question and which ones may be developmentally correlated. Furthermore, these traits have to be measurable in a large number of individuals in order to be considered in a typical plasticity study, which may be difficult for physiological characters.

Several other studies have also provided empirical evidence for the Bradshaw-Sultan effect or in general for the complexity of the relationship between plasticities of a series of traits and reproductive fitness. For example, Lechowicz and Blais (1988) studied the response of genotypes of *Xanthium strumarium* (cocklebur) to different levels of nutrients and water. They found that the relative contribution of different traits to fitness varied with the specific environment. Therefore the plasticity integration of these traits with fitness was altering the relative importance of the traits themselves as seen by natural selection. This has enormous consequences for studies of selection, because it implies that the Lande-Arnold regression approach for estimating the type and strength of selection may depend on plasticity integration. Furthermore, Lechowicz and Blais's results show that as the resource status of the garden went from poor to rich, many selection coefficients decreased dramatically. This is another manifestation of the silver spoon effect, since if conditions are good, the genetic constitution of an individual and the phenotype it expresses matter less and less.[3]

Are There Costs and Limits to Plasticity?

The final topics I would like to discuss concerning the ecology of and selection on phenotypic plasticity are the twin problems of costs and limits and how to measure them. As with any other trait, one can legitimately ask what limits the evolution of phenotypic plasticity. The basic reasoning is simple: If there were no cost or limitation to plasticity, then one would expect the evolution of a *Darwinian monster,* an organism able to perfectly adapt by plasticity to any and all environmental conditions. Quite obviously, Darwinian monsters do not exist, so the question is not only legitimate, but fundamental for our understanding of the limits of adaptive evolution in general. A related way to consider this subject matter, of course, is to think of the generalists-specialists continuum, as we did at the beginning of this chapter.

One of the first attempts to distinguish costs of plasticity from other kinds of effects on fitness was a paper published by Newman (1992) in the context of amphibian metamorphosis. He described four categories of what he called limits to plasticity:

1. *Deficient sensory capabilities.* In order to respond in an adaptive fashion, the organism has to be able to perceive the change in environmental conditions. Such perception, however, might be hampered either by an actual randomness of the environmental fluctuations or by limits in the sensory capabilities of the organisms themselves.
2. *Inability to respond.* Even if the environmental cue can be perceived, there may be an inability to respond appropriately, or simply a time-lag between the stimulus and the response. If the time-lag is long enough,

it will make the response useless, if not maladaptive (because the environmental conditions may have changed in the meantime).

3. *Actual costs of plasticity.* Newman proposed a definition of this category along the following lines: "The cost of plasticity in trait X is defined as a decreased contribution to fitness of other traits (Y) in all or some environments, resulting from a trade-off between plasticity in trait X and the average phenotypic value or level of plasticity of trait Y (and therefore the effect of trait Y on fitness)." He went on to say that the trade-off is presumably created by the functional integration of the traits considered. However, functional integration sounds more like a constraint than a cost, especially if the integration is underlined by genetic correlations among traits. Newman correctly suggested, however, that costs of plasticity, wherever they arise from, could best be measured by comparing plastic and nonplastic genotypes in an environment in which they produce a similar phenotype. Any decrement in fitness of the plastic genotypes would reveal a hidden cost of plasticity.
4. *Lack of genetic variation.* This is a quite obvious limitation on the evolution of any trait, and plasticity is no exception. From a quantitative genetic standpoint, lack of genetic variation can be due to reduced or near-zero heritability (of plasticity, in this case) or to strong genetic correlations between plasticity and other traits. Physiologically, pleiotropy and epistasis could underlie a reduced variation for plasticity and therefore limit its evolutionary response.

Newman's discussion provided a starting point for a more complete theoretical treatment of costs and limits of plasticity, although his approach mixes ecological and evolutionary aspects. The current standard for this discussion is provided by the contribution of DeWitt et al. (1998), who start out by distinguishing between costs sensu stricto and limitations. They then present a list of five categories of costs and four categories of limits as shown in Table 7.2.

I think this classification is a jumping-off point for further research, although as DeWitt et al. themselves freely acknowledge, it may be very difficult to experimentally disentangle all (or even most) of these categories. However, there are some caveats that I would like to note about several classes of costs and limits before we turn to a few examples. It is not at all intuitive that there will ever be production costs *as distinct from the ones incurred by a fixed phenotype.* The authors admit that some people may not consider production costs as plasticity-related, since even a nonplastic genotype will still have to produce the same structures. Furthermore, even if one could demonstrate that the same morphologies are more costly if produced by a plastic genotype, the underlying cost would most likely fall into one of the other categories, especially maintenance and genetic. As for the developmental instability cost, it is a theoretical possibility, but as we saw

TABLE 7.2
Classification of Costs and Limits of Plasticity According to De Witt et al. (1998)

Costs of plasticity	Limits of plasticity
Maintenance: energetic costs of sensory and regulatory mechanisms	*Information reliability:* the environmental cues may be unreliable or changing too rapidly
Production: excess cost of producing structures plastically (when compared to the same structures produced through fixed genetic responses)	*Lag time:* the response may start too late compared to the time schedule of the environmental change, leading to maladaptive plasticity
Information acquisition: energy expenditures for sampling the environment, including energy/time not used for other activities (e.g., mating, foraging)	*Developmental range:* plastic genotypes may not be able to express a range of phenotypes equivalent to that typical of a polytypic population of specialists
Developmental instability: plasticity may imply reduced canalization of development within each environment, or developmental "imprecision"	*Epiphenotype problem:* the plastic responses could have evolved very recently and function more like an "add-on" to the basic developmental machinery than an integrated unit; as such, its performance may be reduced
Genetic: due to deleterious effects of plasticity genes through linkage, pleiotropy, or epistasis with other genes	

in Chapter 4, there is little—if any—evidence that there is any relationship between plasticity and developmental instability. Among the limits described by DeWitt et al., I would like to comment on the last two in particular. The developmental range limit is a matter of empirical investigation, but it will probably turn out to be a straw man. Of course natural selection can produce fixed-response genotypes which, taken together, will show a broader developmental range than any *single* plastic genotype. But this is a rather odd comparison, juxtaposing a population to an individual. If the comparison is made between the subpopulations of plastic and nonplastic genotypes, then the outcome may be in either direction, and this may as well be called a limit of nonplasticity. The epiphenotype problem is perhaps one of the most intriguing from an evolutionary perspective. The degree to which a plastic response is an epiphenotype as opposed to an integrated one will depend on the particular evolutionary trajectory followed by the population or species in recent time (Jacob 1977). Furthermore, a plastic response may start as an epiphenotype and subsequently be integrated by natural selection attempting to reduce the costs and limits of the new response. Indeed, this may be a major venue of evolution for new types of phenotypic plasticity or for phenotypic novelties in general.

Once we have a good working definition of costs and limits, how do we actually measure them? The first to provide an answer to this question within a more general treatment of generalists and specialists was van Tienderen (1991). He reasoned that the standard quantitative genetic approach to the study of selection could be easily extended to determine costs of plastic responses. The current version of this approach is due to further elaboration by Scheiner and Berrigan (1998) and DeWitt et al. (1998). The basic idea is to regress plasticity and the trait mean within one environment against a measure of fitness. The coefficient attached to the within-environment term will tell us if the trait is under selection, as in the classical Lande-Arnold (1983) model. The coefficient for the plasticity term would function analogously, but since it is used in concert with the within-environment term, it actually tells us if there is an advantage (positive coefficient) or a cost (negative coefficient) to plasticity once selection within one environment is accounted for. There are several obvious limitations to this approach, primarily the fact that it works best for two-environment situations (at least in its current incarnation): Plasticity has to be measured by one number if it is to be used in the regression model, and this is easy to do if it is measured across two environments (see Chapter 2). While this approach could be extended to multiple environments by making use of alternative measures of plasticity or by conducting the analyses across two environments at a time, I am concerned here with the general ideas behind the technique and not with the details of its implementation.

To date, the regression approach to measuring costs of plasticity has been used in only two instances, and in both cases with largely negative results. Scheiner and Berrigan (1998) investigated costs of plasticity elicited by predator cues in a genetically variable sample of *Daphnia pulex* by measuring changes in three traits summarizing body size and shape in response to a chemical signal associated with the predator. Their analyses not only failed to reveal any cost of plasticity, but they did not even show a direct cost of juvenile structures in *Daphnia* for which other authors had previously detected costs. The authors speculated that the failure to observe costs of plasticity in their system was due to a lower metabolic rate in *Daphnia* when it is exposed to the predator cue. The idea is that the decrease in metabolic level could conceivably compensate for the expected costs of plasticity.

DeWitt (1998) obtained similar results with another predator-induced system: the change in morphology of freshwater snails of the species *Physa heterostropha* in response to predation by either sunfish or crayfish. He examined plasticity in two traits across twenty-nine families and found only one potential cost of plasticity, namely a negative selection gradient for growth rate associated with plasticity of shell shape. However, he also acknowledged that the particular type of plasticity studied restricts feeding, thereby lowering growth rate. Since the reduced growth rate owing to plasticity may have fitness benefits, the observed negative relationship represents only ambiguous evidence of costs.

When this book was in press, a paper by van Klausen et al. (2000) showed significant costs of plasticity in stolon internode length in the presence of competition in the clonal plant *Ranunculus reptans*. Interestingly, this study was successful probably because of the use of a large number of genotypes—102 in this case.

More effective than these statistical approaches have been studies based on an actual understanding of the developmental aspects and physiological mechanisms potentially underlying both the costs and limits of plasticity. A well-documented example of costs of plasticity is the result of detailed analyses of the heat-shock response in *Drosophila melanogaster*. Krebs and colleagues (Krebs and Loeschcke 1994a; Kreps and Simon 1997) were able to clearly document both the costs and benefits of this response. They found that flies conditioned to high temperature (and hence able to turn on the heat-shock response) survived severe heat stress significantly better than flies that were not acclimated. On the other hand, females that received the conditioning treatment but were then not exposed to stress produced fewer offspring than nonconditioned females. Furthermore, the difference between the two groups increased with the frequency of conditioning treatments. Apparently, turning on the heat-shock machinery when it is unnecessary is extremely deleterious (which also explains the observations of Kohler et al. [2000] that selection favors low levels of the heat-shock protein Hsp70 in chronically metal-stressed arthropods). So far, this is a demonstration of the fact that expressing the wrong phenotype in a given environment is harmful, which is indeed a prerequisite for the evolution of adaptive plasticity. But Krebs and Feder (1997a) went further in exploring the genetic and physiological bases of the conditional costs of heat shock by studying the effects of overexpression of the Hsp70 protein in *D. melanogaster*. They compared extra-copy and excision strains, and found that the latter grew more rapidly and were characterized by a higher survival rate, indicating a metabolic cost of heat-shock overexpression. This cost was evident in the long run even though the extra copies provided a temporary additional protection to the flies when they were exposed to heat stress.

A very different usage of the word *cost* in relation to plasticity is found in studies of reproduction costs and sex allocation in heterogeneous environments. While this is not to be confused with the main thread of our discussion, it does represent an interesting and related area of inquiry. For example, Dunn et al. (1993) studied the cost of environmental sex determination in the crustacean *Gammarus duebeni* and found that plasticity of sex expression to photoperiod produces a very definite cost to the population in the form of intersexuality. Intersexuality is in fact a particularly intriguing side effect of plasticity of sex expression: Intersexual individuals are functionally both males and females, but suffer fitness costs for this double phenotype; at the same time, intersexuality is heritable, in that intersexuals produce more intersexual progeny than do true females. Galen (2000) found that large flowers exact a water cost in *Polemonium viscosum* under dry conditions, leading to an interesting environmental effect on sex allocation:

Under drought, viability selection favors either small-flowered plants characterized by female-biased reproduction or large-flowered individuals with male-biased reproduction. This was because drought survivors (unlike the individuals selected against) showed a negative relationship between flower size and fruit production (so that only small-flowered ones allocated enough to fruit production, the others investing in pollen-making instead).

The literature on costs associated with plasticity is also further marred by the confusion between actual costs of plasticity and two other, entirely distinct categories of fitness decrements: the consequences of expressing the wrong phenotype in a given environment (DeWitt et al. 1998, see Table 7.2), and the detriment incurred by organisms expressing the same phenotype under different conditions (lack of plasticity). If an organism is plastic, but the phenotype expressed in a certain environmental range has low fitness, it is obviously not a cost of plasticity, but simply a case of non- or maladaptive plasticity. On the other hand, if an organism is not plastic and it is exposed to heterogeneous environments, the same phenotype may not be adaptive under all conditions. In fact, part of this literature could be used as evidence for the costs of *not* being plastic!

An especially well-studied category of costs of this type (i.e., not costs of plasticity, but low fitness associated with the expression of a fixed phenotype when the environment is heterogeneous) includes those linked to resistance to herbivores and diseases in plants. For example, Bergelson (1994) has used a path-analytical approach to explore the environmental dependency of costs of resistance to leaf root aphids and downy mildew in two cultivars of lettuce, a case considered here as an instance of reduction in fitness owing to lack of plasticity. The results pointed to environmental effects on as well as genetic specificity of the costs: One of the two cultivars suffered a greater cost of resistance when grown under low rather than high nutrients; the second cultivar showed no such difference. Also, Bergelson suggested that additional trade-offs might occur in these cultivars because of their resistance to two agents, a characteristic that is genetically linked in these near-isogenic lines, although this may not be the case in natural populations. In another study, Simms and Triplett (1994) did not find costs for resistance to fungal isolates in *Ipomoea purpurea* when resistance was measured either as inversely proportional to the total leaf area lost to infection or as the complement of the proportion of leaves infected. However, these authors found a significant fitness reduction (decreased seed and pollen output) owing to *tolerance,* defined as the ability to partially compensate for fitness detriments caused by the disease. They proposed that the observed cost of tolerance could in fact obscure the detection of a similar cost of resistance. Mitchell-Olds and Bradley (1996) found a clear cost of resistance, measured by decreases in growth rate, of *Brassica rapa* to a fungal pathogen, but did not uncover a similar cost for the resistance to another pathogen. This raises the possibility that an initially costly adaptive phenotype may evolve toward a less costly one by natural selection. While this hypothesis was not directly tested in the paper mentioned, it is interesting to note that the fungus for which no costs were detected is specific to the genus *Brassica,* while the cost-inducing one is a more generic parasite of crucifers. Response to herbivory can also be influenced (and limited) by other types of plasticity. For example, Meyer (2000) found that at high soil fertility, *Brassica nigra* maintained a high leaf growth rate in the face of herbivore damage induced by caterpillars of *Pieris rapae.* However, in soils with low fertility, the leaf growth rate was dramatically reduced, with the result that the plants suffered twice the relative amount of damage from the herbivores.

In conclusion, we need to go back to the concern expressed by DeWitt et al. (1998) about the ability of biologists to disentangle so many potential sources of costs and limits to plasticity, a concern backed by DeWitt's (1998) and Scheiner and Berrigan's (1998) inability to find evidence of costs, possibly because of multiple layers of causality underlying the phenotypes they studied. Our discussion so far and the examples of empirical studies of costs suggest that our best hopes reside—not surprisingly—in a multiplicity of approaches, rather than in a single "magic bullet." I submit that van Tienderen's statistical method can be applied as a first screen whenever the data are appropriate. Failure to detect costs in this way, however, should not be considered positive evidence of their absence because of many potentially confounding variables, the situation being somewhat analogous to the difficulty of uncovering trade-offs in empirical studies even though they are expected in theory (van Noordwijk and Jong 1986; Houle 1991). More direct approaches, be they experimental manipulations of phenotypes under controlled environments or investigations of the physiological and mechanistic bases of alleged trade-offs, as in the example of heat shock in *Drosophila* discussed above, will ultimately provide the best answers to the problem.

Conceptual Summary

- All major strategies adopted by sedentary organisms (ruderal, competitive, and stress tolerant) can be associated with particular types of adaptive plasticity.
- Genotypes can be classified ecologically depending on whether they are adapted to a restricted or broader set of environments and based on the presence or absence of plasticity for a focal trait (meaning a trait under selection). The four resulting combinations elucidate the relationships among plasticity, ecotypes, generalists, and specialists.

- The adaptive significance of a plastic response can be studied by one (or a combination) of three approaches: historical analyses (comparative method), optimality analysis (functional ecology), and fitness-correlates methods (Lande-Arnold–type regressions).
- Despite the technical and conceptual difficulties of studying plasticity as an adaptation, several published works can be used as methodological benchmarks for this purpose.
- In the Bradshaw-Sultan effect, homeostasis in one trait (e.g., reproductive fitness) is achieved by plasticity in other, correlated, traits. This can provide a key to understanding apparently puzzling patterns of plasticity and to relating the concept of plasticity integration discussed earlier to the general theory of adaptive evolution.
- Several classifications of costs of and limits to plasticity have been proposed, the major ones being those of Newman and of DeWitt et al. As informative as these may be from a theoretical standpoint, it is unlikely that all the proposed sources of costs can be distinguished empirically, and certainly this is a task of utmost difficulty within a single experiment.
- No convincing statistical evidence of costs has been published so far, but mechanistic studies of the molecular machinery potentially underlying costs of and limits to plasticity have been successfully employed toward this end.

8

Behavior and Phenotypic Plasticity

Habit is a second nature that destroys the first. But what is nature? Why is habit not natural? I am very much afraid that nature itself is only a first habit, just as habit is a second nature.

—Blaise Pascal (1623–1662)

CHAPTER OBJECTIVES

To explore the special case of behavior as a form of phenotypic plasticity. A theoretical framework is provided by which behavior and physiological acclimation are types of genotype-environment interactions that can have macroevolutionary consequences. Various case studies are presented to highlight the similarities between classical developmental phenotypic plasticity and behavioral plasticity, including learning. The chapter concludes with a discussion of the evolutionary lability of behavioral traits and some considerations of the relationship between behavior and plasticity.

Behavioral ecology is an extensive field of research that touches on evolution, physiology, development, and molecular biology, much in the same way as plasticity studies (Boake 1994). I explained in the preface why I have included a brief chapter on behavior in this book while neglecting to do the same for physiology. It is certainly not my intent here to review or discuss in any detail a topic that requires (and has been the subject of) volumes upon volumes of material. However, some aspects of behavioral research have been directly connected to phenotypic plasticity—sometimes in an insightful way and other times more or less vaguely. My aim in this chapter is to elucidate some of the main ideas at the interface between behavior and plasticity while considering a few representative examples from the vertebrate and invertebrate literature. The case studies discussed below all have two underlying themes. First, the environment can be a direct modifier of behavior; hence behavioral ecology belongs to the general field of study of genotype-environment interactions. Second, behavior in turn may affect morphology, thereby linking the two more intimately in an ecological and evolutionary feedback of potentially considerable complexity.

Theoretical Framework

Perhaps one of the earliest and least known theoretical works to explicitly articulate the link between plasticity and behavior was published by Cavalli-Sforza (1974).[1] He proposed treating behavioral plasticity as similar and an addition to physiological plasticity—the latter manifested, for example, in plants' alterations of photosynthetic rates in response to light availability and mammals' changes of respiration rate or production of red cells in response to alterations in oxygen availability. Physiological plasticity is characterized by very short time scales and reversibility, as opposed to morphological/developmental plasticity, which is usually irreversible, at least at the organ level, and unfolds over a longer time span. Also, physiological plasticity is typically possible throughout the life span of an individual, while developmental plasticity depends on *windows of availability* (see Chapter 6). From this standpoint, behavioral plasticity is another kind of plasticity, typical of animals (see the discussions of types of plasticity in the preface and Chapter 2). It should be clear then that, intuitively appealing as the idea may be, plasticity is *not* the plant equivalent of animal behavior: The two are related but distinct phenomena. To further drive this point home, it should be noted that animals show *both* plasticity and behavior (so the two cannot be synonymous). Moreover, some kinds of behavior are highly plastic (learning) and others are much less so ("innate" behavior). Hence the widespread use of the term *behavioral plasticity* to differentiate it from behavior per se. In the following, behavior will be considered as any other phenotypic (i.e., morphologic or physiologic) character that can show a flat reaction norm (no plasticity) or a more or less complex response function to given environmental conditions. Conceivably, an informative parallel can be drawn between behavioral and other kinds of plasticity in the following manner. On the one hand, we can think of learning as similar to developmental plasticity. Both unfold more or less monotonically during the ontogeny of an organism. On the other hand, most every-day behavior is flexible and reversible on a very short time scale, thereby mimicking physiological plasticity. Of course, the whole issue is more complicated in animals by the fact that this kingdom presents us with all of the above categories, and all are somewhat intertwined.

West-Eberhard (1992) is perhaps the author who has most explicitly explored the relationships between plasticity and behavior. She submits that "a fundamental quality shared by behavior and development is condition-sensitivity: both are partly directed by circumstances." Given this fundamental communality, she attempts to frame the entire question of the evolution of phenotypic novelties (Chapter 9) in terms of a chicken-and-egg problem. Are behavioral modifications driving the evolution of new structures, or is it the emergence of new structures in response to environmental changes that leads to alterations in behavioral patterns? This quandary was first considered by Goldschmidt (1940), who offered early valuable insights into the matter. We can easily find examples that fit either end of the spectrum. Presumably, the evolution of land vertebrates was aided by changes in the behavior of fishes living close to the shore and exposed to the air during part of the tidal cycle. We can still observe this in extant species and even induce terrestrial phenotypes in obligatory aquatic amphibians by altering their hormonal balance, hence affecting their behavior (Goldschmidt 1940). On the other hand, West-Eberhard's example of the range of behavioral flexibility in Darwin's finches as a result of morphological adaptation to variation in seed size may represent a case in which morphology drives behavior (though without comparative behavioral and morphological studies any bet is possible as to the directionality of the process). West-Eberhard links this view of behavior-morphology evolution to macroevolution. In her model, major alternative morphologies and/or behaviors can evolve intraspecifically as a result of adaptive responses to distinct environments. In her words, "developmental and behavioral switches divide the phenotype into subunits that evolve semi-independently." That is, plasticity facilitates the partitioning of the phenotype (sensu Wagner and Altenberg 1996), and therefore its adaptive differentiation, without having to limit the appearance of phenotypic novelties to the time around speciation events. Accordingly, macroevolution becomes causally decoupled from speciation, a view that has the potential to finally free us from decades of haggling about punctuated equilibria (Eldredge and Gould 1972; Stebbins and Ayala 1981; Thomson 1992).

A less general but more direct argument for the connection among behavioral, physiological, and morphological plasticities was advanced by Piersma and Lindstrom (1997). They reviewed the literature on the measurements of organ size in vertebrates and noted that the assumption that these measurements are stable and can be used as population or even species averages does not hold. This is of course not surprising in light of the fact that organ size depends in part on environmental conditions and therefore falls under the general category of morphological plasticity. But what Piersma and Lindstrom pointed out is that organ size can change significantly and reversibly over relatively short periods of time, measured in hours or weeks. Examples include snake guts (Fig. 8.1), organ adjustments during lactation, and migration-related changes in the digestive system in

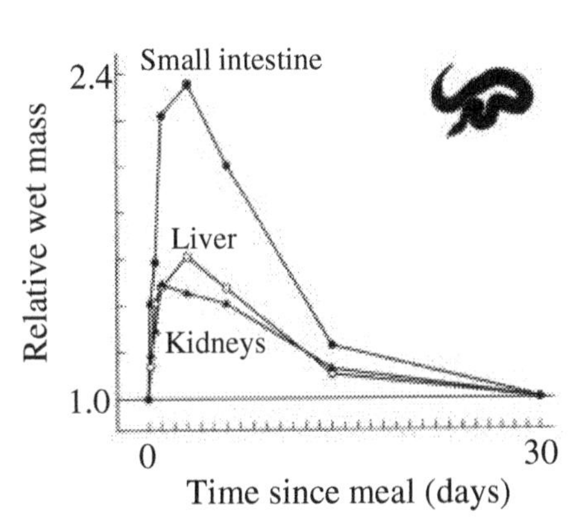

FIG. 8.1. Changes in the anterior part of the small intestine, liver, and kidneys in Burmese pythons after a meal of rats or mice equivalent to 25% of the snakes' body mass. The figure illustrates one type of connection among behavioral, physiological, and morphological plasticities (see text for details). (From Piersma and Lindstrom 1997, based on data from Secor and Diamond 1995.)

birds. All these alterations in morphological plasticity are not only reversible (thereby blurring the distinction proposed by Cavalli-Sforza) but also the result of physiological and behavioral processes, hence inextricably linking the three as facets of a fundamentally similar phenomenon. As I noted in Chapter 2, it is perhaps best to refer to phenotypic plasticity as the general phenomenon and to behavior, physiology, morphology, and development as its different manifestations.

Komers (1997) discussed in some detail scenarios for the evolution of behavioral plasticity in heterogeneous environments, closely following similar optimization approaches used in modeling morphological plasticity (Chapter 10). The general prediction is of course that the range of behavioral responses should increase with environmental variability. More interestingly, however, Komers focused on situations in which, even though alternate environments are similar in terms of resource availability, the costs associated with exploiting such resources may differ substantially.[2] In this case, he found that animals that can better assess the costs would be those characterized by increased behavioral plasticity, if they decrease the costs and increase the benefits of exploiting a particular resource. Consequently, behavioral plasticity should increase with learning capacity, even though the latter is not a prerequisite for the former, although, as I noted earlier, we could consider learning as being akin to developmental plasticity and therefore a kind of behavioral plasticity. Komers also highlighted another important point generally neglected in the empirical literature. Since trade-offs change throughout the life cycle of an animal, its behavioral plasticity should change accordingly. This is analogous to windows for plasticity throughout the development of both plants and animals (Chapter 6, Fig. 8.2), and it suggests yet another fundamental similarity between behavioral and other types of plasticity.

A very different kind of behavioral plasticity was considered by Cheverud and Moore (1994) in their discussion of the quantitative genetics

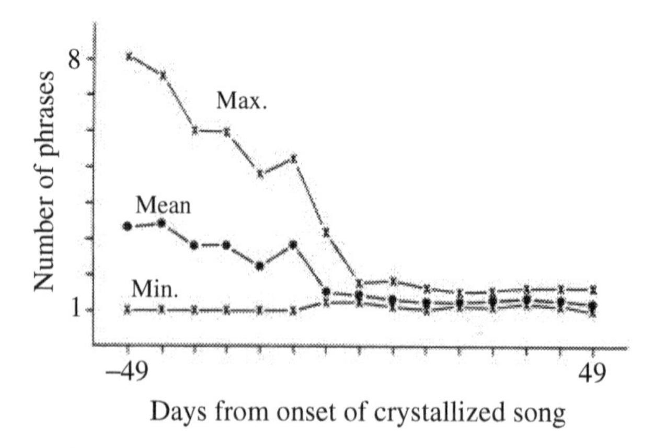

FIG. 8.2. A window for behavioral plasticity in action-based learning of a song in sparrows. The number of phrases included in the song is on the *y*-axis. Note how a population of sixteen birds shows a high degree of variation (the mean and range is reported over the time interval in days) until the window closes and the song becomes canalized. (From Marler and Nelson 1993.)

of parental effects on behavior. In essence, they proposed that the environments experienced by the parents or other relatives of an individual may alter the behavioral pattern of the latter with important, sometimes counterintuitive outcomes, such as a response to selection in the opposite direction of the applied pressure. This is similar to the increasingly well-studied phenomenon of across-generation plasticity in plants (and some animals [Agrawal et al. 1999]), where the phenotypes of the offspring respond in an apparently adaptive fashion to the environments encountered by their parents (Schmitt et al. 1992; Galloway 1995; Sills and Nienhuis 1995; Lacey 1996; Mazer and Gorchov 1996; Fox et al. 1997). The advantage of applying quantitative genetic approaches to this problem is that quantitative genetics by definition allows for a distinction between selection pressure on and selection response of a population.

Some of the following examples should help clarify the conceptual issues raised above concerning the interplay of plasticity and behavior and, in particular, the multifarious effects of environment on behavior and of behavior on morphology.

Case Studies

Behavioral plasticity has perhaps been studied most in insects, for the excellent reasons that the behavioral repertoire is limited, hence more easily quantifiable, and the organisms themselves are easily manipulated. For example, differences in courtship songs are implicated in the reproductive isolation of three morphs of *Chrysoperla* (green lacewings). Wells and Henry (1992) analyzed five features of the song (duration, interval, initial frequency, middle frequency, and end frequency), and found consistent differences among the three morphs (especially in sympatry), which they interpreted as compatible with the idea that the morphs are actually reproductively isolated species. From the point of view of plasticity, it is interesting that variation in the features of the song was dramatically dependent on temperature, with the same feature showing almost no response to the environment in one morph and a marked positive or negative response to increased temperature in another. If—as the authors suggest—variation in the characteristics of the song can cause reproductive isolation and therefore speciation, it is also tempting to consider the possibility that these variations may originate as a plastic response to simple environmental factors such as temperature. This would link plasticity, behavior, reproductive morphology, and macroevolution very much in the fashion envisaged by West-Eberhard and discussed above.

Ciceran et al. (1994) demonstrated a similar effect of temperature on the song of the field cricket *Gryllus pennsylvanicus,* where higher temperatures increase the pulse rate while decreasing interchirp interval and chirp duration. However, the time of the day (even when corrected for temperature) and male spatial density also influenced these characteristics of the song, pointing to the possibility of multiple environmental effects being integrated in the reaction norm for calling song in this species. Interestingly, almost identical results were found by Olvido and Mousseau (1995) in the striped ground cricket, *Allonemobius fasciatus.* Again, chirp rate was positively correlated with temperature, while chirp duration and interchirp interval were negatively related to the same environmental variable. This across-genera consistency points toward fundamental physiological processes being affected in like manner by temperature (perhaps not surprisingly in a poikilotherm), although quantitative differences in reaction norms of full sibs seemed apparent in the latter study.

Plasticity of song production is an aspect of behavioral plasticity that is not confined to invertebrates. Marler and Nelson (1993) studied the distinction between memory-based and action-based song learning in three species of sparrow (*Melospiza georgiana, Spizella pusilla,* and *Zonotrichia leucophrys*). Memory-based learning is the classic mechanism by which a bird learns a song by hearing it, memorizing it, and then matching it via auditory feedback. In the action-based mode, however, the young bird can already produce a variety of songs, from which it will select one that will become canalized during life as an adult. This selection from an initially vast repertoire is made possible because of the social environment in which the bird lives: Social interactions with conspecifics reinforce some forms of the song but not others (Fig. 8.2). This is again equivalent to the concept of windows for plasticity, in that there is a specific ontogenetic phase during which the behavior (or morphology) is plastic, followed by canalization of the phenotype.

Song-learning behavior is indeed a textbook example of nature-nurture interaction, and thereby a prime instance of the complex relationship between behavior and plasticity. Marler (1970) summarized a classical series of studies in the white-crowned sparrow that started with the discovery of the existence of a series of local "dialects" in California and the consequent possibility of genetic differentiation for song attributes. Several cunning manipulative experiments demonstrated that the situation is far more intriguing. Experimenters compared the development of the song in birds exposed to their normal social environment (i.e., hearing their "dialect" song from other birds), birds maintained in acoustic isolation throughout their development, and birds allowed to hear the correct song for the first part of their growth, but then surgically deafened afterward. The results clearly showed that in the early stages of development the birds produce a subsong that is similar to the final form, but it lacks several distinct elements and cannot be identified as any particular dialect. The subsong is apparently genetically determined, and the environment has no influence on its structure. However, the dialects (full songs) develop only if the bird is exposed early on to the full song of its geographical district, thereby demonstrating plasticity for the pattern of maturation of the song.

A series of studies on behavioral plasticity of the fire ant *Solenopsis invicta* in response to the social environment was published by Keller and Ross (1993a,b, 1995). This species is characterized by two types of social organization: monogynous colonies (with one queen) and polygynous colonies (several reproductive queens). Early in development, the queens of the two types of colonies do not differ significantly in aspects of their phenotype such as weight and fat content. However, when the queens assume the winged form later on, monogynous queens are heavier and fattier than polygynous ones. Keller and Ross were able to identify a single locus, *Pgm-3,* that either directly affects these phenotypes or is tightly linked to a locus that does. However, a major cause of the differential phenotypes is actually to be found in the type of colony in which a given queen is raised. That is, the social environment in which the insects mature determines the physiology and phenotype of the adult form. In turn, these differences actually affect the reproductive history of the queens themselves, and therefore the social organization of the colony. In the end, the latter is perpetuated because of the phenotypic effects of the social environment to which the queens are exposed. Keller and Ross call this a case of plasticity affecting "cultural transmission," a situation very similar to the one proposed by Cavalli-Sforza (1974).

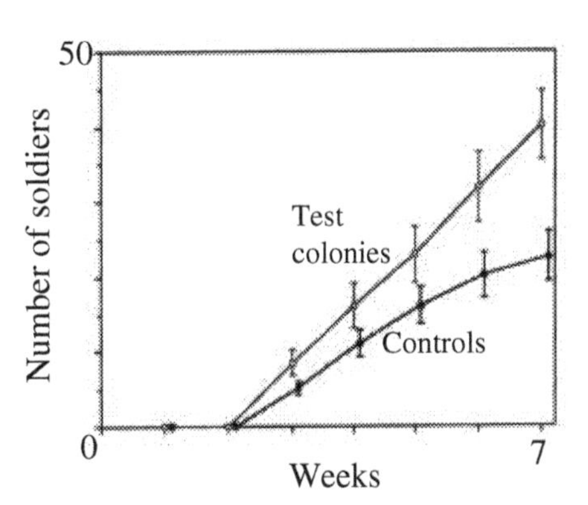

FIG. 8.3. Effect of contact between indigenous workers and foreign workers (open symbols) on the production of soldiers in colonies of the ant *Pheidole pallidula*. Solid symbols indicate the control, where no contact between native and foreign workers was allowed. (From Passera et al. 1996.)

Working with a different model system, Keller and collaborators (Passera et al. 1996) also provided a clear demonstration of adaptive phenotypic plasticity induced by social interactions (Fig. 8.3). Optimization theory predicts that caste ratios in social insects should evolve a reaction norm that is a function of a variety of environmental factors, such as food availability, predation, and competition. However, it is very difficult to alter these variables in an ecologically informative fashion and to study the consequences on colony dynamics. To address the problem, these authors manipulated the exposure of workers of *Pheidole pallidula* to foreign workers and observed the effects on caste ratios. Colonies whose workers had been exposed produced significantly more soldiers within the span of three weeks from contact, again demonstrating a direct effect of the social environment on colony dynamics.

The ease of manipulation of arthropods also allows reciprocal transplant experiments such as the one carried out by Riechert and Hall (2000) to investigate the relative importance of behavioral flexibility and genetic differentiation in the survival of desert spiders living in contrasting environments. *Agelenopsis aperta* living in environments with low prey abundance and competition for sites that provide shelter from the heat are favored if they exhibit aggressive behavior, both toward conspecifics and toward prey. However, some populations occupy riparian habitats, where they experience release from competition for prey and foraging sites. Under these circumstances, low levels of aggression are favored. After the transplant experiments, riparian transplants did very poorly in the novel environment, while the spider obtained from a dry woodland maintained a relatively constant fitness in both native and novel conditions. Interestingly, the distribution of behaviors of the transplants was a continuum between the two "typical" ones, indicating some degree of plasticity. Since the behaviors inappropriate to either habitat were selected against, one would expect the evolution of behavioral plasticity if the rate of migration between the two habitats is significant, as indicated by previous studies published by Riechert (references in Riechert and Hall 2000).

Sometimes it is another organism's plasticity that can elicit adaptive behavioral responses in a given species. Plants of *Nicotiana attenuata* can facultatively produce methyl jasmonate, which increases nicotine production, which in turn substantially reduces the fitness of larvae of a specialist herbivore, *Manduca sexta.* Methyl jasmonate has a strong effect on the tendency of the larvae to leave the host plant and seek an individual that does not produce the chemical (van Dam et al. 2000). This is a situation in which selection favors a kind of physiological plasticity (in the form of inducible defenses) in the plant, which itself favors the evolution of the herbivore's avoidance behavior. Absence of the latter creates a disadvantageous situation for the animal in which plasticity for fitness is caused by the presence in the population of plants with and without inducible defenses (which are in turn costly [Baldwin 1998]). Predator-induced defenses can also provide good examples of interactions between morphological and behavioral plasticity, as in the case of tadpoles of *Hyla versicolor* (van Buskirk and McCollum 2000). The presence of larval *Anax* dragonflies triggers morphological (elongated and shallow bodies and tails) and behavioral (rapid swimming bursts) responses. This combination of plasticity at two levels effectively decreased predation risk to about half. Interestingly, the two responses have to be coordinated, because augmented activity actually increases the risk of predation if it is not accompanied by the morphological changes in the depth of the body and tail.

Considering another classical model system for plasticity study, De Meester (1993) investigated phototactic behavior in *Daphnia magna.* These crustaceans show genetic variation among clones for their response to light stimuli, but such a response becomes much more evident after exposure to the presence of a fish predator in the pond (Fig. 8.4). Furthermore, genetic variation for phototactic behavior is greatly enhanced under the "fish" environment, with some clones showing little, albeit significant, plasticity, and others responding dramatically to the presence of the predator. Interestingly, the plasticity for phototactic behavior in *D. magna* does not actually depend on the immediate vicinity of the predator, but is instead stimulated by a fish-released chemical. Since the chemical is released by fishes at very low density, presumably before they become a serious threat to the *Daphnia* population, this makes it a case of adaptive anticipatory plasticity (Chapter 7). The temporal dynamics of the response were indicative of the underlying mechanics of the phenomenon: While it took less than 24 h after the exposure to the chemical to enhance the phototactic behavior, the response returned to normal only several days after the stimulus was removed. This delay may be caused by the time-lag involved in activating and tran-

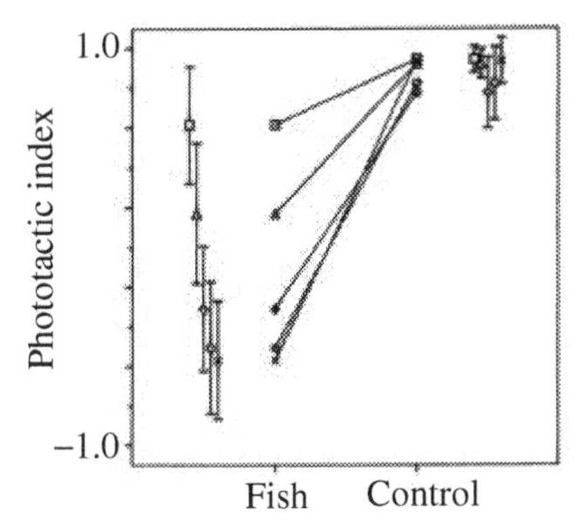

FIG. 8.4. Behavioral plasticity for phototaxis in *Daphnia magna.* Crustaceans not exposed to a predator converge on a similar pattern of behavior, but display a wide range of behaviors if a chemical associated with the fish predator is released in the water. (From De Meester 1993.)

scribing the appropriate gene products at the onset of the response and in turning off the same genes while catabolizing the mRNAs and their products at the offset, although the time frame involved seems too long for a simple explanation in terms of biochemical dynamics. *Daphnia* are also capable of plasticity of behavior that affects habitat selection. Leibold et al. (1994) investigated the effects of genotype, ontogeny, and acclimatization on the habitat preferences of *D. pulicaria.* While both heritability and ontogeny had significant effects on habitat choice, by far the most important determinant was the habitat to which the clone had been exposed. Clones acclimatized to a particular habitat showed a very strong tendency to choose the same habitat. However, the same clones could be acclimatized to a variety of habitats, and then develop a preference for those as well, demonstrating ample flexibility in behavioral patterns. De Meester and collaborators (1995) addressed the combination of antipredator behavior and habitat choice considered by the last two studies. By growing hybrid *Daphnia* in large indoor mesocosms, they could expose them to the fish-released chemical and track the response in terms of distribution of animals in the vertical column. The results confirmed that migration to greater depths is indeed a predator-avoidance behavior, but this is influenced by the genotype, partly through ontogenetic differences resulting in variation in body size (the largest clones tend to dwell at the greatest depth during the day).

Landmark-based learning in honeybees is a situation that presents an interesting example of the communality between behavior and plasticity. The phenomenon is perhaps more closely allied to physiological acclimation: The bees have the capacity to alter their foraging behavior based on the features of the surrounding landscape, gradually becoming more acclimated to those particular features through the use of landmark cues. Collett and Rees (1997), for example, showed that bees use a distinct set of three navigational strategies during the final approach to the target, regardless of the specific configuration of the territory. The bee first aims at a convenient landmark, using it as a reference point to home in on the area where the food resource is located. Then the learned retinal appearance of the landmark is associated with a particular flight trajectory that brings the bee sufficiently close to the destination. Finally, it orients itself so as to move the landmark to a standard retinal position (image matching), sort of like a radar locking onto a target. All of this is not influenced by the environment as such (except in the limited sense of the geographical characteristics of the territory). However, the environment can dramatically affect the landmark-based behavior of honeybees, as demonstrated in a study by Chittka and Geiger (1995). They showed that when the bees occupy a territory with few landmarks, they prefer a sun compass to the landmark method. This alternative behavior again switches to landmark-based targeting if the meteorological conditions are such that the sun cannot provide a reliable reference point. A simple (and temporary) change in the physical environment induces a major (plastic) shift in behavioral strategies.

The intricacies of the relationship between morphological and behavioral plasticity are spectacularly evident in some fishes, for example, in cichlids responding to differences in diet (Meyer 1987). In the threespine stickleback (*Gasterosteus* sp.) foraging efficiency is a function of diet-induced morphological and behavioral plasticity (Day and McPhail 1996), with plasticity improving foraging efficiency for up to 72 days after exposure to a particular type of prey, a time scale matching the one typical of diet variability under natural conditions. Day and McPhail attempted to disentangle the effects of morphological and behavioral plasticity and concluded that the former affects the efficiency of handling of the prey, whereas the latter is more important for improving searching efficiency. The combination, and presumably co-evolution, of the two characterizes ecological variation in freshwater sticklebacks.

A study of garter snakes' (*Thamnophis sirtalis*) foraging behavior provides another example of the relationship between learning and plasticity (Burghardt and Krause 1999). These authors exposed neonatal individuals to three types of diet: fish, leafworms, and mixed. They then measured approach latency (time that the snake takes to reach a dish on which the food resides), capture time, handling time, swallowing time, and total consumption time (the last being the sum of all of the above). Measurements were taken during the first feeding, that is, the first time the animals were exposed to the food, and then again after eleven or twelve meals to see if experience had affected any of the parameters. It turned out that approach latency did decrease significantly with learning if the snakes were exposed to either pure diet, but did not change for animals fed a mixed diet (Fig. 8.5). Also,

FIG. 8.5. Behavioral plasticity of garter snakes in response to fish, worm, or mixed diets. Animals' approach latency (see text) was measured when the neonatal individuals were first fed (left) and then again after eleven or twelve feeding episodes (right). Asterisks indicate significant differences: Note that while learning did improve approach latency in response to either pure diet, it did not affect response to a mixed diet. (From Burghardt and Krause 1999.)

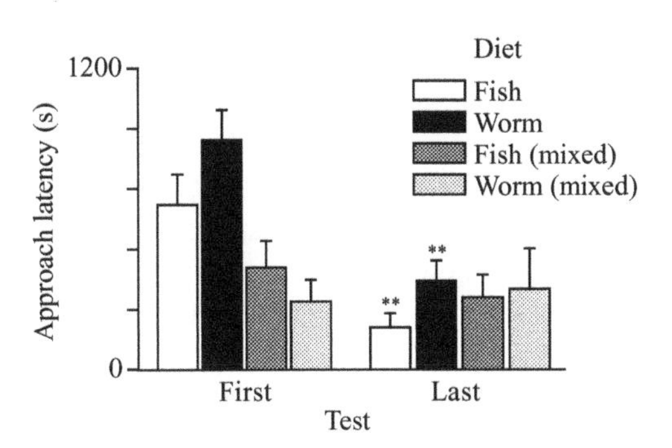

the degree to which different components of feeding time decreased was markedly affected by the specific diet, which, in turn, had little effect on the duration of prey handling. The authors concluded that the snakes showed adaptive behavioral plasticity, but that this was limited by the kind of diet available during the trials.

Nest-building behavior in response to temperature in the house mouse (*Mus domesticus*) is probably one of the best examples of the interaction between selection and constraints in shaping reaction norms for behavioral traits in mammals. Lynch (1994) investigated nesting ability, measured as grams of material accumulated in the nest, at high (21°C) and low (4° or 5°C) temperatures in five natural populations from a latitudinal gradient spanning from Maine to Florida, as well as in replicated lines artificially selected for high or low nesting ability at 21°C and in four standard inbred strains. All genotypes, regardless of provenance, showed significant plasticity and genetic variation for nesting ability, with higher scores at the lowest temperature and no tendency for crossing of the reaction norms. In other words, genotypes performed differently, but maintained the same rank with respect to each other across environments, although the natural populations and the selected lines showed more variation under low than under high temperature. This pattern is consistent with the idea that an increase in nest building with a decrease in temperature is adaptive, and with the possibility that the tight co-evolution of nesting ability at the two temperatures (as evident, e.g., in the selection lines) represents a constraint against future decoupling of the two traits.

In a separate study, Bult et al. (1992) investigated the effects of selection for nest building on the neuronal structure of the brain in the same species. These authors immunocytochemically labeled the hypothalamus of the selected mice and the control line for arginine-vasopressin (AVP), a homeostatic modulator controlling body temperature. They monitored AVP production in two areas of the hypothalamus, the suprachiasmatic nuclei (SCN) and the paraventricular nuclei (PVN). While they did not find any significant difference in the neuronal activity in the PVN, the low-selected line showed a 1.5-fold increase in AVP-reactive neurons compared to the other two lines, which clearly demonstrates a specific neuroanatomical basis to the behavioral response to selection. However, the experimental data left unaddressed what alteration in the brain produced the high-selected line, opening up the possibility that the two responses to selection are underscored by two distinct changes in patterns of neuronal growth and activity.

A final area of study concerning behavioral plasticity involves reproductive behavior. As in the case of *Daphnia* discussed above, *Clethrionomys glareolus* (bank voles) do not respond directly to the presence of a predator—the stoat *Mustela erminea*—but to the odor emanating from it (Ylonen and Ronkainen 1994); this is another case of anticipatory plasticity because the odor can be perceived *before* the predator is close enough to attack. Female voles react to the predator's odor by suppressing breeding behavior, regardless of their age. The females "actively avoided copulations under high predation risk," and therefore "breeding suppression is mediated by a change in female mating behavior." This change in behavior is also accompanied by a marked decrease in body weight, which contrasts with the complete lack of response—both phenotypically and behaviorally—of the male voles. Ylonen and Ronkainen proposed that the selective advantage of this form of behavioral plasticity lies in the fact that shy females avoid the population crash and find themselves the following season in an environment largely free from predators (stoat populations oscillate periodically), ready to take advantage of the favorable conditions after having weathered the storm. Mating behavior also shows adaptive plasticity in female collared flycatchers (*Ficedula albicollis*), according to a study by Qvarnstrom et al. (2000). These females prefer males with a large forehead patch only late in the breeding season, and indeed long-term data confirm that a positive relationship between female reproductive success and male patch size is found only in late breeders (although the reason for this is not clear). The behavioral plasticity is therefore adaptive, regardless of what actually causes the relationship between fitness and sexually selected traits in these animals.

I could cite many other examples of this sort, but by now the reader should realize that behavioral plasticity presents essentially the same range of biological phenomena, complexity, and ecological consequences that we saw earlier in the book when we looked at other kinds of plasticity. Hence, one can feel justified in concluding that behavioral plasticity, while it provides us with a panoply of fascinating and intricate specific examples, does not represent an entirely distinct category of genotype-environment interactions.

Behavior as a Stable Phenotype

Perhaps because of the bewildering flexibility of our own and our closest kins' behavior, we tend to think of behavior as always being an extremely labile character, both on an evolutionary scale (Gittleman and Decker 1994) and in response to environmental conditions (Komers 1997). However, recent work has cast serious doubts on both propositions. Dequeiroz and Wimberger (1993) compared the apparent degree of convergent evolution of phylogenetic hypotheses constructed using behavioral traits with those derived from morphological data. In two different analyses including a variety of data sets they found no significant difference.[3] Similarly, Paterson et al. (1995) compared phylogenetic trees obtained with behavioral and life history traits with trees derived using molecular markers (enzymes and mtDNA). The result was again that behavioral characters are as phylogenetically informative as any other data set.

As for environmental lability, our discussion at the beginning of the chapter and the few examples selected over a wide range of taxa suggest that behavioral characteristics can be treated as much as stable phenotypic traits as morphologies and physiological characters. In the same way we can find invariant morphologies so that behavior can be canalized, while examples of extremely flexible behaviors can certainly be matched by some of the classical cases of morphological plasticity mentioned throughout the book.

In part, our expectations may be the result of a confusion between the lability of a response and the temporal scale of action of the same response. Since behaviors (and physiology) can be affected by environmental changes and respond over very short time scales (minutes to days), it is easy to imagine them as more "variable" (in a broad sense) than morphologies, which change over time scales of days to months. But a physiological (or behavioral) response such as the degree of photosynthesis (or landmark learning) can be under very strict genotypic and environmental control, as much as any morphological character is usually imagined to be. Do we then have only one type of plasticity, representing a fundamental unity across scales of phenotypic responses? Perhaps. The real problem is that for all our studies of plasticity at different levels we still know little about its mechanisms and especially its evolutionary dynamics. Until we know more, I will stick to the idea proposed in Chapter 2 that plasticity can be used to underscore the similarities among different aspects of the phenotype, be they morphological, behavioral, or physiological. A grand unified theory of phenotypic reactions to environmental conditions is still a bit further down the road.

Conceptual Summary

- Behavior per se is not synonymous with phenotypic plasticity, as indicated by the otherwise redundant phrase *behavioral plasticity.*
- A theoretical framework for understanding the relationship between behavior and plasticity has been laid out by several researchers. Cavalli-Sforza suggested that behavioral plasticity is analogous to physiological acclimation, and that they are both related to classical morphological plasticity.
- Certain kinds of behaviors are fixed (i.e., they are not plastic); others change depending on environmental conditions over a brief period of time (i.e., they are similar to physiological plasticity); whereas others—such as learning—have an ontogenetic component analogous to developmental plasticity.
- West-Eberhard has been the chief proponent of a tight relationship between morphological and behavioral plasticity and of a joint role of these two in explaining macroevolutionary phenomena such as the evolution of phenotypic novelties.
- As in the case of morphological plasticity, behavior can be characterized by *windows of plasticity* and by the influence of maternal effects.
- The many examples of behavioral plasticity (or of the effects of plasticity on behavior) present a varied and intricate phenomenology, but one that is not significantly distinct from analogous patterns found when studying classical morphological plasticity.
- Contrary to a general perception, behavioral traits are as phylogenetically stable as morphological or even molecular ones, so that the idea that behavior evolves too fast to be of any use in macroevolutionary studies appears to be unfounded.
- While a grand unified theory of genotype-environment interactions may still be a long way off, indubitably behavioral, physiological, and morphological plasticity has to be included within it as different manifestations of essentially similar biological phenomena.

9

Evolution of and by Phenotypic Plasticity

Not a single one of your ancestors died young. They all copulated at least once.

—Richard Dawkins (b. 1941)

CHAPTER OBJECTIVES

To discuss the evolution of phenotypic plasticity above the species level. The conditions favoring the evolution of plasticity or other strategies to cope with environmental heterogeneity are summarized, and some examples of interspecific divergence in reaction norms are presented. Experimental evolution in heterogeneous environments using selected model systems is discussed as a major novel way to study the long-term evolution of plasticity. The potential role of plasticity in the evolution of phenotypic novelties and speciation is examined. The chapter concludes with a list of new areas of research incorporating the concept of phenotypic plasticity as a primary player in macroevolutionary studies.

Phenotypic Plasticity as a Trait That Sometimes Evolves

Phenotypic plasticity can evolve by natural selection as a trait in its own right. But it does not have to. These two points need not be belabored here. I discussed the ecology of and natural selection on phenotypic plasticity in Chapter 7 and showed that we have clear examples of adaptive plastic responses (which clearly did evolve by natural selection, but see Winn 1999 for a sobering counterexample). At the same time, it is equally clear that some forms of plasticity can be neutral or maladaptive, so that just documenting plasticity is far from being tantamount to showing adaptation. This chapter discusses several aspects of the evolution of plasticity from conceptual and empirical standpoints, and Chapter 10 deals with mathematical modeling of the same.

Adaptive plasticity certainly does not evolve and does not play any role in the evolution of organisms when the environment is constant. This is hardly a limitation, as constant environments only occur in computer models. Environmental heterogeneity, however, is only a necessary but not a sufficient condition. Whenever one or both of two further conditions are present, the evolution of plasticity is more aptly compared to the alternatives of genetic polymorphism or adaptive coin flipping, the random production of alternative phenotypes in the face of unpredictable environmental conditions (Gavrilets 1986; Forbes 1991; van Tienderen 1991; Berrigan and Koella 1994; van Tienderen and Koelewijn 1994; de Jong 1995; Sibly 1995; Scheiner 1998)[1]: first, if the environmental change is predictable by means of reliable cues (anticipatory plasticity, see the discussion in Schlichting and Pigliucci 1995; e.g., Lively 1999), so that the organisms' phenotype will match the environmental requirements with a minimum lag-time; second, if the environmental heterogeneity is experienced by a single individual or by its offspring (see Scheiner 1993a; Agrawal et al. 1999; Fox et al. 1999). Of course, even under these conditions the evolution of adaptive plasticity still depends on the presence and amount of costs associated with it (Chapter 7 and references therein), as well as on the fact that selection must favor different phenotypic states in different environments.

A reliable environmental cue augments the likelihood of the evolution of adaptive plasticity (first condition) because the organism has a way to match the phenotype with the forthcoming external conditions (e.g., cyclomorphosis in *Daphnia* or leaf shedding in deciduous trees). If this cue is not present, plasticity can only be a reaction to conditions that are already in place. If the effectiveness of this plasticity is significantly reduced by lag-times, it may not be useful to the organism. In the latter case, adaptive coin flipping (Kaplan and Cooper 1984; Clauss and Venable 2000; Menu et al. 2000)—wherein the organism produces different kinds of progeny in proportion to the *stochastic* probability of each type of environment—could be favored. Nonanticipatory plasticity can increase fitness in situations in which the effect of lag-time between the environmental change and the plastic response is not crucial, and the evolution of anticipatory versus nonanticipatory plasticity will depend on the circumstances. For example, one can imagine that the relatively slow development of a large body size in an insect exposed to cold conditions will be advantageous even if the low temperatures had set in at the beginning of the life cycle. On the other hand, there may be no time to develop an antipredator device after the predator has attacked, making the reliance on indirect cues vital for the survival of the organism.

The second condition—that either parents or their offspring (or a combination of the two generations) actually experience environmental heterogeneity—is based on our understanding of environments as spatially or temporally variable, and of being coarse- versus fine-grained from the point of view of the organism. The simplest situation leading to the evolution of adaptive plasticity is of course that in which a single individual experiences different environments. Heterophylly in plants (Wells and Pigliucci 2000) is a classic example of this occurrence. If we assume that the environmental change is predictable and that there are limited costs or lag-times, it is advantageous to produce two or more morphologies or behaviors, appropriate to each major class of conditions encountered (e.g., below or above water). The other common predicament that can lead to adaptive plasticity is when the progeny, not necessarily the parent, is likely to experience a different environment, as in seasonal polyphenism in animals (e.g., Kingsolver 1995; McCollum and van Buskirk 1996; Roskam and Brakefield 1996). Because the parent's fitness depends on the survival and well-being of the immediate descendants, and since such descendants will share a large part of the parental genotype (and therefore a similar reaction norm), an environmentally inducible flexibility in the resultant phenotype will be an advantage. Some of the possible phenotypic outcomes of evolution in heterogeneous environments discussed in the references cited above (as well as in Levene 1953; Zonta and Jayakar 1988; Pigliucci 1992; Whitlock 1996) are summarized in Table 9.1. Of course, this list is not exhaustive (note, e.g., the absence of entries characterized by combined temporal and spatial fluctuation), and some of these possibilities require more empirical and or theoretical (modeling) support. However, it will serve as a reference point for discussion.

Knowing that plasticity can evolve or has evolved is only the beginning of the quest. Determining *how* plasticity evolves—which patterns it follows—is a much more daunting task. The answer lies at least in part in comparing the plasticity of closely related species. Surprisingly, however, comparative studies of plasticity are not common, and very few include an explicitly phylogenetic-comparative method approach (but see Roskam and Brakefield 1996; Pigliucci et al. 1999; Pollard et al. 2001). Most of the published studies, and virtually all of those reviewed in the literature, deal with variation for plasticity within species or within populations. This focus may in part be the result of the quintessential fascination of biologists with genetic variation (Levins and Lewontin 1985). However, as Doughty (1995) pointed out, to understand the evolution of plasticity we have to compare reaction norms of extant taxa to the reaction norms of their ancestors, not to the statistical null hypothesis of no plasticity, which is actually the logical one when it comes to ecological and optimality questions. Examples of interspecific comparisons of plastic responses show that indeed plasticity can evolve quite rapidly, since it can be very different in closely related species.

Schlichting and Levin (1986b) reviewed several comparative studies on plasticity in plants. They considered published data on three species of

TABLE 9.1
Possible Phenotypic Outcomes of Evolution in Heterogeneous Environments

Anticipatory environmental cue?	Temporal fluctuation?	Spatial fluctuation?	Evolutionary outcome
No	Yes, shorter than generation time	No	Uniform phenotype physiological plasticity
No	Yes, longer than or equal to generation time, random change but semi-constant proportions	No	Adaptive coin flipping or genetic polymorphism
Yes	Yes, longer than or equal to generation time, regular (seasonal) succession of environments	No	Plasticity by seasonal or generational forms (e.g., cyclomorphosis)
Yes	Yes, longer or shorter than generation time, irregular succession of environments, distinct phenotypes favored at different times	No	Phenotypic plasticity by developmental conversion
No	No	Yes, coarse-grain, environments follow random distribution	Adaptive coin flipping or genetic polymorphism
No	No	Yes, fine-grain phenotype is a continuous function of the environment, lag-time not important	Phenotypic plasticity by modulation
Yes	No	Yes, fine-grain, distinct phenotypes favored in different micro-environments, lag-time important	Phenotypic plasticity by developmental conversion

Phlox, three of *Portulaca,* two of *Sesbania,* three of *Sisymbrium,* and four of *Echinochloa.* Their main conclusions were that not only can plasticity diverge dramatically among closely related species, but that this divergence is not associated with the evolution of across-environment character means. In the terminology discussed in Chapter 1, this indicates that different attributes of the reaction norm (e.g., height and slope) tend to evolve independently of each other, again demonstrating an at least partially different genetic basis of these components of the reaction norm. However, the information summarized in the paper was insufficient to establish whether the divergence was due to response to selective pressures or to nondeterministic phenomena such as genetic drift. Bell and Sultan (1999) showed a close similarity in pattern but a divergence in the amount of plasticity between two closely related species of *Polygonum.* They concluded that the observed differences were consistent with the fact that one of the species experienced a broader niche and was exposed to a more temporally variable environmental regime.

An example of differences in reaction norms among closely related animal species was provided by the work of Morin et al. (1997), which compared the circumtropical *Drosophila ananassae* with the cosmopolitan *D. melanogaster.* The authors studied the reaction norms of these species in response to temperature by using an unusually large number of environments (seven), thereby obtaining data that they could analyze by a modification of the polynomial regression approach described in Chapter 1. One of their main findings was that the two species differed in the temperature of maximum wing and thorax length and ovariole number. In particular, that descriptor of the reaction norm shifted toward lower temperatures in *D. melanogaster,* which originally evolved in a warm climate. Furthermore, an afro-tropical population of *D. melanogaster* displayed a very similar reaction norm to that of *D. ananassae* collected in a similar environment (Fig. 9.1). The authors therefore concluded that evolution of flies occupying warmer environments was accompanied by a modification of the

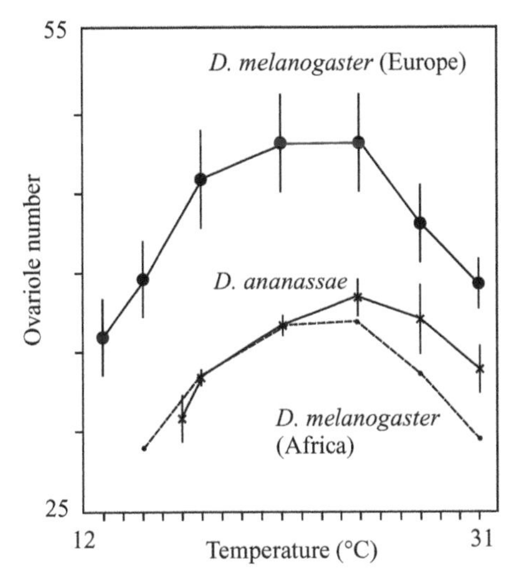

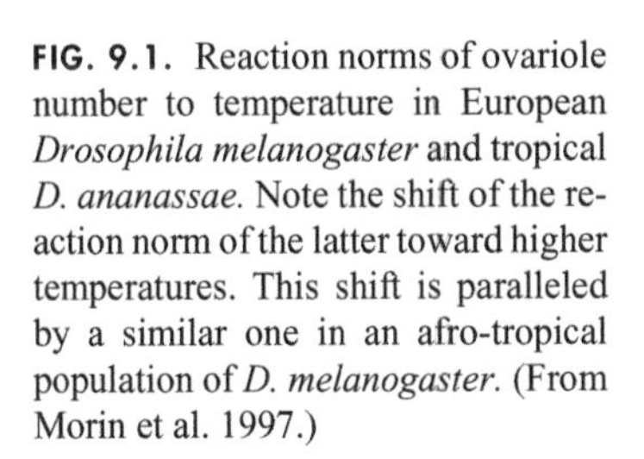

FIG. 9.1. Reaction norms of ovariole number to temperature in European *Drosophila melanogaster* and tropical *D. ananassae.* Note the shift of the reaction norm of the latter toward higher temperatures. This shift is paralleled by a similar one in an afro-tropical population of *D. melanogaster.* (From Morin et al. 1997.)

entire shape of the reaction norm to temperature, in accordance with the adaptive plasticity hypothesis. Other comparative studies of plasticity have been described in various other chapters in this book, depending on whether the emphasis was on ecological, developmental, or genetic aspects of the problem. Given the increasing sophistication of the phylogenetic comparative methods now available (Ackerly 2000a,b; Martins 2000), the time has come for this approach to genotype-environment interactions to contribute to our general understanding of adaptive evolution in a historical context.

Experimental Evolution in Heterogeneous Environments

A rather novel and promising approach to determine what happens to phenotypes, and plasticity in particular, when evolution occurs under changing environmental conditions, is to actually run the course of evolution in a laboratory. Obviously, this cannot be done with most organisms, owing to limitations of space or lifespan (of the experimenter). However, several groups of researchers have successfully employed some "model organisms" such as *Escherichia coli, Drosophila,* or *Chlamydomonas* to this effect. I discuss this research as it refers to four distinct theoretical problems: short- versus long-term evolutionary response, temporally varying environments, spatially varying environments, and novel environments.

Is evolution of plasticity over the short term different from what can occur over the long term? The answer would seem to be yes, in the sense that long-term evolutionary trajectories are affected by factors, such as mutation rates, that do not enter into shaping short-term responses to selection. Therefore, Bell's (1997b) work with *Chlamydomonas* can be considered an indication of how evolution in heterogeneous environments can proceed without the infusion of novel genetic variation. Bell studied the response of this unicellular chlorophyte in a spatially heterogeneous environment consisting of eight different growth media. The base population was obtained by crossing a single mating-type *plus* isolate with four mating-type *minus* isolates, yielding a mixture of forty-eight spores. In each generation the base population was allocated at random among the eight environmental conditions, and aliquots from each habitat were combined to provide individuals for the next base population. Under these conditions, adaptive coin flipping or genetic polymorphism should be the evolving strategies (see Table 9.1). Bell found that a large amount of genetic variance was indeed maintained in the heterogeneous environments when compared to a control homogeneous situation (Fig. 9.2). His interpretation of the results was that directional selection (which was very effective in the homogeneous treatment) was weakened appreciably by the random occurrence of the eight environments. In other words, neither plasticity nor genetic specialization

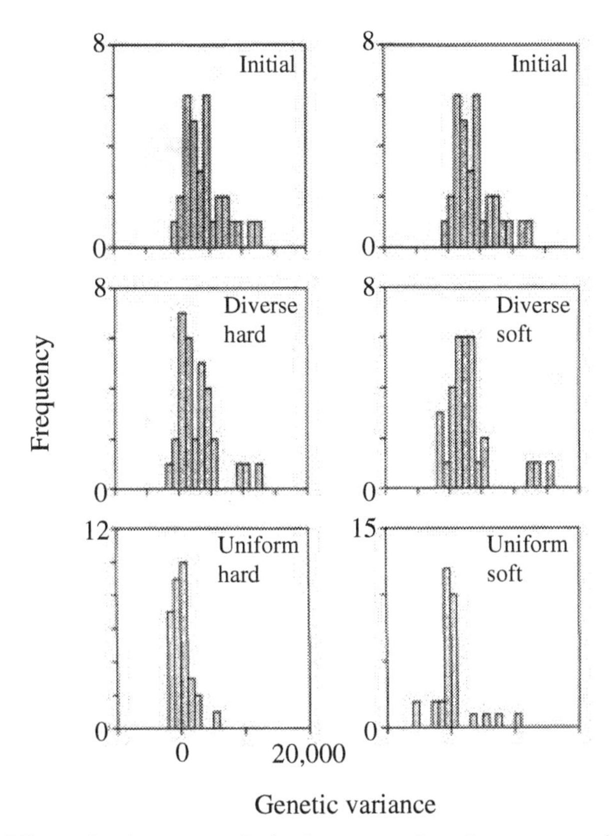

FIG. 9.2. Effects of various types of selection on genetic variance expressed by lines of *Chlamydomonas reinhardtii.* The upper row shows the initial condition, with the frequency of lines characterized by a given amount of variance. Both soft (right) and hard (left) selection in heterogeneous environments (middle row) maintained a substantial amount of genetic variance. Selection under uniform conditions, however, dramatically reduced the genetic variance after fifty generations (bottom row). (From Bell 1997b.)

evolved, in accordance with the predictions of our table. Instead, a genetic polymorphism was maintained for the fifty generations because of the inability of the genetic system to track the changes in environmental conditions under a fluctuating selection regime.

Does temporal environmental fluctuation lead to the evolution of plasticity in the long term? The answer (Table 9.1) should be affirmative only if the environmental change is of longer duration than the generation time and can be predicted to some extent. Leroi et al. (1994) used six lines of *E. coli* to study long-term evolutionary trajectories for 2,000 generations. The environmental heterogeneity was imposed by alternating days (i.e., a time interval spanning multiple generations) at 32°C with days at 42°C.

The transition between the two temperatures, upward or downward, was then effected very rapidly. The derived lines were significantly competitively superior to their ancestor under this environmental regime. The authors, however, wanted to know if this was due to the evolution of better adaptation to the two temperatures independently of one another, to adaptation to the transition, or to both. The answer was that the selected lines were no better than the base population at coping with the transition between temperatures (presumably because such within-generation environmental heterogeneity was not encountered frequently enough to generate selection pressure for plasticity). Instead, they apparently evolved the ability to survive better at constant regimes of *both* high and low temperatures (Fig. 9.3). This finding calls into question the conventional wisdom that "a Jack-of-all-trades is a master of none." Obviously, it was possible to develop strains of *E. coli* that managed to master both environments equally well. This is remarkable in view of the fact that the ancestral population was adapted to an intermediate temperature (37°C), and that therefore this represents an example of simultaneous evolution in response to two novel environments! From the point of view of our theoretical predictions, it is clear why plasticity did not evolve: The environmental change was not predictable and it occurred on too long a temporal scale. On the other hand, genotypes adapted to both environments simultaneously evolved rather than the predicted outcomes of polymorphism or coin flipping (unless either of the latter two actually occurred but was not detectable given the conditions of the experiment and the impossibility of tracking single individuals in the population).

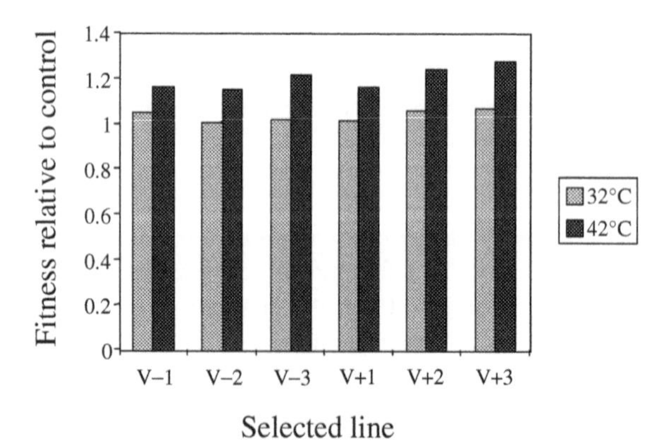

FIG. 9.3. Relative fitness of lines of *Escherichia coli* selected for performance at 32°C (V) or at 42°C (V+) compared to the control line acclimated at 37°C. Note that all selected lines had equal or better fitness than the baseline, having evolved as temperature generalists. However, the same selected lines were no better than the control at coping with the transition between 32° and 42°C or vice versa (data not shown). (From data in Leroi et al. 1994.)

Under what conditions of spatial or temporal heterogeneity, then, does plasticity evolve? The predicted answer here depends on how such heterogeneity actually occurs. That is, what is the spatial or temporal scale of environmental change, how predictable is it, and what is the frequency distribution of the environments (see Table 9.1)? In a very elegant experiment, Reboud and Bell (1997) explored precisely this question, again using *Chlamydomonas,* but this time subjecting their populations to several hundred generations of selection for growth under either light or dark conditions. *Chlamydomonas* can survive in both situations because it is only facultatively photosynthetic. Control lines were grown under a constant environment (either dark or light), and experimental lines were exposed to spatial and temporal variation for light availability. As expected, the controls evolved improved performance in the environment of selection, while simultaneously decreasing performance in the alternate environment. This is a rare direct observation of the generation of a negative genetic correlation between expressions of the same trait in two environments as a consequence of selection. Clearly, this sort of plasticity (of fitness) was the byproduct of evolution in separate environments. The experimental lines, however, displayed a much more complex pattern of evolution. In the *spatial variation regime,* two lines that had previously been maintained under dark or light were mixed and used to inoculate the two environments. This mixture and inoculation process was then repeated each generation. Because the environments were experienced at random and there was no available clue to predict the change, my tabulation of the possible outcomes (above) would indicate that no plasticity will evolve unless the lag-time between the environmental change and the plastic response is not an important factor, which is unlikely for a short-lived organism such as *Chlamydomonas.* Instead, one would see the appearance of adaptive coin flipping or the maintenance of genetic polymorphism. Reboud and Bell found that spatial variation in environmental conditions does in fact lead to maintenance of genetic variation (Fig. 9.4). *Temporal variation* during the experiment was obtained by transferring light lines to dark for one cycle, then back to light, and so on in alternation. A similar procedure was applied to the dark lines. It should be noted that in this protocol, the temporal cycle is exactly coincident with the generation cycle, and our prediction is that plasticity should evolve (Table 9.1), which is what Rebound and Bell actually observed. Furthermore, the above-mentioned negative genetic correlation between levels of performance in the two environments turned out to be the result of mutation accumulation, not antagonistic pleiotropy. Therefore, the authors had selected for a plastic genotype that was well adapted to both environments. Again, experimental evolution provides food for thought about the limitations and constraints on the evolution of plasticity. Also, in this case at least there was no evidence of costs of plasticity (Chapter 7), or one

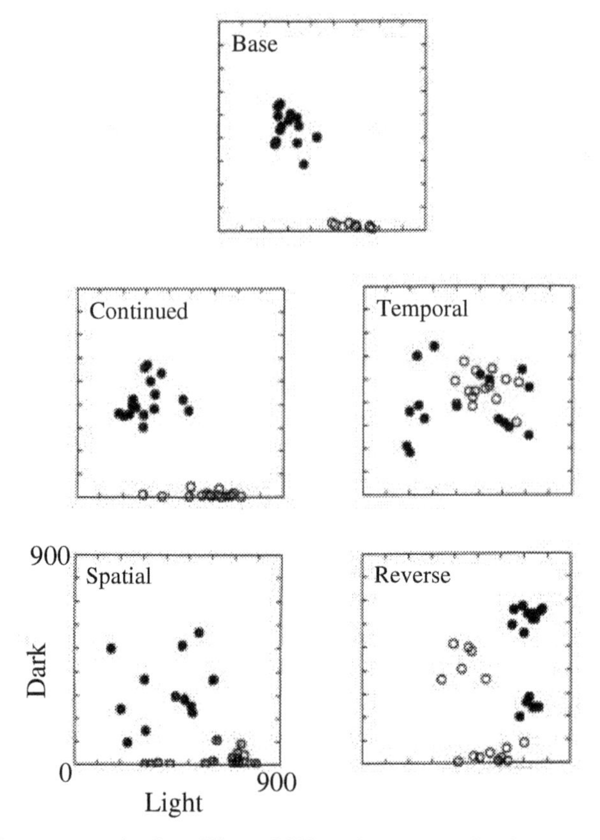

FIG. 9.4. Response to selection of lines of *Chlamydomonas reinhardtii.* Open circles represent lines raised under light, and solid circles are lines in the dark. The base population shows a negative correlation between performance in the two environments. When the lines were maintained in the original environments ("continued") the overall result was no different, and the correlation did not change significantly under a regime of spatial heterogeneity. However, both temporal heterogeneity and reverse selection (in which the dark lines evolved under light and the light lines under dark) reduced the negative interenvironment correlation, generating plastic genotypes in the case of temporal variation. (From Reboud and Bell 1997.)

might conclude that the benefits outweighed whatever costs may have been present.

Finally, does plasticity evolve in response to novel environments? Logic suggests that the answer is no. One would not expect the sudden appearance of adaptive plasticity if a portion of the reaction norm had never before been exposed to the new environment. On the other hand, plasticity may facilitate persistence in a novel environment, if some of the genotypes in the population happen to have reaction norms leading to functional, though possibly suboptimal, phenotypes in the new environment (for an example, see Losos et al. 2000). Indeed, as we shall discuss at the end of this chapter, such a mechanism may be the most important connection between plasticity and macroevolution (also see Chapter 8). Partially to address this question, Travisano et al. (1995b) propagated twelve lines from a single clone of *E. coli* for 2,000 generations in identical glucose-limited environments. They observed a marked increase in fitness in all lines when compared to the initial clone under glucose-limited conditions, whereas the descendants did not differ greatly among each other. They then exposed these independently derived genotypes to two novel environments, in which glucose was replaced by maltose or lactose. The observed genetic variation among lines in response to both novel environments was a hundredfold that observed in the selected environment, supporting the idea that heritability increases in novel environments (Chapter 4). This is presumably because of the sudden exposure of a previously unselected portion of the reaction norm of genotypes that had accumulated mutations that were neutral or quasi-neutral under the original environment. Even more interestingly, genotypes adapted to low glucose turned out to be relatively well adapted to lactose, but not to maltose. This is an example of the simultaneous evolution of maladaptive plasticity to a novel environment (maltose) and of fortuitously adaptive plasticity to another (lactose). In the lactose environment, these genotypes were able to survive simply because their reaction norms are in a sense *preadapted* to the novel conditions.

The above examples, together with the increasing amount of literature on *experimental evolution* (Travisano et al. 1995a; Rose et al. 1996; Elena and Lenski 1997; Ebert 1998; Stowe 1998; Bettencourt et al. 1999), show that model systems can be very powerful tools for bridging the gulf between evolutionary theory and empirical investigation. The problem so far has been that long-term evolution has been inaccessible to the experimenter, so that empirical biology has had to limit itself to mostly indirect approaches, such as comparative analyses. Now classical comparative studies can be complemented by the direct observation of long-term evolution, albeit only within a limited taxonomic range. As far as the evolution of plasticity is concerned, the data support our general theoretical expectations, but they also raise compelling questions about the limits and constraints on the evolution of reaction norms. Such limits must exist (Chapter 7), but so far they have turned out to be more difficult to pinpoint empirically than was expected.

Evolution of Life Histories, Phenotypic Novelties, and Speciation

In this section I discuss the theory and some of the (as yet scarce) empirical evidence that links phenotypic plasticity to three major problems in evolutionary biology: the evolution of life histories, the appearance of phenotypic novelties, and the morphological and behavioral changes associated with speciation. My hope is that should something like this book be written twenty years from now each of these topics will warrant a whole chapter, since a consideration of reaction norms is, I believe, the key to unraveling these and similarly stubborn problems plaguing the modern Neo-Darwinian synthesis.

That evolutionary biology needs to depart from a gene-centered paradigm to account for the complexity of gene-environment interactions is, of course, not a novel thought. Indeed, it was ironically proposed in vague form much in advance of the establishment of the gene-centered paradigm itself (Baldwin 1896). As we saw in Chapter 3, some giants of midcentury biology certainly strove toward such an end, including Goldschmidt, Waddington, and Schmalhausen. Most recently, such calls have been repeated and elaborated upon by various authors (Levin 1988; West-Eberhard 1989; Sultan 1992; Schlichting and Pigliucci 1998). But what does it *mean* to "incorporate reaction norms into macroevolutionary theory"? This is the kind of general statement that sounds as good in principle as it is difficult to fulfill in practice. Below I first cite some selected examples of how phenotypic plasticity can help us understand macroevolutionary phenomena and then summarize my own and other authors' ideas on the topic into a somewhat coherent yet tentative conceptual framework. A more general recent treatment of phenotypic evolution as a whole (i.e., not limited to the contribution of plasticity) is to be found in Schlichting and Pigliucci (1998).

One of the key ideas linking plasticity to macroevolution is the possibility that a plastic reaction norm may be a major (or even sometimes the only) initiator of macroevolutionary change. A potential example of this is found in the work of Leclaire and Brandle (1994) on phytophagous insects. They studied the European rose-hip fruit fly, *Rhagoletis alternata,* as it colonizes two species of *Rosa.* The original host is the common rose, *Rosa canina,* which is widely distributed in Europe. However, the fly infests hips of another species, *R. rugosa,* which is now widespread in Europe but was introduced only about a century ago from East Asia. Plasticity allowed *Rh. alternata* to expand the host range to include the novel environment provided by *R. rugosa,* even though the growth rate of the insect is substantially reduced when it infests the introduced species. Interestingly, although the flies that colonize the two hosts differ somewhat in their morphologies, they are indistinguishable when examined at the level of allozymes. Given that the two hosts co-occur at the same locations, and that there seems to be widespread cross-colonization by the flies, we may be observing the very initial stages of an evolutionary process of divergence. From here, two outcomes are possible: *Rh. alternata* will either evolve into a generalist or split into two daughter species. Speciation through evolution of host races is in fact a currently debated evolutionary outcome in other species of *Rhago-*

letis, and may represent cases of speciation initiated by a purely nongenetic host shift catalyzed by preexisting phenotypic plasticity. In the particular case of *Rh. alternata,* however, Leclaire and Brandle suggest that the alternative outcome of evolution of a generalist is more likely for at least two reasons. First, *Rh. alternata* is characterized by high vagility, which thoroughly mixes individuals growing on each host, thereby precluding any long-term reproductive isolation. Second, and equally important, the two hosts are themselves closely related, so much so that there may be little opportunity for a host recognition system to evolve. Another aspect of this system that has to be considered here relates directly to the role of plasticity in the evolution of life histories. The two hosts have hips that follow different ripening processes and yield different qualities of diet. In fact, *Rh. alternata* shows plasticity of its growth trajectory to diet quality. As a result, the fly develops distinct adult body sizes and reproductive output when it grows on the alternate host: Females that develop on *R. rugosa* are bigger, have more eggs, and may live longer. Furthermore, flies that develop on the novel host have to cope with an alteration in the original pattern of predation: Since the larvae must complete their development before birds take the fruits, they leave the hips early. This means that they are exposed to a higher risk of predation in the soil. Thus, a "simple" host shift mediated by existing plasticity has resulted in a suite of complex *simultaneous* changes in morphology, behavior, and selective pressures. Had such a change been documented in the fossil record instead of in extant populations, one would have applied the standard Neo-Darwinian model and concluded that a long, gradual process of *mosaic* evolution had occurred. The alternative would have been to imagine an unlikely series of mutations occurring simultaneously, thereby yielding a *hopeful monster.* This example outlines a third, hitherto virtually ignored possibility: A shift caused by a change in the environment (the appearance of the new host) that was made possible by preexisting plasticity resulted in an apparently coordinated suite of alterations in the morphology and life history of the organism. No initial genetic change was required (Chapter 8), although standard processes of selection on genetic variation to improve adaptation to the new host and/or reproductive isolation leading to speciation may be operating.

Perhaps one of the most intriguing examples of the macroevolutionary relevance of phenotypic plasticity is provided by temperature-dependent sex determination (TSD) in reptiles and other vertebrates and its relationship with the more common genotypic sex determination (GSD). I discussed how TSD works in some detail in Chapter 6 (see also Fig. 9.5). As for its macroevolutionary significance, Crews (1994) reviewed the evidence supporting the possibility that TSD may actually have been the ancestral condition in vertebrate evolution and that genotypic sex determination evolved later in several groups. Currently, TSD is found among the most

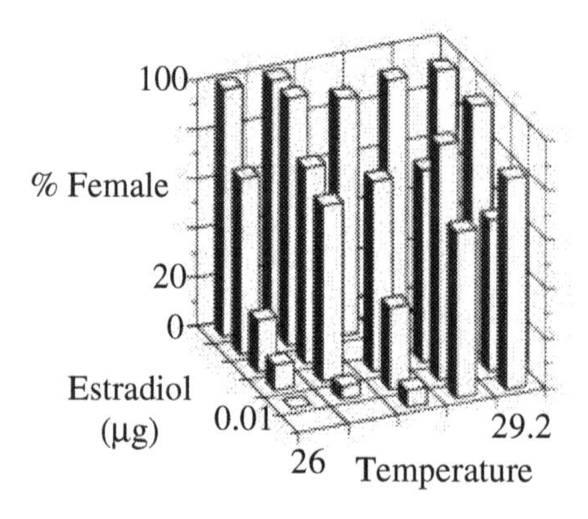

FIG. 9.5. Joint effect of temperature and estradiol on the production of females in the red-eared slider turtle, a clear example of environmental sex determination mediated by hormonal balances. (From Crews 1994.)

ancient reptiles (e.g., turtles), as well as within derived groups such as snakes, lizards, and especially crocodilians (the closest relatives to birds). A single Y-linked gene termed *SRY,* which encodes the testis-determining factor, mediates GSD in mammals. Since *SRY* is part of a large gene family, some members of which are not sex-linked, Crews suggested a scenario in which sex was initially determined by steroid hormones (as in TSD), with *SRY*-like genes being later co-opted for the evolution of GSD. A particularly convincing piece of evidence in this regard is that steroid hormones can determine sex under laboratory conditions in animals with GSD, such as chickens and parthenogenetic whiptail lizards. Apparently, developmental machineries based on genetic sex determination are still responsive to the internal (hormonal) environmental stimulus typical of species with temperature-dependent sex determination.

Phenotypic plasticity has also been implicated in an ingenious alternative hypothesis for the evolution of viviparity. Retention of eggs in the uterus by reptiles as well as other vertebrates and invertebrates has classically been attributed to the hypothesized advantages of protecting the eggs or early hatching. Shine (1995), however, proposed that the higher temperature inside the mother's body may per se have beneficial effects on the phenotype of the offspring (separately and perhaps in addition to the protective effect). Seen from this point of view, viviparity becomes just one way of exposing eggs to higher temperatures. Shine tested his hypothesis by incubating eggs of two species of montane scincid lizards from Australia at controlled temperatures reflecting nest and uterine conditions. He found that a higher temperature has a positive effect on the hatchlings' morphology, behavior, and running speed. This example illustrates how plasticity can provide previously unforeseen alternative explanations for the evolution of novel phenotypes and/or life histories. Genotype-environment interactions may have played a much more fundamental role in macroevolution than has been assumed in the past.

Another reason to think that phenotypic plasticity may be commonly involved in macroevolutionary changes comes from the study of trophic polymorphisms in vertebrates. Wimberger (1994) reviewed several cases of this phenomenon in fishes, amphibians, reptiles, birds, and mammals, though most of his examples come from fishes. He suggests that a major reason for the widespread occurrence of plastic effects underlying the differences between various trophic forms of vertebrates may be that vertebrate bones are known to be responsive to local growth conditions. For example, researchers have observed many dramatic alterations of jaw morphology in animals subjected to different diets. Wimberger calls into question the popular explanation that a genetically controlled developmental switch causes these morphological changes. He suggests the more parsimonious scenario that bone plasticity is a continuous phenomenon, perhaps polygenically controlled, whereas the trophic conditions may be dichotomous. So, the fish feeds on either one kind of prey or on another, leading to an externally imposed discontinuity that yields the appearance of a bimodal distribution of phenotypes (or of disruptive selection) out of what is a continuous plastic response. One of the striking general points emerging from his review is that it is often quite difficult to distinguish between a "true" genetic polymorphism in the classic sense and the presence of alternate morphs caused by plasticity. Wimberger therefore speculates that what was thought (by default) to be an example of the former may actually be an instance of the latter. Furthermore, in many cases *both* plasticity and genetic polymorphism shaped the occurrence and distribution of multiple morphs, which is exactly what one would expect if plasticity is a bridge toward genetic differentiation and, eventually, speciation. Wimberger acutely points out that "plasticity is often invoked only as a last resort to explain polymorphisms." Therefore, it is not surprising that the recent surge in plasticity studies has uncovered several new examples that had hitherto gone unnoticed. Wimberger concludes his review with a discussion of the connections between plasticity and speciation. Phenotypic plasticity cannot cause speciation per se, but it can be associated with assortative mating, which can lead to speciation under certain conditions. For example, it is common for a plastic response to both alter the morphology of feeding structures and simultaneously affect body shape, size, and coloration, as in several cases of fishes (and in particular cichlids) discussed by Wimberger. The latter aspects of morphology are essential components of mate choice, and one can easily envision a diet-induced morphology leading to altered sexual signals and responses, and therefore to assortative mating. This in turn can cause reproductive isolation and speciation, perhaps sympatrically by female mating preferences as has been described in recent models (Gavrilets et al. 1998).

We have seen how plasticity can be invoked in explanations of speciation, modifications of life histories, and the appearance of phenotypic novelties. Perhaps the most astounding example of the latter that I have found in the literature is reported in West-Eberhard's (1989) review on plasticity and evolution. This is the case of the behavioral and morphological changes affecting a mutant goat originally described by E. J. Slijper (1942), and it is well worth summarizing here. The animal was born with very short front legs, but was otherwise normal for all other morphological features. It displayed a remarkable behavioral shift early on by starting to walk in an upright posture. This caused a cascade of morphological alterations during development, which culminated in longer hind legs, a large neck, modified muscle insertions, and an altered shape of the thorax. The point is that if we had found such an animal in the fossil record we would have thought of a complex case of mosaic evolution (Futuyma 1998). Instead, a simple case of behavioral plasticity had catalyzed a composite series of morphological reactions through the correlations connecting different aspects of morphological development. It is easy to see how natural selection could rapidly fine-tune such a promising starting point. Perhaps a similar phenomenon guided the very rapid evolution of the cetacean body plan in mammals (Thewissen et al. 1994) or even the origin of terrestrial vertebrates from fish exposed to tidal conditions.

The Role of Phenotypic Plasticity in Evolution

Sultan (1991), after considering the historical relationship between plasticity studies and the Neo-Darwinian synthesis, concluded that "Phenotypic plasticity must be recognized as central to evolution rather than viewed as a minor phenomenon, secondary to 'real,' genetic, adaptation." Yet, what this central role actually is has been very much the focus of debate. Levin (1988) (see also Wright 1932), for example, suggests that canalization (buffering the phenotype from fluctuations in the environment) and plasticity (environmentally induced changes in the phenotype) may both reduce the coupling between genotype and phenotype. This would make the genotype-phenotype mapping function more fuzzy, and therefore "may reduce the impact of selection . . . and promote evolutionary stasis." It should be clear from reading the present book so far that thinking of plasticity as messing up the genotype-phenotype mapping function is misleading. As plasticity is a property of the genotype, if anything its study *clarifies* the structure of that mapping function and allows us to think more realistically about evolution as a continuous interaction between genotypes and environments. In fact, several of the examples described above may indicate that plasticity can greatly speed up evolution under certain circumstances, to some extent bypassing the need for mosaic evolution or for some intermediate forms. If any one thing may slow down evolution and be considered an internal mechanism producing stasis (as opposed to the external mechanism of stabilizing selection), it is genetic redundancy (Goldstein and Holsinger 1992; Pickett and Meeks-Wagner 1995). We discussed this phenomenon—by which different genotypes may yield very similar phenotypes at least along some portions of their reaction norms—in the light of phenotypic evolution in Chapter 9 of Schlichting and Pigliucci (1998).

What, then, could the role(s) of plasticity be in macroevolutionary phenomena? Below I summarize the ideas of many thinkers, some proposed before and during the Neo-Darwinian synthesis (Baldwin 1896; Wright 1932; Goldschmidt 1940; Waddington 1942; Schmalhausen 1949) and some developed much more recently (Matsuda 1982; West-Eberhard 1989; Sultan 1992):

1. Plastic reaction norms may allow a population to persist under temporarily stressful situations.
2. Plasticity could allow persistence of the population under novel environmental conditions, leaving more time for mutation, recombination, and selection to fine-tune the level of adaptation.
3. Variation among reaction norms in a population may slow down selection (stasis) if the pattern of genotype-environment interactions is such that the reaction norms of different genotypes tend to yield similar phenotypes under the prevailing environmental conditions.
4. Variation among reaction norms in a population may accelerate selection (punctuated evolution) if the environmental range is such that the reaction norms of different genotypes tend to yield highly dissimilar phenotypes.
5. Plasticity may generate phenotypic novelties as a secondary effect of behavioral changes.
6. Plasticity may generate phenotypic novelties as a secondary effect of environmental changes, as when a new portion of the reaction norm is exposed to selection.
7. Plastic reaction norms can provide the intermediate step of Waddington's genetic assimilation. If observed at a coarse temporal scale, it will look like conventional evolution by allelic substitution because the plastic phase may be very short. See Schlichting and Pigliucci (1998) and Chapter 10 for a broader discussion of the role of genetic assimilation in phenotypic evolution.
8. If the plasticities of different traits are correlated, the appearance of a novel environment or a behavioral change may lead to coordinated changes at the whole phenotype level, which would look like mosaic evolution if observed at a coarse temporal scale.
9. Plasticity at the cellular or molecular level (e.g., enzyme reaction curves) may play a major role in the evolution of development, if the

internal environment to which cells are exposed changes. This process may have started the evolution of differentiation in multicellular organisms.

10. Phenotypic plasticity should probably be considered the default state of organic systems (whole organisms or their components), because of the inherent physical-chemical properties of biomolecules, which tend to alter their properties when some aspects of their environment change. Any lack of plasticity (homeostasis) is then to be considered the result of canalizing selection and as the derived, presumably adaptive, state.
11. Plastic reaction norms can be the target of selection (through their effects on fitness) and yield flexible strategies under the conditions of temporal or spatial heterogeneity described above.

Given this long list, evolutionary biologists are faced with two novel tasks. First, we have to acknowledge this new spectrum of evolutionary mechanisms and incorporate it into our thinking and modeling of phenotypic evolution. Second, we need to design research projects and empirical approaches that can demonstrate or rule out the role of plasticity in macroevolutionary phenomena. Given the rapidity and temporary nature of variation for reaction norms, we may run into a paradox. Our usual complaint that most macroevolutionary phenomena are too slow to be observed and dissected directly may turn into the opposite difficulty: Things may happen so fast under the appropriate conditions that the real cause (e.g., a transitory phase of plasticity) will be difficult to pin down. This should be a welcome challenge for people interested in the real stuff of evolution.

Conceptual Summary

- Adaptive phenotypic plasticity can certainly evolve as a strategy to cope with environmental heterogeneity; however, it does not represent the only possible outcome of evolution in changing environments.
- To determine which strategy is more likely to evolve in response to environmental fluctuations, one has to consider if it is possible for natural selection to exploit indirect cues signaling the change in conditions (as opposed to responding directly to the change itself) and the scale of temporal or spatial fluctuations in the environment.
- The use of model organisms with fast generation time is now making it possible to follow macroevolutionary dynamics under laboratory conditions; this approach can be used to test fundamental ideas about the evolution of phenotypic plasticity and the maintenance of genetic variation in natural populations.

- Theoretical considerations as well as intriguing empirical results suggest a hitherto unforeseen role for phenotypic plasticity in the evolution of life histories, the appearance of phenotypic novelties, and the speciation process itself.
- Genotype-environment interactions have the potential to play a much more central role in evolutionary theory than has been granted so far, and entirely new research programs can be derived from a thoughtful consideration of this possibility.

—10—

The Theoretical Biology of Phenotypic Plasticity

In order to shake a hypothesis, it is sometimes not necessary to do anything more than push it as far as it will go.

—Denis Diderot (1713–1784)

CHAPTER OBJECTIVES

To introduce different approaches to the modeling of genotype-environment interactions and discuss their strengths and limitations. Several classes of models are considered: quantitative genetic, optimization theory, gametic, empirically informed, and situation-specific models. The chapter ends with a general discussion of the utility of modeling in organismal biology.

While Diderot was certainly not thinking of modern theoretical evolutionary biology, it can be argued that that is exactly the point of producing a body of theoretical work in a discipline as complex as ours. The idea is to stretch the assumptions and see how far they will go before breaking against the evidence gathered from the real world. When this happens, one goes back to the drawing board and fashions a better model, based on assumptions that are more realistic, and so on, in a continuous feedback between empirical and theoretical research. In this chapter, I summarize the status of theoretical studies on the evolution of phenotypic plasticity, ending with a general discussion of the benefits and limitations of mathematical modeling in evolutionary biology (as opposed to, e.g., particle physics). For convenience, and in part following a tentative conceptual classification, I have divided the kinds of models published so far into four categories:

1. *Quantitative genetic models,* based on the statistical theory of multi-locus evolution of quantitative characters and ostensibly the most common modeling approach to plasticity.
2. *Optimization models,* usually lacking genetic details (or reducing them to a quantitative genetic approach), but focusing on trade-offs and phenotypic-level analyses. These were the first models ever applied to the study of plasticity.
3. *Gametic (population genetics) and individual-based models,* attempting to directly incorporate complex genetic effects such as pleiotropy and epistasis and usually—but not necessarily—limited to the study of the behavior of a limited number of loci.
4. *Empirically informed and situation-specific models,* a heterogeneous category including attempts to model specific situations (such as nitrate and phosphate uptake in patchy soil) or ones based on an empirical parameterization of crucial features of the model.

My discussion here focuses on the relationships among these approaches, their limitations, and the special perspectives they bring to the study of phenotypic plasticity. As with any other chapter in this book, this is by no means a complete review of the published studies, but rather a wide-angle survey guided by what I think are the conceptually most important advances published in recent years.

Quantitative Genetic Modeling of Phenotypic Plasticity

The opening salvo of the modern quantitative genetic theoretical approach to the study of phenotypic plasticity was a paper published in the mid-1980s by Via and Lande (1985), which followed up on Via's (1984a,b) empirical investigation of the quantitative genetics of plasticity in herbivore-plant interactions. I described that work in Chapter 3, so I will concentrate here on work done by theoretical quantitative geneticists simultaneously with or subsequent to the Via-Lande paper.

Gavrilets (1986) independently published a paper very similar to that of Via and Lande (which is not as widely recognized because it is in Russian). In many respects, the two papers are similar, but Gavrilets went further on several points. First, he considered evolution of plasticity in environments that are varying stochastically in time. He then derived equations describing not only the evolutionary trajectories of mean values but also the genetic variances, and studied them analytically. He explored the possibility of levels of genetic variation for both trait means and plasticity being maintained by mutation, and predicted that populations with plastic genotypes evolve to a state of maximum geometric fitness. Last, he also verified that the quasi-linear reaction norm model is well supported by experimental data. As we shall see below, later papers confirmed several of these conclusions.

Another little-known paper by Gavrilets (1988) also anticipates many of the later findings of other researchers. There, he significantly extended the approach in the 1986 work, using the quasi-linear model:

$$z = g_1 + g_2 + e \quad (10.1)$$

where z is a variable characterizing the effects of the environment on the phenotype, g_1 is the genetic contribution to the across-environment mean of the trait, g_2 is the genetic contribution to the plasticity of the trait, and e is the error. After having derived equations for the mean g_1 and g_2 values and for the variance-covariance matrix relating the two quantities, he solved the system analytically and derived equilibrium values for the relevant quantitative genetic parameters. An example is the estimate for the mean plasticity, that is, the mean of g_2, is cov(z, θ) var(z), where θ is the optimum phenotype. Analogously, the genetic variance for plasticity is $\sqrt{[MV \text{var}(z)]}$, where M is the mutational variance and V is a measure of the strength of stabilizing selection. An interesting prediction of this model, as yet experimentally untested, is that the level of genetic variation for plasticity should be higher in stable environments than in more variable ones. Another prediction is that the average plasticity should increase with the variability of the environment of selection and go to zero if the developmental window of expression of plasticity and the action of selection become increasingly separated in time.

A very different view of the relationship between phenotypic plasticity and maintenance of local genetic variation was published a few years later by Gillespie and Turelli (1989). I discussed their work and Gimelfarb's (1990) critique of the same in Chapter 4.

De Jong (1990b) was the first to explore the effects of genotype-environment interactions on the structure of the genetic variance-covariance matrix, the fundamental player in all quantitative genetic models of evolution. First, she demonstrated that the presence of a significant interaction term produces a quadratic relationship between a given character and the environments in which that character is expressed. Importantly, this holds even if the allelic effects on each trait are linear functions of the environment. She then discussed two alternative scenarios. In the first, independence of polygenic effects (i.e., absence of pleiotropy) leads to a change in the sign of the additive genetic covariance from one environment to another. In the second case, pleiotropy may maintain the same sign of the covariance regardless of the environment. This is a key conceptual distinction that links the idea of developmental constraints to the structure of the pleiotropy underlying a certain phenotype and to the reaction norms expressed in a given environmental range.

Van Tienderen (1991) further investigated the effect of soft versus hard selection on plasticity, along similar lines to Via and Lande, but he was also the first to explicitly consider the consequences of adding costs to plasticity (Chapter 7) in a quantitative genetics model. He found that under soft selection the equilibrium reaction norm is a compromise between the optimal phenotype across environments and the lowest cost incurred by the genotype because of plasticity. The outcome, as in the Via and Lande model, is always one final equilibrium. In the case of hard selection, however, van Tienderen confirmed the possibility of multiple peaks, up to three in his case, with populations becoming specialized to one environment, being equally adapted to two environments, or compromising between these two extremes. Costs play a fundamental role here: If they are absent or negligible, the equilibrium reaction norm is close to the optimum, and the evolutionary outcome is indistinguishable from the one obtained under soft selection. However, whenever costs (assumed to be associated with the difference between habitats in average response) increase, the presence of multiple peaks with different regions of attraction makes the final outcome dependent on the initial conditions. This means that historical accidents may play a fundamental role in the evolution of adaptive phenotypic plasticity. These theoretical expectations also constitute the reason for the current intensive activity in search of empirical demonstrations of costs of plasticity (Chapter 7).

A major conceptual advance in modeling reaction norms, albeit one that has enjoyed little empirical follow-up so far, was published by Gomulkiewicz and Kirkpatrick (1992). They argued that reaction norms (like developmental trajectories [Kirkpatrick and Lofsvold 1992]) are best thought of as continuous quantitative characters and not as discrete points in low-dimensional space. When reaction norms are considered in this light, it becomes necessary to adopt a different mathematical approach, which represents them as continuous functions rather than as a finite set of genetic correlations. This is known as the infinite-dimensional character approach, and it allows the same calculations of evolutionarily relevant quantities such as genetic variances and covariances that are possible with the classical methods, although the mathematics involved is more sophisticated and requires special software. (In essence, discrete data points are fitted by any of a number of smoothing functions, such as Legendre's polynomials.) Gomulkiewicz and Kirkpatrick considered both spatial (hard versus soft selection) and temporal (within- and between-generations) variation in the characteristics of the environment, confirming that different evolutionary trajectories may result even if the same optimum is favored in all cases.

Along the same lines of previous authors, they also concluded that the presence of genetic constraints may direct the population onto different equilibria, not necessarily close to the specified target of selection.

A similar attempt at generalizing the mathematical treatment of reaction norms by fitting data to continuous functions was published by Gavrilets and Scheiner (1993), who studied the evolution of the shape of reaction norms under a variety of circumstances. They focused on three special cases: plasticity expressed in two environments (the most common empirical approach published to date), populations characterized by linear reaction norms, and populations characterized by a mixture of linear and quadratic reaction norms. In the case of temporal environmental variation, they found that selection favors, as is intuitive, the genotype with the highest geometric mean fitness across environments. More interestingly, Gavrilets and Scheiner were able to demonstrate a theoretical link between the environment in which an individual develops and the environment in which selection occurs (which, of course, can be different). In this case, selection favors the linear reaction norm whose slope (i.e., amount of plasticity) is proportional to the covariation between the two environments. These authors found that linear reaction norms are always favored, even when nonlinear ones are possible and are encountered in the population, but they admit that there is only limited empirical support for this result. In particular, nonlinear reaction norms are fairly common (see, e.g., Fig. 9.1) and are expected on the basis of simple biochemical and functional ecological considerations (e.g., reaction curves of enzyme activity to monotonic changes in environmental parameters such as temperature or pH). Part of the problem may reside in the fact that Gavrilets and Scheiner's model deals with genetics at the zygotic level, thereby assuming additive and Gaussian effects. The mathematics of gametic nonlinear models incorporating realistic details such as epistasis is much more complex.

More recent papers on the quantitative genetics of reaction norms have appeared from members of the "Dutch school," especially van Tienderen and de Jong (van Tienderen and Koelewijn 1994; de Jong 1995). These works discuss the theoretical underpinning of the debate between Via (1993), on the one hand, and Scheiner (1993b) and Schlichting and Pigliucci (1993, 1995), on the other, as to if and when phenotypic plasticity is the target of selection. Van Tienderen, de Jong, and their colleagues compare the reaction norm and the character-state methods of describing and modeling plasticity data (Chapter 1). They conclude that the two approaches are mathematically related, and even provide methods to switch from one to the other. However, they also conclude that the kind of biological interpretation one is liable to make does indeed depend on how one looks at plasticity, and de Jong states that "the reaction norm model seems to be simpler and to give rise to more predictions" (p. 493). Nonetheless, she adds that "the list of predictions shows that the character-state model and the reaction-norm model allow separate predictions and are therefore both indispensable" (p. 506).

Quantitative genetic models of the evolution of plasticity have evolved along predictable yet fundamentally important trajectories during the last fifteen years. Starting with Via and Lande's and Gavrilets' novel treatment of the subject, we have seen the exploration of the effects of the type of selection (soft versus hard), the kind of environmental heterogeneity (temporal versus spatial; within versus across generations), the presence of costs, and the genetic constraints arising directly from the pleiotropic structure of the genetic variance-covariance matrix. Furthermore, reaction norms themselves have been modeled in all possible fashions, from a discrete set of points in sharply defined environments to continuous functions applicable in a variety of circumstances. It seems to me that most of the results are in good qualitative agreement with the known empirical results, and when they are not it is clear that the model is addressing too restricted a set of circumstances or is based on biological assumptions that are too simplistic. Some of the predictions of these models, of course, are hardly testable, since it is not feasible to conduct experimental evolution trials that are sufficiently long-term to cover the necessary span of several hundreds to thousands of generations. This may be possible with very special model organisms such as *Escherichia coli* or *Chlamydomonas* (see Chapter 9), but these violate other assumptions typical of quantitative genetic models, such as diploidy or random mating.

Perhaps more important, the quantitative genetic models proposed so far make the simplifying assumption of constancy or proportionality of genetic parameters (variance-covariance matrices) over long periods of time.[1] This is untenable on theoretical grounds, since **G**-matrices are affected by mutations and selection, as well as by other forces capable of altering allelic frequencies (Turelli 1988; Roff 2000). It is also questionable in light of at least some empirical results documenting cases of changes in the variance-covariance matrices of different species or populations. However, the actual data provide a mixed picture with variation or stability of **G** at different taxonomic ranks and for different groups of organisms. (For a discussion of several examples and references, see Roff and Mousseau 1999.) Such varied outcomes should perhaps not be surprising; after all, some combinations of traits are known to be more strongly linked in some groups than in others, and certain arrangements of traits are bound to be more stable than others on functional or developmental grounds. However, the evidence for evolutionary changes in **G** is sufficient to compel us to include it in our models of evolutionary trajectories (see Shaw et al. 1995; Pigliucci and Schlichting 1997). This can, in principle, be done by modeling the dynamics of the **G**-matrices themselves, although this is not a mathematically trivial task. I suspect that the best way around this obstacle is to move to empirically informed models similar to some of those described below in a different context. This has been the trend in ecological models addressing community dynamics (Cain et al. 1995; Kareiva et al. 1996) and is an approach that comes with at least one distinct advantage and one major disadvantage. The advantage is that the model is much more realistic and makes quantitative predictions that are directly testable. The disadvantage is that the results of the theoretical treatment are applicable to a more specific set of circumstances than those that usually concern theoretical population biologists. Is this a temporary impasse, or does it reveal a fundamental limitation imposed by the complexity of biological systems on our abilities to produce relatively simple yet general mathematical models of the problems we are interested in?

Optimization Models of Reaction Norm Evolution

Nongenetic models of phenotypic plasticity actually have a much longer history than their quantitative genetics counterparts, as the former go back at least to the attempts of Levins (1963) to describe fitness sets under different ecological scenarios. The basic idea is to set up a situation based on reasonable biological assumptions about the organism and the environment, explicitly incorporate trade-offs among alternative allocations of resources or costs, and then ask if there are stable or dynamic equilibria and where they lie. These models are nongenetic because they do not incorporate any specific parameter summarizing the genetic underpinnings of the traits studied. The necessary available genetic variation is simply assumed to be there. However, Charlesworth (1990) pointed out that by and large these models can be thought of as equivalent to quantitative genetic models with additive polygenic inheritance.

Although the pioneering work on optimization modeling of phenotypic plasticity published by Lynch and Gabriel (1987) is more properly seen as a model for the evolution of niche width, they tied it in with phenotypic plasticity for fitness. These authors developed a model to examine the response of an asexual population to density-independent gradients of environmental variation. They found that the plasticity of a genotype for fitness (which they called *environmental tolerance*) depends on the environmental optimum, its developmental variance between individuals, and the expected genetic contribution to the breadth of adaptation. They also concluded that temporal heterogeneity selects for genotypes with broader adaptations and that the within-generation component of environmental variation contributes the most to this outcome. In their model, spatial variation also selects for broader adaptation of a genotype, if it occurs in conjunction with within-generation temporal environmental variation.

A major area of research in optimization modeling of phenotypic plasticity is the relationship between age and size at maturity, a classical problem of evolutionary ecology. Perhaps one of the most influential papers in this field is Stearns and Koella's (1986) simulation of a series of scenarios and their comparisons with a large set of empirical data from various sources. Their basic model assumes that fecundity increases with size and that offspring's juvenile mortality decreases with an increase in the age at maturity of the parents. In other words, a higher investment of resources translates into a better quality of the offspring. The environment-dependent curves describing the relationship between age and size at maturity are obtained by varying the growth rate of the organisms and calculating the optimum combination of the two life history parameters. The results converge on four basic types of curves, each favored by a specific subset of environmental conditions (Fig. 10.1). If mortality does not depend on the growth rate, an L-shaped curve is obtained (No. 1 in Fig. 10.1). When juvenile mortality increases as an inverse function of growth rate, the resulting curve has a sigmoid (No. 2) or a paraboloid (No. 3) shape, depending on a parameter describing the power to which the inverse of growth is raised. If adult mortality increases as an inverse function of growth rate the curve is L-shaped again, while if both juvenile and adult mortality increase when growth rate decreases (a combination of the previous two cases), the curve should be keel-shaped (No. 4). The most general result of the model is that plasticity in age-size at maturity is favored, since the optimal situation is for an organism to slide along a continuum of age-size combinations, not to settle on a specific pair of values of the two. Stearns and Koella applied their model to data from nineteen fish populations, as well as to other organisms such as fruit flies, red deer, and humans. In the case of the fish populations, there was a very good quantitative agreement between the predictions and the observations (Fig. 10.2), which is more than can be claimed for many models in population biology. In other cases, predictions agreed in a qualitative fashion with the actual situation.

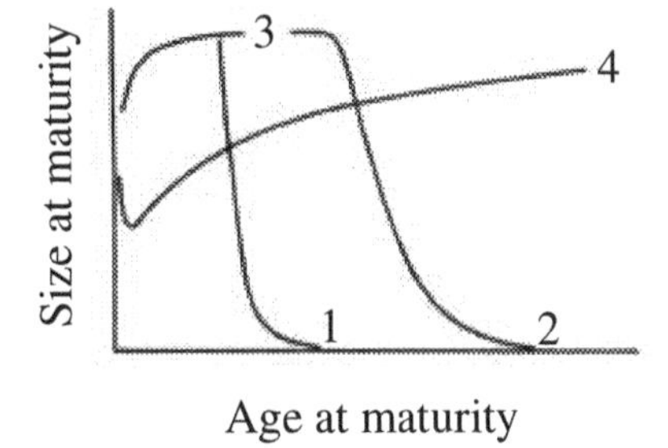

FIG. 10.1. Four shapes of the environment-dependent relationship between age and size at maturity according to different optimization models explored by Stearns and Koella (1986). Numbers refer to the types of curves: (1) L-shaped, (2) sigmoid, (3) paraboloid, and (4) keel-shaped. See text for details.

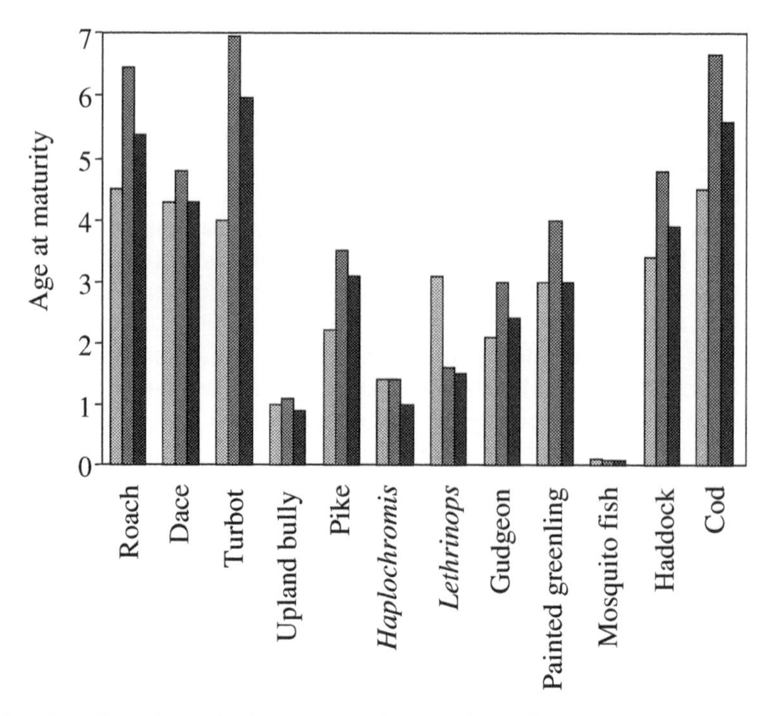

FIG. 10.2. General quantitative agreement between the predictions for age at maturity of Stearns and Koella's (1986) model and observations for a variety of fish species. Observed values are on the left in each cluster of bars. The other two bars are generated by the model for different values of a parameter measuring the effect of age at maturity on juvenile mortality. (From data in the original paper.)

Houston and McNamara (1992) followed up on Stearns and Koella by considering plasticity as a state-dependent life history *decision* that an animal (or plant) has to make. Their general conclusion is that in almost all situations it does not make sense to talk about fitness as a distinct measure in each possible habitat, because the optimal solution to the problem of environmental heterogeneity depends on what the organism does in all ecologically relevant habitats. Houston and McNamara conceded that in special circumstances one can ignore the outcomes in alternate habitats, in which case their model yields the same results as that of Stearns and Koella. A limitation of Houston and McNamara's attempt is that they assume *infinite plasticity,* that is, no restrictions on the expression of a trait in alternate environments (i.e., neither costs nor constraints to plasticity).

An additional component of the size-age problem related to the approaches of both Stearns and Koella and Houston and McNamara is the relationship between the size and the *number* of offspring of a given parental genotype. Forbes (1991) generalized a classical solution found by Smith and Fretwell to the case of variable environments and in particular of environments varying temporally between generations. This is equivalent to asking what the effect is of variable offspring survival on the optimal allocation of parental resources. From the parent's point of view, it is a matter of plastically allocating resources as a function of the environment likely to be encountered by the progeny. Forbes found that with environmental heterogeneity the optimal size of the offspring is greater than the one predicted by the classical model. This implies that the optimal number of offspring goes down correspondingly owing to an assumed trade-off between the two quantities. However, Forbes also concluded that the importance of the environmentally induced variance in offspring survival is dramatically reduced for very high offspring numbers. This may suggest that plasticity in life history traits is more relevant to K- rather than r-selected populations.

Stearns and Kawecki (1994) published a follow-up to the 1986 study on size and age at maturity by generalizing the model to a spatially patchy environment in which the fitness in one patch is not independent of the fitness at other patches (contra to what was assumed by Stearns and Koella [1986]). This is because the new model considers a situation in which good habitats function as sources for poor habitats (sinks), with a net flow from the former to the latter. Contrary to the findings of Stearns and Koella, the favored reaction norm is not obtained simply by connecting the local optima, since now both quality and frequency of the environments play a significant role. Kawecki and Stearns (1993) found that selection is expected to be stronger in good and frequent habitats rather than in poor and rare ones. This is probably because fewer offspring are produced in poor environments, and consequently genes favored in these environments constitute a smaller proportion of the long-term gene pool than those favored in good environments, leading to a reduced impact of bad-environment genes on the evolutionary trajectory of a population. Of course, all of this actually depends on the frequency of "good" and "bad" environments or, more realistically, on the shape of the continuous function describing the distribution of environmental quality. For a further elaboration of the Kawecki-Stearns model, see Sibly (1995).

Yet another set of divergent results on the never-ending saga of age and size at maturity was published by Berrigan and Koella (1994). Their model also predicts a variety of shapes for reaction norms, and their data compare well to actual observations in *Drosophila* when the model is parameterized accordingly. A major result of this model is that a correlation between growth rate and juvenile mortality, a biologically reasonable circumstance, can change the shape of the optimal norm of reaction, thereby linking directly phenotypic plasticity, life history parameters, and environmental quality. Why are the shapes of reaction norms found by Berrigan and Koella *qualitatively* different from those predicted by previous models? In part, this is a result of the fact that they considered a different set of assumptions, especially maximization of net reproductive rate and independence of mortality rates from age and size at maturity. Perhaps more important, however, their measure of fitness was different. This is part of a long discussion among evolutionary biologists: What is the best measure of fitness for reaction norms? The answer, which I have hinted at in several chapters of this book, depends on the type of environmental heterogeneity as well as on the selection regime on the reaction norms, which in turn also depends on the type of environmental heterogeneity. For example, Gomulkiewicz and Kirkpatrick (1992) used the geometric mean fitness under soft selection and the arithmetic mean fitness under hard selection in their quantitative genetics infinite-dimensional model.

Optimization modeling has been used recently to address the problem of limits and costs of plasticity (Chapter 7). For example, Tufto (2000) investigated the possibility that adaptive plasticity depends on the availability of an imperfect environmental cue correlated with the phenotypic optimum. The results show that the amount of plasticity at equilibrium reaches a maximum equal to the squared correlation between the environmental cue and the phenotypic optimum. The author also discusses the role of costs of acquiring pertinent information about the environment, as well as the evolution of sensory systems in general.

Overall, even a cursory survey of this literature indicates that the results of optimization models are much more varied (sometimes even inconsistent) than those obtained with quantitative genetic models. There may be several reasons for this state of affairs, but I think the major one is that the parameter space encompassed in typical optimization models is much more complex than the one considered in quantitative genetic models. There simply are too many parameters that can be used to describe the ecology of a population, and the more realistic one wants the model to be, the more aspects of the environment and of the phenotypes of the organisms one has to consider. There is almost no end to the types and combinations of trade-offs we can assume are underlying a particular plastic response. In contrast, most quantitative genetic models have to deal "only" with the **G**-matrix and the type and intensity of selection (though they could be complicated ad infinitum if we were to add details of the genetic architecture, finite population sizes, mutation, migration, and so on).

On the positive side, the optimization literature has been characterized by more direct and quantitative attempts to actually test the models. Probably the main reason for this is that the authors of these models are generally more interested in evolutionary equilibria than in the dynamics that lead to them. Therefore, if one assumes that natural populations are in fact at or near equilibrium, one does not need prohibitively long data series to test the models. Of course, it is very possible that populations are not at equilibrium, depending on a variety of factors including migration rates and breeding systems (Hartl and Clark 1989), in which case the validity of the empirical tests referred to above is conditional on reasonable evidence of attained equilibrium in the populations investigated.

Where is optimization modeling going? I submit that two major directions will be of utmost interest in the future. On the one hand, we would like to see hybrid ecological *and* genetic models (Charlesworth 1990) instead of the two parallel and mostly mutually exclusive (or at least not explicitly connected) sets of assumptions and problems so far tackled by quantitative geneticists and theoretical ecologists. The second direction addresses the problem alluded to earlier that real situations may be more dynamic and complex than is usually assumed. Computer-intensive dynamic models based on the idea of state variables (i.e., variables that can change their value throughout the simulation, in part in response to the current state of the model) are becoming increasingly popular and have been hailed as the tool that will allow the unification of life history and behavioral ecology theories (Clark 1993). Unfortunately, a typical limitation of even this approach is that the dimensionality of a problem can still easily overwhelm even the most sophisticated computer technology available to date, not to mention the less aesthetic appeal of "brute force" computation over simple and elegant analytical solutions.

Gametic, Population Genetic, and Individual Models of Plasticity

An alternative to the quantitative genetic approach to modeling the genetic aspects of the evolution of a character—and of plasticity as a special case—is what is known as a *gametic model.* I include in this category all models that actually comprise direct information on allelic or genotypic frequencies at specific loci as well as models simulating the fate of single individuals in a population. This is an extension of the classical approach of population genetics (Hartl and Clark 1989), and it is intellectually particularly satisfying because it can provide analytical, exact solutions to the problem at hand while incorporating a wealth of mechanistic detail. On the other hand, its main disadvantage is that it is impractical or even impossible to handle the computational difficulties of solving gametic problems for situations in which more than a few loci are involved—which led to the onset of quantitative genetics as an alternative approach based on statistical methods in the first place.

One of the first attempts at modeling plasticity by recourse to a gametic model was published by Zonta and Jayakar (1988) and Lorenzi et al. (1989), who addressed the behavior of two diallelic loci (nine genotypes) in response to temporal environmental variation. The two models analyzed correspond to two loci displaying additive effects on the character mean or to one locus affecting the mean and the other the variance (plasticity). These authors concluded that heterogeneous environmental conditions can (but not necessarily do) maintain phenotypic variance in the population. However, the relationship between plasticity and genetic variation in this model is not straightforward, and it depends on the specific contributions of alleles at both loci to the phenotypic variance as well as on the specific selection regime.

An attempt to bridge the gap between population (analytical) and quantitative (statistical) theory was published by de Jong (1990a). She explored the way in which a multilocus population genetic model translates into population-level statistical parameters such as genetic variances, covariances, and correlations. One of the most surprising (and largely ignored) results is that an additive genetic correlation of 1 between the expressions of the same trait in two environments does not necessarily imply the absence of genotype-environment interactions.[2] This is because if the same loci affect the expression of the trait in both environments and the allelic sensitivities are such that the effects remain proportional within each environment, the additive genetic correlation will remain unaltered, even though the reaction norms will differ in slope. In general, however, it is very difficult to truly bridge the population to quantitative genetic gap because if the number of loci increases, the possibility of exploring the results of specific genetic phenomena such as pleiotropy and epistasis from a mechanistic standpoint (one of the advantages of population genetic models) rapidly fades owing to the increasing analytical intractability of the problem.

An example of how complex things can get—even with simple, one-locus–two-allele population genetic models—is offered by a paper in which I explored the linear and nonlinear effects of dominance, drift, selection, and probability of environmental change (Pigliucci 1992). While the linear effect of selection clearly dominates the evolutionary dynamics, many interaction terms turn out to affect it significantly, and the total variance explained by nonlinear terms in the model is very strong (20%). Contour plots of fitness surfaces can illustrate the complex and sometimes counterintuitive outcomes of the evolutionary process when so many forces are at work simultaneously and with different strengths (Fig. 10.3).[3] I then used the same model to further explore the relationship among population genetics, quantitative genetics (interenvironment genetic correlation), and reaction norm approaches to model the evolution of plasticity (Pigliucci 1996b). The main conclusion was that neither interenvironment genetic correlations nor analyses of variance describing variation in reaction norms are necessarily good predictors of the evolutionary equilibrium. Only knowledge of the ac-

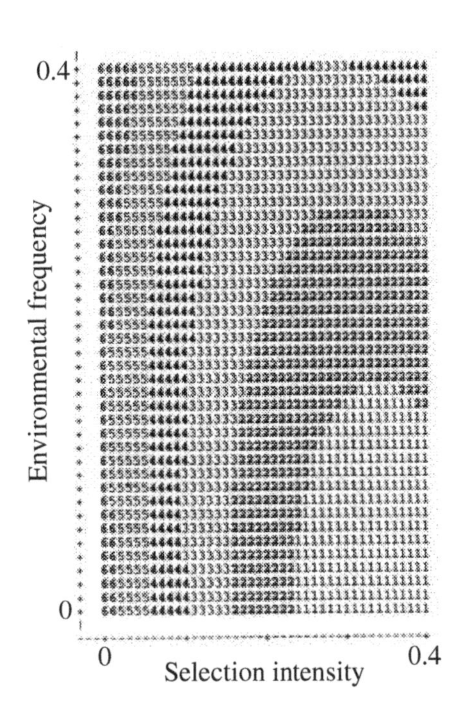

FIG. 10.3. Population genetic model (one locus, two alleles) of the evolution of phenotypic plasticity. The model explores the evolution of reaction norms when the frequency of occurrence of two environments (on the *y*-axis, summarized as frequency of the environment favoring the A allele), the selection intensity (on the *x*-axis), the population size (here fixed at $N_e = 255$), and the degree of dominance of the A allele (here fixed at $h = 0.25$) are accounted for. Numbers label quasi-equilibrium frequencies of the A allele, with the highest range indicating a frequency between 0.4 and 0.5, and the lowest range indicating the extinction of the allele from the population. Note how the frequency isoclines are markedly nonlinear as a result of a higher-order interaction between selection and environmental frequency. (From Pigliucci 1992.)

tual genetic basis of the trait (i.e., of the effects of the alleles at the relevant locus) leads to a good understanding of the system. This, of course, is not surprising in light of the difference between population and quantitative genetic theories highlighted above: A statistical approach is used only as a surrogate, not as a true alternative, to a mechanistic/analytical solution.

Another way around the complexity obstacle for population genetic models is the one pursued by Zhivotovsky et al. (1996) when they modeled evolution of plasticity in spatially heterogeneous environments in haploid populations. Their model accounts for mutation and weak selection, with the intensity of selection being variable among niches. They concluded that for relatively simple population genetic models it is possible to calculate the equilibrium reaction norms, although—in agreement with other studies—these are a complex function of both the type of environmental heterogeneity and the genetic properties of the system being investigated. The same authors then considered the evolution of plasticity when changes in gene frequencies, environmental effects, and linkage disequilibria simultaneously affect the evolutionary trajectory (Zhivotovsky et al. 1996). In this case, the population was mating randomly, and at each generation the individuals colonized different niches, each again subjected to weak Gaussian selection. While initially trait variability is attributable to the effect of additive loci and to homeostasis induced by heterozygosis, the dynamics and the forces underlying it change with time. Plasticity in this system evolves by invasion of genetic modifiers that affect the trait epistatically, altering the niche-specific optima as well as the strength of selection. While phenotypic optima are not achieved, reaction norms evolve despite the fact that the modifiers may affect only the individual rate of dispersal, that is, even when the modifiers do not actually alter the trait values. This paper represents the first clear analysis of the relative contribution of within- versus between-niche effects on the evolution of phenotypic plasticity, a distinction that is implied in several other models but never expressly investigated (but see my later discussion of Scheiner 1998).

An attempt to link population genetic modeling to ecological considerations of the type common in optimization models was published by Whitlock (1996). He noted that the evolution of niche breadth, which he equates with phenotypic plasticity, is usually assumed to be limited by trade-offs between fitness components in different environments (the idea that a Jack-of-all-trades is a master of none). However, Whitlock correctly points out that such trade-offs are rarely observed, and it is difficult to demonstrate them experimentally.[4] He then focuses on other limitations on the evolution of reaction norms and specifically on the idea that a broader niche may entail a slower evolutionary rate. This is another way of stating the hypothesis that plasticity can slow down evolution by buffering natural selection (see the discussion in Chapter 9). Whitlock concludes that a narrower niche is associated with a higher chance of fixing beneficial alleles, because selection is stronger, with a consequently lower probability of fixing deleterious alleles by drift. These conclusions hold, in his model, even in the special case in which fitnesses in different environments are uncorrelated (i.e., no trade-offs are assumed), which is usually thought of as the best scenario for the evolution of generalists. According to this author, narrower niches will also evolve because genes favored in local habitats may be associated with habitat selection, assortative mating, or simply reduced migration. This would increase the effectiveness of those genes by limiting their expression to the environments in which they are favored. The general conclusion, therefore, is that the evolution of plastic generalists is a much more complex affair than simply accounting for trade-offs generated by variable fitness in multiple environments.

A model explicitly considering both the effects of genetic architecture and the particular structure of the environmental heterogeneity (in this case, in a spatial sense) is the one published by Scheiner (1998). He assumed the presence of three types of loci affecting the phenotype: those whose expression is not affected by the environment, those that respond to the environment in a linear fashion, and those that respond in a quadratic fashion. Although the biological meaning of the distinction between the last two categories is unclear given how little we know of the genotype-phenotype mapping function generating observable reaction norms, it turns out that assuming a complex genetic architecture dramatically affects the results. For plasticity to evolve in Scheiner's model, the environment must strongly influence gene expression (not surprisingly) and the between-generation changes in the environment experienced by a given individual because of migration between demes have to be predictable (the situation described as *anticipatory plasticity* in Chapter 4). The response to selection is a complex result of the contribution of the different kinds of loci and the demic structure of the population, with "pure" strategies (globally optimal plasticity or local genetic specialization) rarely favored. Instead, the evolutionary outcome is a compromise between plasticity to the immediate range of environments likely to be encountered and specialization to any one of them. The migration rate turns out to be a key variable, with higher rates correlated with a higher occurrence of plasticity, because the progeny of a given individual is more likely to experience environments that are different than the one in which the parental generation was selected. Scheiner also found that selection on plasticity is a function of its global fitness and, therefore, will occur only in structured populations. This, however, may not be an important limiting factor, given that most natural populations are structured to some extent because of ecological, demographic, or genetic factors (Hartl and Clark 1989). De Jong (1999) further explored Scheiner's model, deriving analytical solutions for the system. She found that the evolution of curved reaction norms is the result of unpredictability of selection during development in substructured populations (not of genetic architecture, as suggested by Scheiner). This happens even if the optimum reaction norm is in fact linear, provided that there is genetic variation for curvature and an asymmetry in the frequency distribution of the relevant habitats. She pointed out that the degree of structuring assumed by Scheiner can be obtained in a system in which adult migration is coupled with the absence of zygote dispersal, a situation very different from previous models such as Via and Lande's (1985).

As we have seen, this mixed class of models has the advantage of being able to incorporate more mechanistic details of the evolution of phenotypic plasticity, thereby rendering a biologically more relevant scenario. However, this strength is somewhat limited by the inevitable computational constraints that arise if one attempts to simulate the evolution of a trait affected by many loci, in turn modulated by complex interactions with the genetic and external environments. Part of the appeal of these models is that they can be formulated in a way that merges population and quantitative genetics or genetics and optimality (ecological) considerations. However, in each of these hybrid attempts the complexity of the genetic architecture must be reduced to relatively simple scenarios, again because of the immediate onset of computational limitations. While this is a technological, not a theoretical, obstacle, it is not clear if and when our computers will become sophisticated enough to simulate much more realistic, and therefore informative, situations than the ones briefly reviewed here. It is also far from clear whether our brains will ever be able to process and absorb the outcomes of such realistic simulations.

Empirically Informed and Situation-Specific Models

The last approach to modeling phenotypic plasticity that I consider here is a mix of two kinds of models: (1) attempts that conceptually fall into one of the other categories discussed above, but differ because they are empirically informed, i.e., some parameters of the model are fixed by collecting actual data in natural populations (we saw some examples of these while discussing some of the age-size models); and (2) situation-specific models, in which the mathematical treatment is tailored to a particular kind of plasticity, usually because it corresponds to a phenomenon of interest to physiologists, the mechanics of which are understood well enough to allow the incorporation of many details that would be unthinkable in the more general types of models discussed so far. I consider one example of the second kind and one that can be clearly thought of as a combination of the two.

A typical example of situation-specific models is the one by Sall and Pettersson (1994) addressing photosynthetic acclimation as a special case of phenotypic (physiological) plasticity. Photosynthetic acclimation is the phenomenon by which plants change the temperature optimum for the rate of photosynthesis as a function of temperature during growth. Sall and Pettersson explored two versions of their model, both based on physiological concepts: a stationary model dealing with a fully acclimated plant and a dynamic model based on the idea that the rate of change of acclimation is proportional to the deviation from full acclimation. Both models share the underlying assumption that two curves describe the behavior of the plant: a short-term and a long-term response function. These authors were able to account for the empirically demonstrable fact that plants usually do not fully acclimate to the growth temperature (i.e., the effective growth temperature to which the plant is acclimated does not correspond to the optimal temperature for the same species as determined by physiological assays). The suggested reason is that the optimal strategy focuses on the maximization of carbon uptake and not on temperature per se. In general, a dynamic model was found to be (predictably) more realistic and capable of accounting for more subtle aspects of the reaction norms of plant photosynthesis to temperature as known from the physiological literature.

The work of Jackson and Caldwell (1996) (Fig. 10.4) represents a hybrid between a situation-specific and an empirically informed model. They

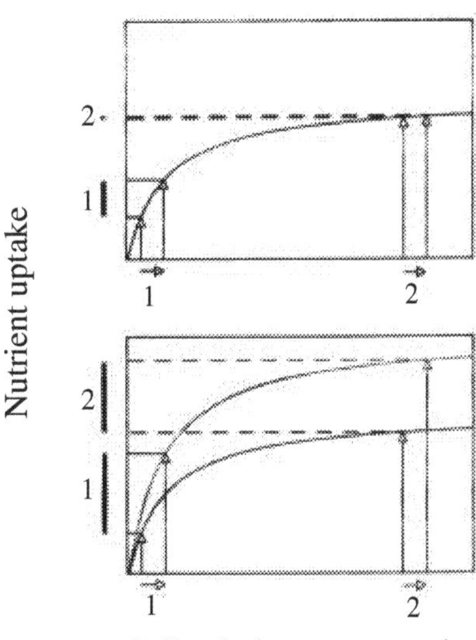

FIG. 10.4. Increased efficiency of nutrient uptake as a function of soil solution concentration in the case of no plasticity (upper diagram) and plasticity (lower diagram): (1) indicates low nutrients and (2) high nutrients. Note that in the absence of plasticity, increased nutrient availability augments uptake only in case 1, whereas with plasticity both cases yield an enhanced uptake, and even if we consider only case 1, plasticity provides a sharper increase in uptake than the absence of plasticity. (From Jackson and Caldwell 1996.)

approached the problem of plasticity to heterogeneity in nutrient resources (namely, phosphate and nitrate) using biochemically based kinetic and flow equations. In particular, they investigated the potential advantage of two components of root plasticity in heterogeneous environments: changes in uptake kinetics and in root proliferation. To set up their model, these authors calibrated the soil characteristics and physiological parameters of the roots based on data from a five-year field study. This led to simulating a range of variability in nutrient availability of threefold for phosphate and twelvefold for nitrate. They found that plasticity accounted for a more efficient uptake of both nutrients, but especially of the more variable nitrate (a seven- to twentyfold increase in efficiency) with plasticity accounting for 50% of the acquired phosphate and 75% of the acquired nitrate. The higher efficiency of plants displaying phenotypic plasticity was equally attributed to augmented uptake kinetics and enhanced root proliferation for phosphate, whereas increased uptake was almost entirely responsible for the more efficient use of nitrate. This highlights the possibility that distinct components of a plastic response may be characterized by different roles and relative importance depending on which aspect of the environment is varying.

Empirically based and situation-specific simulations have distinct advantages and disadvantages. In both cases, one gets a most welcome feeling that the simulation addresses a relevant biological scenario in a realistic fashion. Furthermore, the predictions generated by these models can be compared with actual data in a quantitative, not just qualitative, manner, thereby better fulfilling the promise to "predict" the specifics of a situation.

On the other hand, this higher capability for yielding quantitative predictions and better biological flavor comes at the expense of generality. A model designed to simulate photosynthetic acclimation as a case of plasticity cannot easily be applied to any other situation, since the equations underlying such a model are specific to the physiology of photosynthesis and the physics of reactions to temperature. Furthermore, in many cases we still lack adequate knowledge of the mechanistic details necessary to parameterize these models.

The Future: Of What Use Is a Model, and Do We Need More?

What is the role of modeling in evolutionary biology? Schlichting and I have argued that it is very different from the role of mathematical models in physics, the standard comparison for the "hard" sciences (Pigliucci and Schlichting 1997). In physics—especially particle physics—theory is usually (with occasional spectacular exceptions) far ahead of experimentation. Theoretical scientists do most of the cutting-edge research, with the experimentalists perennially running just to keep up with them by at least partially confirming precise predictions about the intimate structure of matter. There are two fundamental reasons for this state of affairs: On the one hand, the objects of research in theoretical physics are the simplest constituents of matter, which means the least complex systems in the universe; on the other hand, it just happens that doing any meaningful empirical research on such elementary particles requires achieving very high energies under controlled conditions, one of the most expensive propositions in modern empirical science.

Biology, especially at the organismal level, presents an entirely different, almost reversed, picture. Its foci are the most complex systems known in the universe (organisms, populations, communities, and ecosystems), but it is also true that a great deal of sophisticated empirical research can be done on them with comparatively little expenditure, especially at the population to community levels. This combination has put experimental biology into a different relationship with theoretical efforts ever since the molding of the Neo-Darwinian synthesis in the 1930s and 1940s. Unlike the situation in physics, the two enterprises run side by side on parallel tracks, and the history of the field is punctuated by examples of productive feedback between theory and empirical evidence (e.g., the sexual selection debate between runaway selection and good genes). On the other hand, we also have plenty of instances in which theoretical models appear not to be testable under most biologically relevant circumstances (e.g., how will we ever know if a quantitative genetic model of a panmictic diploid population projecting predictions ten thousand generations into the future is actually in agreement with reality?). Given this difference between the physical and biological sciences, one could ask whether we need more models of the evolution of plasticity (or, for that matter, of evolution in general). I would argue for a qualified yes, provided that we escape the trap of physics-envy that always lurks in the hallways of biological departments, exemplified years ago by Fisher (1930), who explicitly declared that his aim in formulating the "fundamental" theorem of natural selection was to provide biology with theoretical tools equivalent to the principles of thermodynamics in physics. All of the categories of models discussed above are more or less useful for furthering our understanding of biological evolution. Moreover, many classes of models complement one other and can be used in various combinations to approach the same fundamental questions. However, these models usually do not embed an accurate or sophisticated view of reality (unlike their counterparts in physics). Instead, they represent rough sketches of what is out there, and, as such, they seldom forge far ahead of empirical research. On the contrary, empirical evidence can be used to discard or improve the performance of these models more by questioning their assumptions than by verifying their specific predictions.

As the Oxford dictionary puts it, a *model* is "a simplified description of a system for calculations," and all models lie between two extremes. On the one hand, they can embody a description that is so simplified that it fails to capture many of the fundamental features of the system. On the other hand, the description can be so realistic that the model itself is too cumbersome and specific and is therefore of little use for the understanding of general principles. I submit that many quantitative and optimization models currently fall close to the first extreme, while empirically informed and situation-specific models are closer to the second. Of course, this is no reason not to further pursue a continuous feedback between empirical and theoretical biology. As Immanuel Kant put it, "Experience without theory is blind, but theory without experience is mere intellectual play." All we need is a better mutual understanding of the limitations intrinsic in each approach and an appreciation of where the strengths of each lie.

Conceptual Summary

- Genotype-environment interactions can be modeled by using a variety of independent, not mutually exclusive, approaches.
- Quantitative genetics theory provides a standard set of tools for modeling the evolution of plasticity, mostly following the pioneering work of Via and Lande and, independently, of Gavrilets.
- Optimization theory has also been used extensively to model aspects of the evolution of plasticity, especially with reference to life history de-

cisions and the relationship between age and size at reproduction. These models do not incorporate specific information about the genetic basis of the traits considered, or otherwise assume a standard many-allele-with-small-effects approach analogous to quantitative genetic modeling.

- Population genetic (gametic) models have the advantage of being capable of incorporating details of the mechanics underlying the trait, but are computationally and analytically limited to the consideration of only a few loci simultaneously.
- Empirically informed and situation-specific models are the most powerful in terms of producing quantitative predictions, but they are also the most restricted in application because by definition they are constructed to deal with very specific situations.
- In general, modeling in evolutionary biology has a different role to play than in other branches of science, for example, particle physics. The objects to be modeled are extremely complex, and the predictions are mostly of a qualitative nature. The role of empirical biologists in this case is focused on checking the assumptions, rather than the output and predictions, of models.

—11—

Phenotypic Plasticity as a Central Concept in Evolutionary Biology

I don't pretend to understand the Universe—it's a great deal bigger than I am.

—Thomas Carlyle (1795–1881)

CHAPTER OBJECTIVES

To tie together the different threads of the book by showing how phenotypic plasticity can usefully be incorporated in all areas of inquiry in evolutionary biology. The chapter uses recent studies on the model system *Arabidopsis thaliana* as a roadmap to the delineation of a research program ranging from microevolution within and among populations to macroevolution and the appearance of phenotypic novelties.

This final chapter departs from the general structure of the book in two ways: First, it addresses several aspects of plasticity research simultaneously; and second, it is based primarily on examples from a single line of research on a single model organism, which has been carried out in my laboratory over the last few years. The reason for these departures is that I wish to explore the relationship between the specific study of phenotypic plasticity and the broader questions with which evolutionary biology is concerned (Carroll 2000). Studying plasticity is all the more justified, and intellectually most satisfying, if we can frame the results in a general context. Even better, plasticity research is of the most interest when its study sheds light onto old and yet unresolved evolutionary problems. In order to explore the connection between phenotypic plasticity and broader evolutionary questions, I have attempted to develop a theoretical framework connecting several fields of evolutionary biology by the way in which they touch upon or are influenced by phenotypic plasticity studies. I weave my argument by way of example—following a venerable tradition in evolutionary biology as well as the structure of the other chapters in this book. In the process, I briefly discuss the kinds of bridges that phenotypic plasticity can help build among the subdisciplines and levels of analysis typical of evolutionary biology.

The central idea I am developing here is that an essential part of evolution deals with adaptation to changing environmental conditions, and that therefore plasticity studies have something to say (to a greater or lesser extent) about population genetics, ecology, microevolution, and macroevolution. I illustrate this principle by studying a single type of plasticity in a single group of plants, in line with the recent explosion of research on *model organisms* across biological disciplines (Kellogg and Shaffer 1993). This is convenient not only because I happen to be studying that particular form of plasticity in those plants but also because I know the details of this story much better than I could master those of other equally fascinating tales. The major reason for my choice, however, is that this example illustrates rather nicely how a relatively simple system can function as a gateway to inquiries about much broader scientific issues. The story that is going to be told in the following pages is, of course, incomplete, and it will very likely always remain so. In fact, some aspects are simply sketched and only give a taste of things to come. On the other hand, my aim in writing this book is not to provide a conclusive set of answers, but rather to analyze how we got the tentative answers we are currently holding on to and, more important, to suggest avenues to explore the many questions that are still out there.

The Many Implications of Phenotypic Plasticity

Evolutionary biology is certainly a large and fascinating field of inquiry. Yet, there are very few parts of it that have not been touched by studies devoted to the understanding of phenotypic plasticity (the only major exception being paleontology, since it is impossible to study reaction norms of extinct species, contra Reyment and Kennedy 1991). I have compiled a list—almost certainly incomplete—to help the reader bring the disparate threads of earlier chapters together into an organic whole:

1. *Adaptogenesis*. Understanding how adaptations come about is one of the most challenging fields of evolutionary studies, dealing with problems ranging from the measurement of selection under natural conditions (Lande and Arnold 1983; Endler 1986) to the distinction between adaptation as the result of a historical process or as a currently conferred advantage (Gould and Vrba 1982), from the definition and measurement of fitness (McNeill Alexander 1990; Burt 1995) to the understanding of what consti-

tutes a character (Pigliucci 2001; Wagner 2001) and therefore the target of selection (Gould and Lewontin 1979; Wagner 1989). Studies of potentially adaptive phenotypic plasticity have explored the same gamut of difficulties and solutions, with the added complication of having to measure traits and fitnesses under several environmental conditions. Tests and discussions of adaptive plasticity hypotheses are abundant (e.g., Cook and Johnson 1968; Takahashi 1975; Khan et al. 1976; Mitchell 1976; Givnish 1982; Baskin and Baskin 1983; Grime et al. 1986; Lively 1986b; Hazel et al. 1987; Newman 1992; Gotthard and Nylin 1995; Dudley and Schmitt 1996; McCabe and Dunn 1997; Schmitt et al. 1999; Chapter 7). The major contribution of the plasticity perspective to this research is that organisms *must* adapt to multiple environmental conditions. It is not really a matter of constant versus changing environments. Rather, it depends on the degree, frequency, and type of environmental heterogeneity that living beings have to face during the course of their lives. This realization may lead simultaneously to the recognition that there can be stricter limits to (Via and Lande 1985) as well as previously unforeseen possibilities of (West-Eberhard 1989) evolution by natural selection (Chapter 8).

2. *Natural variation and its maintenance.* The existence and maintenance of genetic variation in natural populations have been at the center of endless discussions in evolutionary biology, from the early days of the Neo-Darwinian synthesis (summarized in Provine 1981), through the application of electrophoretic techniques (Lewontin and Hubby 1966), to the neutral theory of molecular evolution (Kimura 1954, 1983). Quite clearly, plasticity studies have demonstrated at least one thing: Natural genetic variation for phenotypic traits across varied environments is present—to different degrees for different traits—in virtually every organism that has been looked at (Bradshaw 1965; Sultan 1987). This observation catalyzed renewed efforts to model the circumstances that may lead to the maintenance of genetic variation, and the ubiquitous existence of environmental heterogeneity has been factored into much of the published theory on this topic (e.g., Levene 1953; Kawecki and Stearns 1993; Lande and Shannon 1996; Chapter 4).

3. *Quantitative genetics.* Quantitative genetics is a classical tool in biology (Fisher 1930; see summaries in Falconer 1989 and Roff 1997), and it has seen a considerable renewal of interest within several subfields of evolutionary biology during the last two decades (e.g., Lande 1980; Cheverud et al. 1983; Lande and Arnold 1983; Slatkin 1987; Barton and Turelli 1989; Shaw et al. 1995; Roff 1997). A significant part of this interest has in fact been catalyzed by the natural application of quantitative genetics techniques to studies of phenotypic plasticity (Falconer 1952; Via 1984b). Consideration of plasticity has yielded two major advances in quantitative genetics: the empirical demonstration that heritability is environment-

dependent and the conceptual extension of the idea of a genetic correlation to the general case of multiple environments (Chapter 1).

4. *Modeling.* The twin fields of quantitative genetics and optimality models have been ruling evolutionary theory for the past few decades (e.g., Levene 1953; Levins 1963; Hoekstra 1988; Williams 1992; Gavrilets and Hastings 1994; Doebeli 1996). Modeling evolution in heterogeneous environments is a general extension of these efforts (see Chapter 10). Focusing on reaction norms has actually helped merge the two strategies of theoretical modeling, in that it couples quantitative genetic variation and optimality problems related to allocation of resources in favorable and unfavorable environments or to phases of the life history in which an organism may be exposed to distinct environmental conditions.

5. *Character correlations and constraints.* The study of character correlations has increased in importance in evolutionary biology since the realization that genetic correlations among different traits may either enhance or dramatically retard adaptive evolution (because of antagonistic or synergistic pleiotropy (see, e.g., Berg 1960; Cheverud et al. 1983; Lande and Arnold 1983; Phillips and Arnold 1989; Riska et al. 1989; Arnold 1992; Fry 1993; Houle et al. 1994; Merila et al. 1994; Johnson and Wade 1996; O'Neil 1997). The logical extension of these problems to multiple environments is yielding a general theory of constraints in evolutionary biology as well as compelling empirical examples of the relationship between constraints and natural selection in shaping phenotypic evolution (Via 1987; Cavicchi et al. 1989; van Tienderen and Koelewijn 1994; Simons and Roff 1996). One particular concept that emerges from the study of plasticity is *plasticity integration* (or the coordination of different plasticities rather than simple within-environment character means) in response to environmental changes (Chapter 7). The genetics, ecology, and evolution of plasticity integration are still completely open fields of inquiry.

6. *Gene regulation and its evolution. Regulatory genes* have long been the implied culprits in everything that was difficult to explain in evolutionary biology. The recent proliferation of studies in molecular developmental genetics has afforded us the opportunity of finally gaining a peek into the *black box* of genetic regulation and its relationships to organogenesis and adaptogenesis (Ha and An 1988; Avery and Wasserman 1992; Doyle 1994; Nijhout 1994). Along the same lines, our understanding of the actions of genes is now being expanded to the general case of multiple environments (Chapter 5). In fact, a major contribution of plasticity studies in this field is the realization that many (if not most) regulatory genes are directly involved in context-dependent pathways, and that several respond differentially to an array of environmental stimuli. The field of phenotypic plasticity has also yielded the novel concept of *plasticity genes,* hierarchically high-level regulatory elements that directly detect environmental signals and trigger one

of a series of alternative developmental pathways. Moreover, the study of natural variation of regulatory genes and their products promises to provide one more bridge between micro- and macroevolution (Krebs and Feder 1997a,b; Krebs et al. 1998; Purugganan and Suddith 1998, 1999).

7. *Comparative method—phylogenetic patterns.* The modern comparative method has seen an explosion of research, which has culminated in its establishment as a requirement for any serious investigation into the long-term evolution of any trait (e.g., Felsenstein 1985; Harvey and Pagel 1991; Gittleman and Luh 1992; Hillis et al. 1994; Martins and Hansen 1997; Abouheif 1998; Ackerly and Donoghue 1998; Ackerly 2000a,b; Martins 2000). Plasticity has hardly entered this realm (but see Roskam and Brakefield 1996; Pigliucci et al. 1999; Pollard et al. 2001), mostly because of the cumbersome logistics associated with comparative plasticity studies (Chapter 9). However, plasticity can in principle be considered as any other quantitative trait and in particular as being analogous to animal behavioral traits (Gittleman and Decker 1994), for which comparative studies are certainly possible and quite informative. As I mentioned before, Doughty (1995) correctly pointed out that the null hypothesis for a reaction norm when comparing different species is not one with a zero-slope (i.e., lack of plasticity), but the most likely reaction norm of the common ancestor of the group under consideration. I predict that, given its importance, phylogenetically informed studies of plasticity will become very common within the next few years and will add an essential ecological element to comparative biology, which has been partially missing from the picture.

8. *Macroevolution.* Evolution above the species level and the evolution of phenotypic novelties have been major missing links in the Neo-Darwinian synthesis, and are still the subjects of much debate but little empirical study (Eldredge and Gould 1972; Stanley 1975; Levin 1983; Agur and Kerszberg 1987; Larson 1989; McKinney and McNamara 1991; Coyne 1992; Hall 1992a; Lee 1993; McShea 1994; Gatesy and Dial 1996; Zrsavy and Stys 1997). As I pointed out in Chapters 8 and 9, phenotypic plasticity may play a very important role in explaining the evolution of novel phenotypes as well as in accounting for speciation and colonization of new niches. Should a critical mass of empirical examples accumulate in the near future, it might lead to a major shift in our way of thinking about "familiar" macroevolutionary phenomena such as preadaptation and mosaic evolution.

Evolutionary Biology of Plasticity: Four Levels of Analysis

I think one can (more or less arbitrarily) identify four hierarchical levels of concerns and questions that evolutionary biologists have been asking, each of which is either affected by or has profound implications for phenotypic plasticity studies. First, we have microevolution within populations. Here the questions are of the kind: Is there genetic variation (e.g., for plasticity: Mackay 1981)? What kind of selective forces act on such variation (Weis and Gorman 1990)? Furthermore, what types of constraints limit the response to selection (van Tienderen and Koelewijn 1994)? The second is microevolution among populations. This is the level at which we can see the outcome of past episodes of selection and their interaction with genetic constraints. In other words, the observable patterns of current variation among populations may be thought of as a kind of "fossil" evidence of the recent past evolutionary history of those populations (Jernigan et al. 1994; Armbruster and Schwaegerle 1996). Third is macroevolution at the species level. In this instance, we are interested in uncovering patterns of phenotypic differentiation among species and establishing if and to what extent they are connected to speciation events themselves (Coyne 1992; Grant 1994). As far as plastic responses in particular are concerned, we would like to know how they vary in patterns and type among closely related species. Fourth is macroevolution at higher taxonomic levels. I am not advocating a special status for orders or families of plants and animals, but am referring instead to the fact that we do not yet have a good understanding of how traits and plasticities come into being in the first place. This, in many cases, must have happened long ago, resulting in a wide across-taxa conservation of the mechanisms underlying some types of plasticity. This may be thought of as the problem of the evolution of phenotypic novelties and *bauplane.* It is one (relatively easy) thing to understand how something can be altered by natural selection and historical accidents once it is in place, but much more difficult to produce a testable theory of how the necessary genetic machinery for a character or plasticity emerged in the distant past.

In order to address all four levels of analysis, I make use here of my own research as well as that of colleagues on the model plant *Arabidopsis thaliana* (discussed earlier in these pages) and its close relatives (Alonso-Blanco and Koornneef 2000), focusing in particular on the study of plasticity in response to light perception.

Microevolution within Populations

At the level of microevolution within populations, the two important determinants of adaptive evolutionary processes are selection (and the environments that induce it) and constraints (and the limits on the genetic covariation that determine them). For example, plants commonly show plasticity to the spectral quality of light, which is altered by foliage shade and is therefore an indicator of ensuing competition: the so-called *shade avoidance syndrome* (Givnish 1982; Casal and Smith 1989; Schmitt 1997; Smith and Whitelam 1997; Lee et al. 2000; Weinig 2000a,b). The degree of this response is usually related to the habitat in which they have evolved (with little plasticity in shade-tolerant species and enhanced plasticity in shade-intolerant ones [Bradshaw and Hardwick 1989]). Before any progress can be made in understanding the relative roles of selection and constraints, we have to quantify *two* types of variation: the commonly measured genetic variation within whatever collection of populations we are interested in and the degree and type of *environmental* heterogeneity. This is where the study of plasticity has made painfully clear that we know a good deal more about the genetics than about the ecology of natural populations. It is easy enough nowadays to sample natural populations and get reliable estimates of genetic variation at either the quantitative or the molecular genetic level. But accurate assessments of environmental variation are very difficult and are still rare in the context of plasticity studies, insofar as they represent the equally necessary "other side" of the evolutionary equation.

Several studies have now been published with intensive sampling of the physical conditions of the environment (e.g., Lechowicz and Bell 1991; Cain et al. 1999) or have utilized the plants themselves as indicators of environmental heterogeneity (*phytometers*). The unequivocal results seem to be twofold: On the one hand, there is significant environmental variation among distinct sites (occupied by different populations that often acquire the characteristics of ecotypes [van Tienderen 1990; see also Chapter 4]). On the other hand, within a given site most of the environmental heterogeneity occurs at scales of a few centimeters to a little over a meter (see, e.g., Kalisz 1986; Lechowicz et al. 1988; Stratton 1994; Stratton and Bennington 1996, 1998). These are the spatial scales that one would expect to favor the evolution of both within-plant and whole-organism types of phenotypic plasticity (Chapter 9).[1] Once we know what kind and range of environmental variation we are dealing with, we can then proceed with standard quantitative genetic experiments on reaction norms under controlled conditions to uncover the extent of genetic variation for plasticity present in a given population (Fig. 11.1).

Another component necessary for a full understanding of microevolutionary patterns and processes within populations is a quantification of selection pressures under natural conditions. The quantitative study of natural selection has a well-established theoretical foundation (Lande and Arnold 1983; Manly 1985; Endler 1986), yet our database concerning phenotypic plasticity is in this respect rather sparse. Selection, of course, is particularly important and much more difficult to study in highly heterogeneous conditions, especially those favoring the evolution of plasticity. One approach that has been used in my laboratory is to measure selection coefficients on the same trait at different sites and in different years, and then statistically compare the patterns and intensities of selection across sites/years. For example, we have studied selection on leaf number in *Ara-*

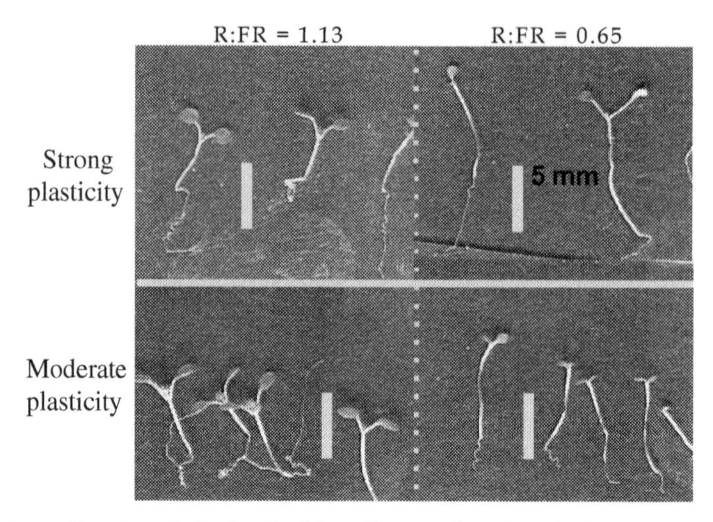

FIG. 11.1. Genetic variation for plasticity of hypocotyl (embryonic shoot) length in *Arabidopsis thaliana* exposed to normal (left) and foliar (right) red:far red light. The upper panels show a family characterized by very strong plasticity and the lower panels display a family with moderate plasticity. (From Callahan and Pigliucci, unpublished.)

bidopsis thaliana and found that at two sites separated by a few miles there is directional selection for increased leaf number (Callahan and Pigliucci, in press). However, the intensity of selection was $\beta = 0.32$ in one case, and $\beta = 0.59$ in the second case. This difference was highly significant, indicating that selection was much stronger at the second site than at the first. Heterogeneity in the selection coefficients was also found between years at the same site, and this may lead to selection for plasticity for leaf production in this plant.[2]

The final component an evolutionist needs in order to understand microevolution at this level is the type and extent of genetic constraints that may limit the response of the population to selective pressures.[3] One type of genetic constraint has already been dealt with implicitly: lack of or reduced genetic variation. A straightforward survey of the genetic variation for reaction norms present in the population will directly quantify that (Scheiner and Lyman 1989). But there is a second major category of genetic constraints (in the sense explained in the preface), which deals with the genetic architecture of the plant or animal: the dominance, pleiotropic, and epistatic relationships within and among the loci affecting the character mean (or plasticity) of interest. One approach to studying the genetic architecture in detail has utilized the characterization of the environmental responsiveness of single, double, and higher-order mutants available owing to the enormous research effort of contemporary developmental and molecular geneticists (Chory 1993; Kenyon et al. 1993; Kondo et al. 1994; Leon-Kloosterziel et al. 1996; Swalla and Jeffery 1996; Schichnes and Freeling 1998; Cohn and Tickle 1999). This information will be particularly useful for evolutionary studies when it is coupled with a knowledge of selective gradients, and I think this angle of attack on the problem will be used extensively in the near future, at least in model organisms. However, a more direct—and somewhat related—method uses the array of techniques known as quantitative trait loci mapping to pinpoint genes that affect the particular plasticity under study *in natural populations,* as opposed to laboratory lines that are better characterized but may be of more evolutionarily dubious usefulness (Chapter 2).

The stage is now set to move on to the next level, the study of microevolutionary patterns among populations, which are the outcomes of the processes we have discussed in this section.

Microevolution among Populations

A common observation in plasticity and in evolutionary studies generally is that natural populations appear to display distinct patterns of character variation and covariation (Chapter 7). It is therefore interesting to inquire into the degree to which these patterns are the result of natural selection or of genetic constraints.

The key to this line of research is the concept of genetic correlation amply discussed throughout this book. Both genetic correlations between two traits expressed in the same environment and genetic correlations between the expressions of the same traits in two (or more) environments can be indicative of selection (*functional correlations*) or constraints (*structural correlations*). As is well known, the study of constraints in general is the focus of a major debate in modern evolutionary theory (Antonovics 1976; Gould 1980; Maynard Smith et al. 1985; Wagner and Altenberg 1996; Phillips 1998; Armbruster et al. 1999; Merila and Bjorklund 1999; Hodin 2000; Pigliucci and Kaplan 2000). The consideration of plasticity has added two very important dimensions to this debate: first, the existence of constraints connecting the expression of the same trait in multiple environments (Via 1987; Andersson and Shaw 1994, see Chapter 2; Hebert et al. 1994); and second, the possibility that constraints themselves can be plastic, that is, that they can be affected by environmental effects. As an example of the latter category, my group studied the constraint that relates the number of leaves and bolting (flowering) time in several populations of *A. thaliana.* This constraint is found in early-flowering populations, but not in late-flowering ones, and it is affected by environmental change (Pigliucci et al. 1995). Figure 11.2 shows that a change in photoperiod (from short to long days) does not alter the slope of the correlation defining the constraint

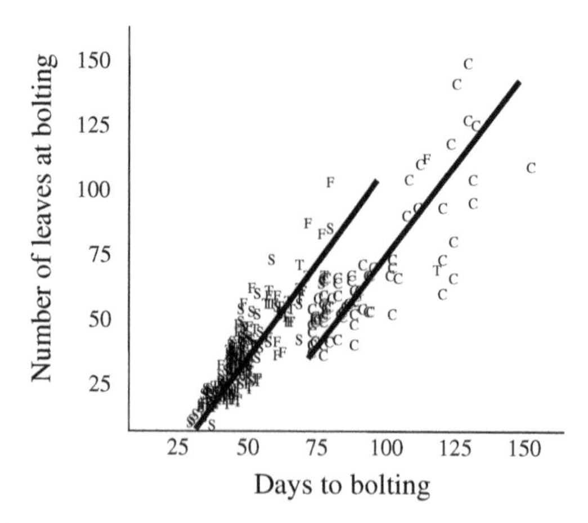

FIG. 11.2. Environmental shift induced by exposure to different photoperiods on the constraint relating bolting time to leaf production in *Arabidopsis thaliana*. The two lines indicate the correlation between the two traits: While the slopes of the lines are identical, a short photoperiod shifts the constraint toward the upper right because it causes later bolting and increased leaf production. Letters indicate different timing of the switch between short and long photoperiods: C = always short days (control); F = switch to long days after emergence of first leaf; T = switch to long days after emergence of third leaf; S = always long days. (From Callahan and Pigliucci [in press].)

but does shift the whole line toward the upper right of the diagram. This means that the two traits are affected by the photoperiod in the sense that plants flower later and with more leaves under short days, but that the overall *relationship* between the two remains constant.

There are at least four approaches that can be used to study constraints on plasticity and, by extension, on any trait experimentally. First, one can induce mutations in a series of genetic backgrounds and look for mutants that lie significantly outside the correlation line defining the constraint. While this general type of experimental attack on the problem of genetic constraints has not been widely used so far, it may prove to be a promising avenue of inquiry that can be generalized to any group of traits that are apparently genetically constrained by one another (see Camara and Pigliucci 1999 and Camara et al. 2000 for examples and references).

A second, related, approach is to characterize the pleiotropic and epistatic effects of known mutants affecting the plasticities of interest. This kind of study is still at an embryonic stage (Schmitt et al. 1995; van Tienderen et al. 1996; Ballaré and Scopel 1997; Callahan et al. 1999; Pigliucci and Schmitt 1999, see Chapter 2), since only a few relevant mutants have been isolated in a limited number of model organisms. However, the method has the potential to yield insights into constraints affecting high-level regulatory genes controlling whole cascades of phenotypic effects.

A third general way of tackling the selection-constraint conundrum is to carry out selection experiments to verify to what extent evolutionary changes in the focal trait really do translate into correlated responses in another trait(s). This has been attempted for evolution within stable environments (see, e.g., Wilkinson et al. 1990; Mueller et al. 1991; Hillesheim and Stearns 1992; Partridge and Fowler 1993), but only a few papers have dealt with environmental heterogeneity (Robertson 1964; Hoffmann and Parsons 1993). Of particular interest is the study of the correlated responses of functionally similar types of plasticity. For example, plants have to react to different aspects of light availability, such as photoperiod and light quality. However, light quality is patchy on a very small spatial scale, thereby possibly leading to the evolution of types of plasticity such as the shade avoidance syndrome. Photoperiod, on the other hand, varies across much larger geographical scales (and seasonally, of course), possibly inducing selection for latitude-specific ecotypes. Since both aspects of light variation are perceived, to some extent, by the same kind of molecules (the phytochromes [Reed et al. 1994; Shinomura et al. 1994; Yanovsky et al. 1995]), one would expect some degree of inevitable constraint relating the two types of responses (Blazquez and Weigel 2000).

The fourth approach to the problem of constraints is represented by applications of the comparative method. For example, the question raised above about the degree of independence of the evolution of shade avoidance and photoperiod responses can be resolved by phylogenetic comparisons. In particular, if selection shaped the shade avoidance response first, adaptive evolution to photoperiod may have followed within the limits imposed by constraints determined by a partial overlap of the biochemical pathways. One would therefore expect this co-evolution between plasticities to be reflected in the sequence of reconstructed character states when mapped onto an intraspecific phylogeny. Studies of this sort using a wide range of ecotypes of *A. thaliana* are currently in progress and should soon yield an answer to this classic chicken-and-egg question (Pollard et al. 2001).

Evolution within- and among-populations has so far accounted for most of the scientific literature on plasticity. However, the next two levels of macroevolution certainly constitute the most intellectually interesting as well as almost entirely open fields of inquiry for the future.

Macroevolution above the Species Level

The use of the term *macroevolution* has often been characterized by widespread ambivalence. Sometimes it is used to mean evolution above the species level, and sometimes as a synonym for speciation. Some researchers consider macroevolution the evolution of higher-level taxa (i.e., above the species or genus), and it is also used to indicate evolution of phenotypic novelties. Here, I make a distinction between evolution immediately above the species level, which includes, but is by no means limited to, speciation, and evolution at higher taxonomic levels, which is characterized by the appearance of many of the most puzzling phenotypic novelties and certainly by the origin of body plans (Hall 1992a; Lee 1993; Raff et al. 1994; Zrsavy and Stys 1997). The first class of phenomena is the subject of this section, and the following section deals with the second class.

Although I treated the evolution of phenotypic plasticity in Chapter 9, let me approach the topic again within the model system defined in this chapter, for ease of continuity with the lower levels of analysis discussed so far. The few investigations carried out on the evolution of shade avoidance in *A. thaliana* (Pigliucci et al. 1999; Pollard et al. 2001) show quite clearly that this syndrome can evolve very rapidly. For example, when we plotted the plasticities of leaf number against a simplified phylogeny of several *A. thaliana*'s ecotypes and some of its relatives, we found that there is a strong response in the distantly related *A. griffithiana–A. pumila* group, but that a closer relative, *A. petraea,* has lost most of this plasticity (Fig. 11.3). Furthermore, plasticity of leaves to shade has been recovered by the Moscow ecotype of *A. thaliana.* The behavior of other plasticities followed similar patterns in the case of most traits associated with shade avoidance, but completely different patterns for traits independent of shade responses. We need more detailed data sets of this kind to acquire a general picture of when and under what circumstances plasticity evolves and at what pace, and these will soon be available through more complete and robust phylogenetic hypotheses based on molecular markers (Koch et al. 2000).

This information can also shed light on the relative roles of genetic/phylogenetic constraints and ecological selective pressures. For example,

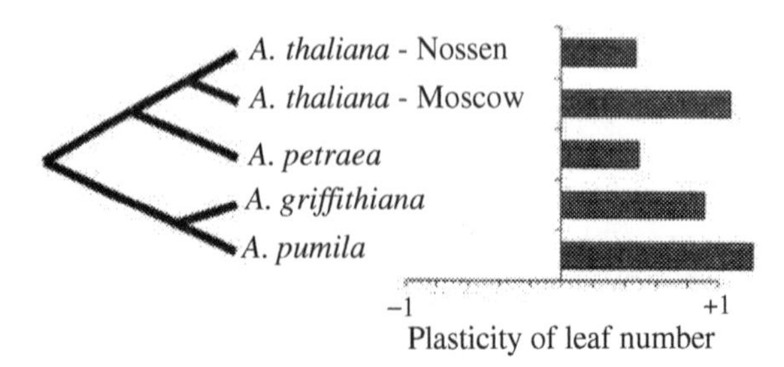

FIG. 11.3. Evolution of phenotypic plasticity of leaf production in response to shade (bars) mapped onto a partial phylogenetic tree of *Arabidopsis thaliana* and close relatives. The plasticity is calculated as the difference between trait mean under higher light and under low red:far red ratios (simulated shade). Note that all plasticities retain the same pattern (the bars fall on the same side of the diagram). (From Pigliucci et al. 1999.)

the fact that many plasticities of traits composing the shade avoidance syndrome tend to show similar phylogenetic patterns can be interpreted in two ways. First, it could be the reflection of strong genetic constraints imposed by the fact that shade avoidance is triggered by one or a few high-level regulatory genes such as the phytochromes. On the other hand, it may well be that there is significant evolutionary flexibility in the system because of many potentially diverging steps in the transduction pathway leading from the receptor of the light stimulus to the observable plasticity. This flexibility may not be reflected in independent plasticities of the shade avoidance traits because of selection to maintain a unified, coherent phenotype under variable light conditions.

An answer to this sort of conundrum is possible only by reaching back into the two previously considered levels of analysis. Studies of genetic variation within populations will tell us if there is variation for the plasticity of shade avoidance traits, thereby falsifying the genetic constraint hypothesis. The alternative scenario predicts selection on plasticity for shade within populations, also an empirically testable hypothesis. Finally, research at the across-population level should highlight genetic variation for shade avoidance plasticity across populations. Such variation would be expected to sort out nonrandomly in the case of the ecological hypothesis (e.g., with early-flowering ecotypes distinct from late-flowering ones), or randomly (following patterns of overall genetic similarity) if the genetic constraint hypothesis is correct.

Macroevolution at Large Scales: Where Does Plasticity Come From?

One can argue that the deep mysteries of evolutionary biology lie outside of what most evolutionary biologists actually do in their daily research. In fact, it was precisely a frank exploration of such a view that got one of the foremost geneticists of the first part of the century, Richard Goldschmidt (1940), into trouble with the mainstream evolutionary community. Nevertheless, it is quite clear that we know a lot more about every nuance of the developmental biology and molecular genetics of fruit flies than we do about how the insects came into existence to begin with (but see Carroll et al. 1995). This same problem applies to the study of phenotypic plasticity. We have a wealth of information on genetic variation for plasticity in natural populations, and we even think we can predict the future evolution of plasticity thousands of generations into the future by using quantitative genetic models. However, when it comes to the question of how any form of plasticity evolved to begin with, we are forced into reverent silence.

Hopefully, things will improve thanks to the combination of two very distinct approaches: on the one hand, the advances of molecular genetics, especially the ability to probe for unknown DNA sequences homologous to known ones; on the other hand, the development of the evolutionary comparative method, in turn dependent on cladistic methodology as well as on molecular techniques. If we need a reason for why it is fruitful to combine mechanistic and organismal biology, I would be hard pressed to provide a better one.

For a practical example concerning our model system, let me illustrate the advances that have been made in our understanding of where shade avoidance came from, and how it became what we observe today. (Good summaries of this story can be found in Clack et al. 1994; Mathews et al. 1995; Quail 1997a,b; Wu and Lagarias 1997.) We currently know of the existence of five phytochromes in *Arabidopsis* (A, B, C, D, and E). The first two, A and B, were discovered by mutagenesis. Probing the *Arabidopsis* genome for similar sequences identified the rest. We are just now beginning to understand what the other three phytochromes are doing: Phytochrome-D may be partially redundant in the shade avoidance response (Aukerman et al. 1997; Devlin et al. 1999), phy-E influences stem elongation and flowering time (Devlin et al. 1998), and phy-C controls vegetative development and flowering time in concert with phy-B (Halliday et al. 1997; Qin et al. 1997).

Researchers soon started looking for phytochrome-like molecules in other organisms, and found them everywhere, including in green algae (Wu and Lagarias 1997). This latter finding gives us a very important glimpse into the remote origins of shade avoidance plasticity. The molecules that make it possible have an ancient history and were possibly originally not even used as photoreceptors in the sense applied by physiologists to modern plants. Moreover, since we know that phytochromes belong to a single gene family, we can obviously derive the conclusion that their functions evolved slowly over a long period of time by repeated gene duplication.

A related and important question can be asked once a gene phylogeny is available: Do similarities in gene sequences foreshadow similarities in the biochemical functionality of the resulting molecules? The answer is rather complex. Phytochromes A and B, which play partially overlapping and partially opposing roles during the development of *A. thaliana* (McCormac et al. 1993; Reed et al. 1994; Shinomura et al. 1994; Yanovsky et al. 1995), are in fact very distinct from each other at the sequence level, defining the two main branches of the phytochrome gene phylogeny (Mathews and Sharrock 1997, Fig. 11.4). On the other hand, the functional redundancy of phy-B and phy-D referred to above is in agreement with their recent common origin. Then again, phy-C works similarly to phy-B (see the references listed previously), yet it clusters phylogenetically with phy-A. It seems that any hope of neatly matching history and functions of gene sequences is unlikely to be realized, which of course is no surprise to evolu-

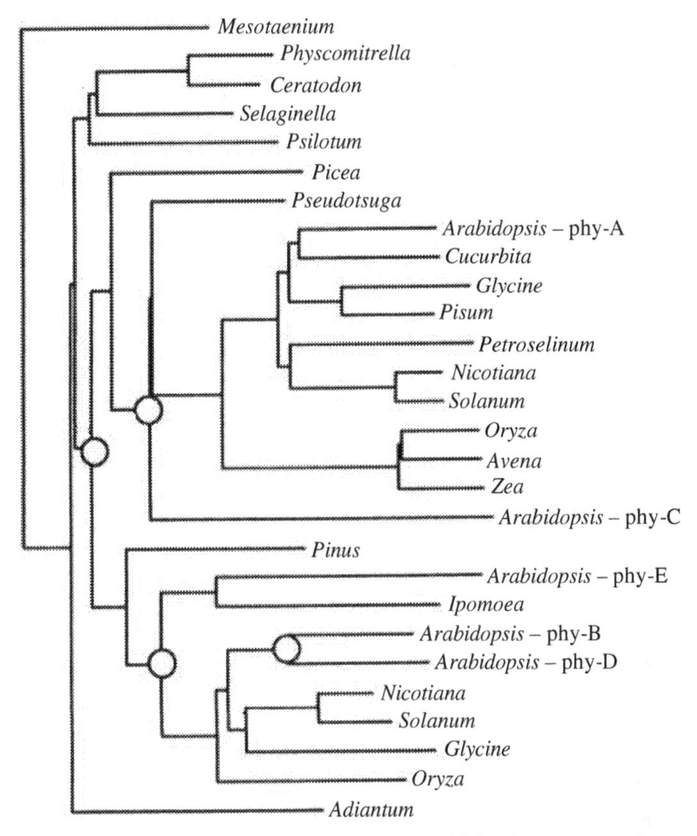

FIG. 11.4. A gene phylogeny of phytochromes across widely divergent taxa, from green algae such as *Mesotaenium* to ferns and fern allies and into gymnosperms, monocots, and dicots. Note the major split between two subfamilies: phy-A and phy-C on one side, and phy-B, phy-D, and phy-E on the other; phy-A and phy-B have partially overlapping and partially opposing functions throughout the development of *Arabidopsis thaliana.* (From Mathews and Sharrock 1997.)

tionary biologists who appreciate the complexity of evolution at various levels between the genotype and the external phenotype. This intricate pattern may be the result of the fact that regulatory gene evolution comes about more by altering the short sequences that determine when and where the gene is expressed than by gross alterations of the gene's main sequence (Ingram et al. 1995; Ganfornina and Sanchez 1999). A much more detailed gene phylogeny of plasticity genes is necessary to further these inquiries. Also, and most important, careful studies of the ecology of species displaying significant differences in the number or regulation of these genes are required to understand the deep macroevolution of phenotypic plasticity. Given the current state of the field, it will be some time before we will be able to put together a coherent scenario, but we at least have a map of the road ahead.

We are getting close to the end of this book, and I devote the epilogue to a field of inquiry that strikes close to home, in that it concerns the genetic and environmental components of human traits. If the study of phenotypic plasticity is important for many aspects of evolutionary biology, its most general meaning is to be found in the fact that it also helps to clarify one of the oldest and most controversial questions vexing human science, philosophy, and ethics: Are phenotypes determined primarily by nature or by nurture? While the answer should be clear from what we have been discussing throughout this book, the remaining pages will delve into some of the details that are most relevant for human biology, including the infamous question of heritability of cognitive traits.

Conceptual Summary

- Phenotypic plasticity is central to the study of many aspects of evolutionary biology for the simple reason that organisms develop in specific environments and that these environments are often labile over short time and spatial scales.
- Examples of areas in which genotype-environment interactions can play a pivotal role in future research include the study of adaptogenesis, the problem of the maintenance of genetic variation in natural populations, quantitative genetics, modeling of evolutionary trajectories, the study of character correlations and constraints, the evolution of gene regulation, comparative phylogenetic research on adaptive evolution, and the study of macroevolution.
- Recent research on the model system *Arabidopsis thaliana* provides examples of how questions in evolutionary biology can be pursued at four levels: microevolution within populations, microevolution among populations, macroevolution above the species level, and evolution of phenotypic novelties.
- The examples presented can be used as a roadmap to integrative inquiries spanning the whole range of methods and questions in modern evolutionary biology.

— Epilogue —
Beyond Nature and Nurture

Men have an extraordinarily erroneous opinion of their position in nature; and the error is ineradicable.

—W. Somerset Maugham (1874–1966)

Now, at the end of a technical book on genotype-environment interactions, is the time to pause and consider the broader implications of research on phenotypic plasticity. What originally motivated me to tackle this field was the fascinating relationships it has with the age-old philosophical question of nature versus nurture. I subsequently followed with passion the debates on the inheritance of human cognitive traits, with their obvious implications for social policy. Long diatribes, political agendas, and much misinformation have obscured the underlying philosophical and applied aspects of these questions. As a scientist-citizen, therefore, I would like to conclude this book on a less technical and more general note by discussing what the research on phenotypic plasticity that we have examined in the last eleven chapters can tell us about the nature-nurture question.

The debate on the relative importance of genetics and environment in determining human traits has been prolonged and acrimonious. Many great minds have engaged in it during the last three hundred years, from philosophers John Locke and Thomas Hobbes, to biologists Stephen J. Gould, Richard Lewontin, and Edward O. Wilson, to social scientists Arthur Jensen, R. J. Herrnstein, and C. Murray, to name but a few. A great number of books and an even greater number of book reviews, articles, and media appearances have been devoted to the problem. Gould's *The Mismeasure of Man* (1996) echoed Lewontin and colleagues' *Not in Our Genes* (1984), while Wilson had previously published *On Human Nature* (1978), and Herrnstein and Murray subsequently wrote *The Bell Curve* (1994). They all provide interesting philosophical speculations, but, in one way or another, hard empirical evidence is sorely lacking in all of them.

This claim may seem harsh; yet, readers who have followed the discussion of phenotypic plasticity that makes up this book will realize that we must draw just such a conclusion from what we know (and do not know) of plasticity. I do not insist that we cannot (in principle) know anything about reaction norms in humans or that we do not know anything about genetic or environmental effects on human characteristics. However, I do submit that we currently know little about the biological basis of human nature that is pertinent to the very public and often rancorous contemporary dialogue. I especially contend that we know very little that can sensibly inform our social policies. The more important question the reader should keep in mind is: Given our current methods of investigation, *can* the debate move forward? As Richard Lewontin (1998) wrote rather uncompromisingly about the study of the evolution of human cognition: "I must say that the best lesson our readers can learn is to give up the childish notion that everything that is interesting about nature can be understood. . . . It might be interesting to know how cognition (whatever that is) arose and spread and changed, but we cannot know. Tough luck."

Nature versus Nurture

Let us start by asking if there is really still a debate between opposing camps. While advocates on both sides of the nature-nurture debate in humans agree in theory that the solution to the enigma falls somewhere in the middle between the two extremes, it is important to ask what exactly does it *mean* to be in the middle in this case. Are we in the middle if we say that 50% of intelligence is determined by genes and 50% by the environment? What about 70% and 30%? Or is it more reasonable to propose that intelligence (or any other complex human trait) is the result of a vague and unspecified *interaction* (or an *emergent property*) between nature and nurture?

The position of Gould, Lewontin, and others is that the environment is the major determinant of human nature. If the cause of intelligence, aggression, or any other aspect of human behavior is not in our genes (as is explicitly affirmed in the 1984 book by Lewontin et al.), it must surely be found in the environment. How can biologists claim that genes have little to do with the human condition? Proponents of the modern nurture school do acknowledge that genes influence human behavior, but they add that whatever such influence may be, it can be overridden by the comparatively much greater effect of environmental conditions, chiefly education and socioeconomic status. For example, Gould (in the revised edition of *The Mismeasure of Man,* p. 355) says that "the *biological* basis of human uniqueness leads us to reject biological determinism" (his emphasis). He is implying here that what we know of the biology of humans argues against the possibility that genes play a major role in shaping human characteristics. The problem is that we do not actually know much about the biological basis of human nature, and such sweeping generalizations are simply unwarranted.

On the other side of the divide, Jensen, Herrnstein, Murray, Wilson (to some extent), and to a lesser degree evolutionary psychologists such as Pinker (1997) are convinced that genetics and natural selection are the major factors shaping the physical as well as mental characteristics of all living beings, including humans. When Murray (1998) suggests (in the title of one of his articles) that "IQ will put you in your place," he is implying that IQ is inscribed in stone in the DNA of each one of us. He, like the nurturists, also acknowledges the opposing camp—by saying that an optimal environment may have some effect. But in his thinking the role of the genes is so powerful that he and Jensen (but not Wilson or Pinker) go on to suggest that governments should not invest time and money to improve education and economic conditions because that will not alter one's place in society as determined by IQ. The bright people are going to stay that way regardless of the conditions, and the others will lag behind regardless of what society does.

It will be instructive to consider two of the major arguments adduced by each school in some detail, to show how apparently clear-cut conclusions tell us more about the limitations of the scientific method than about the question of genotype-environment interactions in humans. One of Gould's major attacks against genetic determinism is based on what he calls the reification fallacy. Simply put, the fact that a statistical correlation exists between two or more variables does not imply that that correlation corresponds to a physical entity or demonstrates the existence of a real phenomenon underlying the correlations (a point originally made by philosopher David Hume in 1748). Genetic determinists interpret the observation that the scores from different types of intelligence tests correlate in multivariate space as an indication of the existence of an underlying factor summarizing general intelligence. Gould objects that this is simply a statistical artifact of the different tests being designed in a similar way, so that of course individuals who score high on one test will score high on another, while individuals who do not do as well on one test will also fail another one. This may be true, but the same reasoning should then apply to more mundane and rather uncontroversial scientific research. For example, Gould (1984, 1989) described the statistical covariation among morphological traits measured in land snails of the genus *Cerion*. In that body of work, he argues that these correlations show the *existence* of several constraints, that is, limitations on the future evolutionary trajectories of these populations of snails, and he goes on to *name* such constraints. This is also a case of reification, and, as scientists, we should agree that it is a valuable, albeit limited, tool that can be used if its shortcomings are well understood. The catch is, we do not get to pick and choose when to use it and when not.

In the opposing camp, naturists have always argued that the most incontrovertible evidence for a strong genetic influence on human cognitive traits comes from rigorous studies of identical twins reared together or apart. The statistical arguments are quite complex, but they boil down to the fact that identical twins have a very high degree of genetic similarity, compared, for example, to fraternal siblings (or twins), or to unrelated individuals—such as adopted children (even identical twins are not necessarily genetically identical, given the possibility of somatic mutations). If identical twins show a higher correlation of, say, their IQ scores than the control groups, this is taken as *prima facie* evidence for genetic determination of intelligence. However, there are some serious problems with this conclusion. First, this approach sidesteps the fact that the real problem is how to determine human reaction norms, not to simply estimate heritabilities. We saw in earlier chapters that even if heritabilities are high in a particular environment, this does not tell us anything about plasticity or genotype-environment interactions (Chapter 1). Second, the identical-twin argument underestimates the fact that the environment in which these or other siblings are raised can play a major role. If identical twins are reared together, their environments are also going to be very similar. In statistical terms, this results in a complete confounding effect of genetics and environment. Furthermore, even separating the identical twins at birth does not guarantee that their environments will be dissimilar, thereby allowing a statistical decomposition of nature and nurture effects. Indeed, most adoption agencies take great care in assigning children to families of social, economic, and cultural status similar to the ones that characterize the biological family (Levins and Lewontin 1985; Lewontin 1992). Attempts at counteracting this difficulty by using some measures of the environment as statistical covariates in the analyses are technically sound, but clearly insufficient. It is hard to believe that just measuring, say, income (Murray 1998) is enough to statistically remove environmental differences when one is studying complex cognitive traits. Identical-twin studies are not very useful for the simple reason that they control for only one of the two factors—the genetics but not the environment. Thus, while they certainly yield evidence of a genetic contribution to human behavior, they do not provide any information on the all-important genotype-environment interactions.

What Do We Actually Know about the Biological Basis of Human Characteristics?

The discussion so far does not have to be taken as a nihilistic statement on the state of our ignorance of genetic and environmental effects on human characteristics. We do indeed know a lot, but we do not know as much as is necessary to claim a good understanding of the problem or to insist that science is now in a position to positively inform social policy.

There is little doubt that genes affect brain development. Since the brain and the peripheral nervous system determine human behavior, it is undeniable that we behave in part because of our genes. A large and fascinating literature on brain damage provides direct and incontrovertible evidence that alterations in the physical structure of the brain (which can be brought about by accident or by mutations) directly affect all sorts of human behaviors, including subtle personality traits (Ramachandran 1996; Gazzaniga 2000).

A recent example of a scientifically sound investigation of the biological basis of complex human behaviors is a study of the peculiarities of Einstein's brain compared to appropriate controls from the general population (Witelson et al. 1999, Fig. E.1). As is well known, Einstein was gifted with an unusual mathematical ability; it is also well known that increased

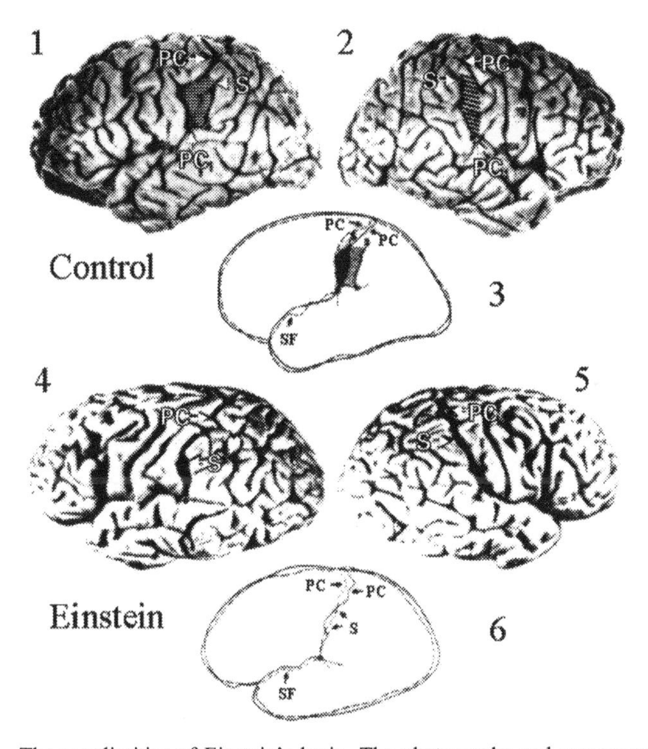

FIG. E.1. The peculiarities of Einstein's brain. The photographs and accompanying diagrams show lateral sections of a typical human male brain (1–3) and Einstein's brain (4–6). Note the relatively anterior position of the sylvian fissure (SF) bifurcation and the associated greater posterior parietal expanse in Einstein's brain. PC = posterior commissure. (From Witelson et al. 1999.)

mathematical ability is associated with an expansion of the inferior parietal region of the brain. Witelson and colleagues found that Einstein's brain was about 15% larger than the controls precisely in that region and therefore concluded that they may have pinpointed at least part of the biological basis of the great physicist's rare ability. This, of course, is not necessarily to say that Einstein's mathematical genius was written in his genes: A change in brain structure can be brought about by mutations or by environmental influences (or both).

One of the best examples of why both nurturist and genetic determinist arguments are partially right, yet cannot resolve the most important questions, is given by the environmentally curable genetic disease known as phenylketonuria (PKU). PKU is caused by a simple *inborn metabolic error,* that is, a mutation that does not allow the formation or proper functioning of an enzyme that metabolizes the amino acid phenylalanine. Often (though not always)[1] this results in an accumulation of the amino acid in the brain during development, which in turn causes a host of phenotypic and behavioral effects, including severe mental retardation (see Kaplan 2000 for an in-depth discussion and references). The link between the genetic level (the mutation and consequent enzymatic defect) and the phenotype/behavior is apparently straightforward, and therefore offers an ideal example of how genes can affect behaviors, although no genetic determinist would go so far as to say that this is a gene *for* normal intellectual development (on the concept of *genes for,* see Kaplan and Pigliucci 2001). On the other hand, a very simple dietary change (i.e., a change in behavior) can completely neutralize the genetic effect: An individual can grow up normally by carefully avoiding phenylalanine, which is why many soda cans carry a warning label for PKU patients. This is the most extreme and one of the few clearly documented instances of a genetic basis of plasticity that one can find in humans. PKU is *both* genetically determined *and* environmentally plastic, but the crucial question remains: What are the reaction norms of different human genotypes for PKU? There surely are several possible phenotypes, not only because the mutation can be caused by different alleles at the main locus, but also because the phenotypic expression of the mutation depends on the other genes that interact with that locus. Furthermore, there may be plenty of other environmental circumstances that affect the degree of occurrence and gravity of PKU symptoms. We simply do not know and—given the obvious limitations on experimental research in human genetics—we will not know any time soon.

Summarizing, we can—and often do—know that the expression of a trait is affected by changes in a person's genetic makeup. Similarly, we know that a given characteristic or behavior can be altered by changes in the environment. What we do not know is whether and how these two effects interact with each other. Unfortunately, this is the most crucial aspect of the nature-nurture controversy in humans.

What Are We to Do? Science versus Policy

What kind of evidence would settle the nature-nurture debate on the biological basis of human behavior? The answer is in fact very simple and involves improvements in experimental design. Such experiments were carried out, for example, by Cooper and Zubek (1958) on rats, a long time ago. These authors compared "intelligence," as measured by the ability to avoid mistakes in running through a maze, in two inbred, genetically distinct lines of rats. One line had been selected for high performance in the maze ("bright" line), the other for particularly low performance ("dull" line). When reared under a standard environment, comparable to the one in which the lines were originally selected, the two strains showed a highly significant difference in their maze-running abilities. Cooper and Zubek, however, also reared individuals of the two lines in two other environments: one in which the cage was entirely devoid of visual and tactile stimuli ("poor" environment) and one in which the developing animals were exposed to brightly colored walls and toys ("enriched" environment). The results were simply stunning (Fig. E.2). Under the poor conditions, the bright rats performed as badly as the dull ones; furthermore, under the enriched environment the dull rats did as well as the bright ones! This translates into a high heritability and marked genetic differences in the standard environment, but no heritability under either extreme environment. The inescapable conclusion is that maze-running ability in rats has a very plastic reaction norm and that different genotypes converge onto similar phenotypes under extreme environmental conditions. I am not suggesting here that this particular experiment is flawless. For one thing, only two genotypes were tested, and they were certainly not a representative sample of natural populations. Furthermore, the cognitive ability being tested was a relatively specific one, and its correlation to generalized intelligence (if there is such thing) remains to be ascertained. Finally, one could easily conceive of other (and more naturalistic) kinds of environments that might yield very different results. However, these are methodological problems that can easily be addressed in the lab and are much less daunting than those associated with direct research on human subjects. The point I want to make is that these are the kinds of data that would go a long way toward providing an empirical answer to questions about heritability and plasticity of human behaviors. Even though extrapolations from rats to humans are obviously dangerous (especially in the case of cognitive abilities), an improved

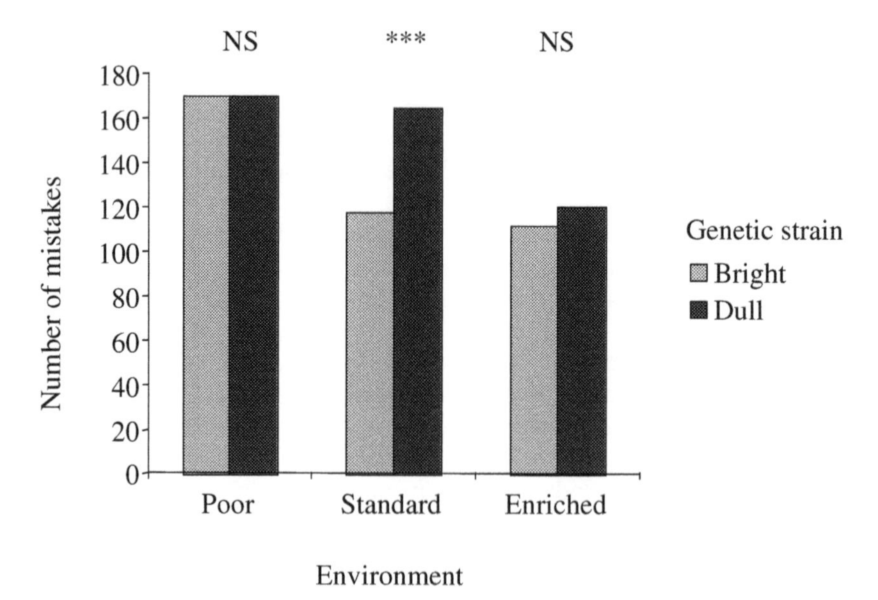

FIG. E.2. Effect of nurture (quality and stimulation properties of the environment) on "intelligence" (ability to avoid mistakes in mazes) in genetically distinct laboratory lines of rats. Note how the genetic differences are significant (i.e., there is a significant heritability) only in the environment in which the lines were selected, but they disappear in the other two environments. (After data from Cooper and Zubek 1958.)

version of this experiment could perhaps be carried out on animals phylogenetically closer to our lineage, for example, some lower primates. Even similar experiments in other vertebrates would go much further toward a partial resolution of the problem than any additional speculation.

So how should scientists answer pressing requests from the public and policy makers that relate to our knowledge of intelligence and other behavioral traits in humans? I believe some answers and suggestions can be given, but only if accompanied by careful statements about the serious limitations imposed by the impossibility of carrying out the proper experimental manipulations. While good science can be done by using only indirect and statistical approaches, we have to remember that, as yet, the only known way to dissect the functionality of complex systems is by direct manipulation.

The first and perhaps most important thing we can safely conclude from simple observation is that human behavioral traits tend to be plastic. Therefore, regardless of any ideologically motivated agenda, education programs that improve the learning environment of an individual do have a good chance of succeeding. In fact, even if human individuals are genetically distinct in their abilities and even if an improvement in environmental conditions will maintain such differences (a very unlikely scenario by comparison with known reaction norms in other organisms), it would still be worth funding education.

The second fact we can state with confidence is that most human traits (including behavioral ones) do have a genetic basis. It simply cannot be otherwise, given that the structure of the brain is dictated in part by genes and considering all we know about the effects of brain anomalies on behavior. Genetic differences between genders and among races and individuals may indeed exist, regardless of how politically incorrect such an admission might be. This acknowledgment, however, does not justify any discrimination or abuse. Variation among genotypes is a fact of nature, completely impervious to our ethical and moral standards. Striving to achieve social equality is a matter of attitudes and policies, both of which can be changed by individuals and governments. To justify draconian policies based on the fact that they allegedly reflect the "natural" state of affairs is to commit what in philosophy is known as the naturalistic fallacy: What *is* does not prescribe what *ought* to be (Hume 1739–40).

What scientists cannot and should not venture to say, however, is that we understand the interactions of genes and environments in humans to a point that that knowledge can safely be used to inform social policies. This is indeed a crucial question because specific educational approaches, for example, may be more or less fruitful depending on the precise shape of human reaction norms. The same can be said for policies concerned with curbing crime or for a host of other fundamental and difficult societal decisions. It should be eminently clear by now that this is where the line must be drawn, and the only honest answer a scientist can give is: *I do not know.* It is astounding to see how difficult it can be to utter these simple words. The reasons to yield to the temptation of saying more than one's data and theories allow are, of course, easy to understand. Sociologists, psychologists, and philosophers of science have long pointed out that personal egos, social prestige, and financial rewards all play into that temptation in a remarkably complex ensemble (Kuhn 1970). There are—and always will be—some questions that science cannot answer. Learning to live with this conclusion actually empowers scientists because, by being humble, they can pursue the many positive aspects of the single most effective tool humans have ever devised for understanding the world: the scientific method. It also is the only decent thing to do.

Notes

2 – Studying and Understanding Plasticity

1. This is also the reason for the persistent unpopularity of studies on maternal effects: to actually estimate maternal effects—especially in multiple environments—requires even larger sample sizes. So, maternal effects, as plasticity for most of the twentieth century, are usually "assumed" to be unimportant, even though we do not know *how* important they actually are (see Roach and Wulff 1987; Schmitt et al. 1992; Galloway 1995; Schmitt 1995; Tollrian 1995; Lacey 1996; Mazer and Delasalle 1996; Rossiter 1996; Watson and Hoffman 1996; Fox et al. 1997; Fox and Savalli 1998). For the specific case of parental effects in clonally propagated material see Schwaegerle et al. (2000).
2. The term *phenocopy* was introduced by Goldschmidt (1940) to indicate environmentally induced phenotypes that look like known mutations. Goldschmidt's idea was that phenocopies give us a clue about the functional genetics underlying a particular trait. The manipulation experiments described here adopt exactly this logic and extend it to tests of costs and adaptation.

3 – A Brief Conceptual History

1. A similar line of reasoning was followed simultaneously by both Goldschmidt (1940) and Waddington (1942, 1960). See Schlichting and Pigliucci (1998) for a more detailed account.
2. For example, see G. G. Simpson's retreat from his idea of *quantum evolution* in the fossil record, resurrected decades later by Gould and Eldredge's *punctuated equilibria* (Eldredge and Gould 1972).
3. Although important experimental work was done by C. H. Waddington in evolutionary biology and by a series of researchers, such as D. S. Falconer, focusing on applied biology.

4 – The Genetics of Phenotypic Plasticity

1. The concept of overdominance dates back to the original work of Lerner (1954) and of Waddington (1961).

2. The problems with interpreting genetic covariances and correlations from a mechanistic standpoint discussed in Chapter 1 obviously pose a dilemma if we want to keep studying constraints in quantitative genetic terms. However, I am not aware of a published solution to this problem, so for now I will continue to use quantitative genetic parameters and interpret them in the classical sense (Cheverud 1984, 1988), with the caveat that we now know better and are awaiting a breakthrough that addresses the valid criticisms of Houle (1991) and Gromko (1995).
3. I realize that the distinction between micro- and macroenvironment is not necessarily clear-cut either. However, we can study the grain of environmental variation *as perceived by the organism* and determine the response to and fitness consequences of either level of variation. In this sense, I think the literature on plasticity shows a distinct biological phenomenon from the literature on developmental noise, as reflected by the fact that plasticity proper is never measured as asymmetry in biological structures.
4. Fisher's expectation that there should not be much genetic variation around was a direct consequence of his fundamental theorem of natural selection. If there is genetic variation for traits related to fitness, it means that selection can improve the fitness of the population. By so doing, it would eliminate the genetic variation. Since—in Fisher's view—most genes have effects on fitness, we should not observe much residual genetic variation in nature. See Hartl and Clark (1989) for a more comprehensive explanation, as well as for a discussion of the many cases in which the "fundamental" theorem does not apply.

5 – The Molecular Biology of Phenotypic Plasticity

1. That genes specifically affecting plasticity exist was implied by Schmalhausen (1949) and by Bradshaw (1965) and more explicitly stated by Scheiner and Lyman (1989).
2. For example, we still by and large think of genetic effects as being "modified" by epigenetic interactions or environmental conditions, not of the three players as being on an equal footing. See Lewontin (2000) for a frank discussion of this issue.

6 – The Developmental Biology of Phenotypic Plasticity

1. The classical thinking about ultimate causes is that they refer to teleological explanations, such as the intentions and actions of a designer. But that is a restrictive anthropomorphic view of ultimate causes: The increase in average fitness of a population owing to natural selection can be considered an example of a nonteleological ultimate cause driving adaptive biological evolution.
2. The plastochron is a measure of development time originally devised by Erickson and Michelini (1957). It is defined as the time interval between the initiation of two successive structures or, more broadly, as the interval between corresponding stages of development of successive structures (initiation, maturity, or any intermediate stage of development can be chosen as reference).
3. Note that what looks like a true limit to plasticity—from a genetic standpoint—was demonstrated by Scheiner and Lyman (1991) when they could not further decrease plasticity of thorax size in *D. melanogaster* in response to temperature. They attributed this failure (all the more intriguing since there was still significant variation for plasticity in the selected lines) to epistatic effects.

7 – The Ecology of Phenotypic Plasticity

1. For some time now people have even questioned the very concept of niche, but I still think it has definite value, though this is not the place to enter into that controversy.
2. Note that this is one example of a directly identifiable link between mechanistic and organismal biology, which tempers the position promulgated by G. Bell and that I partially endorsed at the end of Chapter 5.
3. I wonder if this consideration illuminates somewhat the dark debate between naturists and nurturists about the biological bases of human intelligence and behavior (Bouchard et al. 1990; Feldman and Lewontin 1990). See the epilogue for a possible answer.

8 – Behavior and Phenotypic Plasticity

1. One could argue that I should go back to research on behavior itself, including the ground-breaking works of I. Pavlov and B. F. Skinner, but I am concerned here with individuals who have expressly addressed the similarities between plasticity and behavior.
2. Note that this author refers to costs in general, including reduced growth and reproduction, errors in developing the appropriate response, costs of maintenance of the developmental machinery that underlies the plastic response, and information-acquisition costs (see Chapter 7).
3. Note that if a trait is highly evolutionarily labile its use for phylogenetic reconstruction will likely overestimate the degree of convergence among species, because the same state for the character will evolve several times independently but it will be interpreted by the phylogenetic algorithm as an indication of common descent.

9 – Evolution of and by Phenotypic Plasticity

1. Adaptive coin flipping falls into a general class of models addressing the relevance of variance in a phenotypic trait. The joint effect of means and variances in the evolution of life histories is discussed in Lacey et al. (1983). The relationship between components of phenotypic variance (genetic, environmental) is covered in Bull (1987). The importance of mean-variance models for evolution in stochastic environments is elaborated upon in Real and Ellner (1992).

10 – The Theoretical Biology of Phenotypic Plasticity

1. With the exception of the two papers by Gavrilets (1986, 1988) cited at the beginning of the chapter. Also, within the different context of developmental noise, Gavrilets and Hastings (1994) presented a quantitative genetic framework for analyzing *short-term* effects of stabilizing selection on **G**-matrices.
2. Which is what Via (1987) had found, and what emerges from a graphical comparison of reaction norms and interenvironment genetic correlations such as the one presented in Chapter 1. Those results are still valid because a purely statistical or graphical approach does not tell us anything about the underlying genetic machinery. What de Jong (and Pigliucci 1996b) demonstrated is that knowledge of the

mechanistic details does provide alternative scenarios that cannot be derived through coarser approaches.

3. Gimelfarb (1994) reached a similar conclusion while exploring nonlinear effects in a quantitative genetic model. His results show that the presence of genotype-environment interactions can cause significant nonlinearity in the parent-offspring regression, which in turn can translate into a reverse response to selection. Another odd consequence of this model is that directional selection can actually *increase* heritability.
4. As we saw in Chapter 1, this does not mean that the trade-offs are not there, but simply that the interaction between the genetic architecture of an organism and the environment is more complex than is assumed in most available models of life history evolution (Houle 1991).

11 – Phenotypic Plasticity as a Central Concept in Evolutionary Biology

1. As pointed out earlier, the term *within-organism plasticity* refers to the situation in which different parts of a single individual encounter distinct environments and react accordingly. Typical examples of this kind of plasticity are heterophylly, the production of below- and above-water leaves in semiaquatic plants, and the differentiation between shade and sun leaves on a single tree.
2. The most intuitive scenario leading to plasticity would be directional selection of opposite sign in the two environments. However, if the phenotypic optima in different environments (years) are distinct, it will still lead to the evolution of plasticity, even though the genotypes are currently being pushed in the same general direction in both environments. Note that this scenario would require stabilizing selection around the two optima, which is a likely possibility given that the plant simply cannot keep increasing its leaf production indefinitely.
3. Note that this discussion is couched in terms of the deterministic forces of selection and constraints only. Obviously, mutation, migration, recombination, and random drift will still affect whatever scenario plasticity studies may be able to elucidate.

Epilogue: Beyond Nature and Nurture

1. I am in debt to my philosopher friend Jonathan Kaplan for pointing out to me that the literature on PKU is far more complex than the simple story found in textbooks. One of the reasons there is more variation in the phenotypic expression of the disease than is commonly acknowledged may be that the effects of the mutation depend on the genetic background in which it occurs. Having the same gene for PKU can result in different levels of phenylalanine in the blood, and having the same level of phenylalanine in the blood can result in different levels of physical problems, depending on the individual. In other words, some combinations of other genes interacting with the focal one may greatly reduce, or entirely eliminate, the deleterious effects.

— References —

Abouheif, E. (1998) Random trees and the comparative method: a cautionary tale. *Evolution* 52:1197–1204.

Ackerly, D. D. (2000a) Taxon sampling, correlated evolution, and independent contrasts. *Evolution* 54:1480–1492.

Ackerly, D. D. (2000b) Comparative plant ecology and the role of phylogenetic information, pp. 391–413 in: *Plant Physiological Ecology,* M. C. Press et al. (eds.). Oxford: Blackwell.

Ackerly, D. D., and M. J. Donoghue. (1998) Leaf size, sapling allometry, and Corner's rules: phylogeny and correlated evolution in maples (*Acer*). *American Naturalist* 152:767–791.

Adams, M. W. (1967) Basis of yield component compensation in crop plants with special reference to the field bean, *Phaseolus vulgaris. Crop Science* 7:505–510.

Adams, S. R., Pearson, S., Hadley, P., and Patefield, W. M. (1999) The effects of temperature and light integral on the phases of photoperiod sensitivity in *Petuniaxhybrida. Annals of Botany* 83:263–269.

Agrawal, A. A., Laforsch, C., and Tollrian, R. (1999) Transgenerational induction of defences in animals and plants. *Nature* 401:60–62.

Agur, Z., and Kerszberg, M. (1987) The emergence of phenotypic novelties through progressive genetic change. *American Naturalist* 129:862–875.

Alberch, P. (1991) From genes to phenotypes: dynamical systems and evolvability. *Genetica* 84:5–11.

Alberch, P., Gould, S. J., Oster, G. F., and Wake, D. B. (1979) Size and shape in ontogeny and phylogeny. *Paleobiology* 5:296–317.

Alonso-Blanco, C., and Koornneef, M. (2000) Naturally occurring variation in *Arabidopsis:* an underexploited resource for plant genetics. *Trends in Plant Science* 5:22–29.

Alpha, C. G., Drake, D. R., and Goldstein, G. (1996) Morphological and physiological responses of *Scaevola sericea* (Goodeniaceae) seedlings to salt spray and substrate salinity. *American Journal of Botany* 83:86–92.

Ambros, V., and Moss, E. G. (1994) Heterochronic genes and the temporal control of *C. elegans* development. *Trends in Genetics* 10:123–127.

Andersson, S., and Shaw, R. G. (1994) Phenotypic plasticity in *Crepis tectorum* (Asteraceae): genetic correlations across light regimens. *Heredity* 72:113–125.

Andersson, S., and Widen, B. (1993) Reaction norm variation in a rare plant, *Senecio integrifolius* (Asteraceae). *Heredity* 73:598–607.

Antonovics, J. (1976) The nature of limits to natural selection. *Annals of the Missouri Botanical Gardens* 63:224–247.

Antonovics, J., and Tienderen, P. H. v. (1991) Ontoecogenophyloconstraints? The chaos of constraint terminology. *Trends in Ecology and Evolution* 6:166–168.

Armbruster, W. S., and Schwaegerle, K. E. (1996) Causes of covariation of phenotypic traits among populations. *Journal of Evolutionary Biology* 9:261–277.

Armbruster, W. S., Stilio, V. S. D., Tuxill, J. D., Flores, T. C., and Runk, J. L. V. (1999) Covariance and decoupling of floral and vegetative traits in nine Neotropical plants: a re-evaluation of Berg's correlation-pleiades concept. *American Journal of Botany* 86:39–55.

Arnold, S. J. (1992) Constraints on phenotypic evolution. *American Naturalist* 140: S85–S107.

Asins, M. J., Mestre, P., Garcia, J. E., Dicenta, F., and Carbonell, E. A. (1994) Genotype x environment interaction in QTL analysis of an intervarietal almond cross by means of genetic markers. *Theoretical and Applied Genetics* 89:358–364.

Atchley, W. R., and Hall, B. K. (1991) A model for development and evolution of complex morphological structures. *Biological Review* 66:101–157.

Aukerman, M. J., Hirschfeld, M., Wester, L., Weaver, M., Clack, T., Amasino, R. M., and Sharrock, R. A. (1997) A deletion in the PHYD gene of the *Arabidopsis* Wassilewskija ecotype defines a role for phytochrome D in red/far-red light sensing. *Plant Cell* 9:1317–1326.

Avery, L., and Wasserman, S. (1992) Ordering gene function: the interpretation of epistasis in regulatory hierarchy. *Trends in Genetics* 8:312–316.

Baatz, M., and Wagner, G. P. (1997) Adaptive inertia caused by hidden pleiotropic effects. *Theoretical Population Biology* 51:49–66.

Bagchi, S., and Iyama, S. (1983) Radiation induced developmental instability in *Arabidopsis thaliana. Theoretical and Applied Genetics* 65:85–92.

Baldwin, I. T. (1998) Jasmonate-induced responses are costly but benefit plants under attack in native populations. *Proceedings of the National Academy of Sciences of the USA* 95:8113–8118.

Baldwin, J. M. (1896) A new factor in evolution. *American Naturalist* 30:354–451.

Ballaré, C. L., and Scopel, A. L. (1997) Phytochrome signalling in plant canopies: testing its population-level implications with photoreceptor mutants of *Arabidopsis. Functional Ecology* 11:441–450.

Barton, N. H., and Turelli, M. (1989) Evolutionary quantitative genetics: how little do we know? *Annual Review of Genetics* 23:337–370.

Baskin, J. M., and Baskin, C. C. (1983) Seasonal changes in the germination responses of buried seeds of *Arabidopsis thaliana* and ecological interpretation. *Botanical Gazette* 144:540–543.

Bazzaz, F. A., and Sultan, S. E. (1987) Ecological variation and the maintenance of plant diversity, pp. 69–93 in: *Differentiation Patterns in Higher Plants*. London: Academic Press.

Beacham, T. D., Murray, C. B., and Barner, L. W. (1994) Influence of photoperiod on the timing of reproductive maturation in pink salmon (*Oncorhynchus gorbuscha*) and its application to genetic transfers between odd- and even-year spawning populations. *Canadian Journal of Zoology* 72:826–833.

Becher, H. (1993) Bootstrap hypothesis testing procedures. *Biometrics* 49:1268–1272.

Bell, D. L., and Sultan, S. E. (1999) Dynamic phenotypic plasticity for root growth in *Polygonum:* a comparative study. *American Journal of Botany* 86:807–819.

Bell, G. A. (1997a) *Selection: The Mechanism of Evolution.* New York: Chapman & Hall.

Bell, G. A. (1997b) Experimental evolution in *Chlamydomonas.* I. Short-term selection in uniform and diverse environments. *Heredity* 78:490–497.

Bell, G., and Lechowicz, M. J. (1991) The ecology and genetics of fitness in forest plants. I. Environmental heterogeneity measured by explant trials. *Journal of Ecology* 79:663–685.

Bennington, C. C., and McGraw, J. B. (1996) Environment-dependence of quantitative genetic parameters in *Impatiens pallida. Evolution* 50:1083–1097.

Berg, R. L. (1960) The ecological significance of correlation pleiades. *Evolution* 14:171–180.

Bergelson, J. (1994) The effects of genotype and the environment on costs of resistance in lettuce. *American Naturalist* 143:349–359.

Berrigan, D., and Koella, J. C. (1994) The evolution of reaction norms: simple models for age and size at maturity. *Journal of Evolutionary Biology* 7:549–566.

Bettencourt, B. R., Feder, M. E., and Cavicchi, S. (1999) Experimental evolution of Hsp70 expression and thermotolerance in *Drosophila melanogaster. Evolution* 53:484–492.

Bewley, J. D. (1997) Seed germination and dormancy. *Plant Cell* 9:1055–1066.

Black, S., Eriksson, G., Gustafsson, L., and Lundkvist, K. (1995) Ecological genetics of the rare species *Vicia pisiformis:* quantitative genetic variation and temperature response in biomass and fecundity. *Acta Oecologica* 16:261–275.

Blanckenhorn, W. U. (1991) Life-history differences in adjacent water strider populations: phenotypic plasticity or heritable responses to stream temperature? *Evolution* 45:1520–1524.

Blanckenhorn, W. U. (1998) Adaptive phenotypic plasticity in growth, development, and body size in the yellow dung fly. *Evolution* 52:1394–1407.

Blazquez, M. A., and Weigel, D. (2000) Integration of floral inductive signals in *Arabidopsis. Nature* 404:889–892.

Blouin, M. S. (1992) Comparing bivariate reaction norms among species: time and size at metamorphosis in three species of *Hyla* (Anura: Hylidae). *Oecologia* 90:288–293.

Blows, M. W., and Sokolowski, M. B. (1995) The expression of additive and nonadditive genetic variation under stress. *Genetics* 140:1149–1159.

Boake, C. R. B. (ed.) (1994) *Quantitative Genetic Studies of Behavioral Evolution.* Chicago: University of Chicago Press.

Bouchard, T. J. J., Lykken, D. T., McGue, M., Segal, N. L., and Tellegen, A. (1990) Sources of human psychological differences: the Minnesota study of twins reared apart. *Science* 250:223–228.

Bourguet, D., and Raymond, M. (1998) The molecular basis of dominance relationships: the case of some recent adaptive genes. *Journal of Evolutionary Biology* 11:103–122.

Bradshaw, A. D. (1965) Evolutionary significance of phenotypic plasticity in plants. *Advances in Genetics* 13:115–155.

Bradshaw, A. D., and Hardwick, K. (1989) Evolution and stress—genotypic and phenotypic components. *Biological Journal of the Linnean Society* 37:137–155.

Brakefield, P. M. (1997) Phenotypic plasticity and fluctuating asymmetry as responses to environmental stress in the butterfly *Bicyclus anynana,* pp. 65–78 in: *Environmental Stress, Adaptation, and Evolution,* R. Bijlsma and V. Loeschcke (eds.). Basel: Birkhauser Verlag.

Brakefield, P. M., and Breuker, C. J. (1996) The genetical basis of fluctuating asymmetry for developmentally integrated traits in a butterfly eyespot pattern. *Proceedings of the Royal Society of London* B263:1557–1563.

Brakefield, P. M., and Kesbeke, F. (1997) Genotype-environment interactions for insect growth in constant and fluctuating temperature regimes. *Proceedings of the Royal Society of London* B264:717–723.

Brakefield, P. M., Gates, J., Keys, D., Kesbeke, F., Wijngaarden, P. J., Monteiro, A., French, V., and Carroll, S. B. (1996) Development, plasticity and evolution of butterfly eyespot patterns. *Nature* 384:236–242.

Breto, M. P., Asins, M. J., and Carbonell, E. A. (1994) Salt tolerance in *Lycopersicon* species. III. Detection of quantitative trait loci by means of molecular markers. *Theoretical and Applied Genetics* 88:395–401.

Brodie, E. D., III, and Janzen, F. J. (1996) On the assignment of fitness values in statistical analyses of selection. *Evolution* 50:437–442.

Bronmark, C., and Miner, J. G. (1992) Predator-induced phenotypical change in body morphology in crucian carp. *Science* 258:1348–1350.

Brumpton, R. J., Boughey, H., and Jinks, J. L. (1977) Joint selection for both extremes of mean performance and of sensitivity to a macroenvironmental variable. *Heredity* 38:219–226.

Bubli, O. A., Imasheva, A. G., and Loeschcke, V. (1998) Selection for knockdown resistance to heat in *Drosophila melanogaster* at high and low larval densities. *Evolution* 52:619–625.

Bull, J. J. (1987) Evolution of phenotypic variance. *Evolution* 41:303–315.

Bult, A., Zee, E. A. v. d., Compaan, J. C., and Lynch, C. B. (1992) Differences in the number of arginine-vasopressin-immunoreactive neurons exist in the suprachiasmatic nuclei of house mice selected for differences in nest-building behavior. *Brain Research* 578:335–338.

Burghardt, G. M., and Krause, M. A. (1999) Plasticity of foraging behavior in garter snakes (*Thamnophis sirtalis*) reared on different diets. *Journal of Comparative Psychology* 113:277–285.

Burks, R. L., Jeppesen, E., and Lodge, D. M. (2000) Macrophyte and fish chemicals suppress *Daphnia* growth and alter life-history traits. *Oikos* 88:139–147.

Burt, A. (1995) The evolution of fitness. *Evolution* 49:1–8.

Cain, M. L., Pacala, S. W., Silander, J. A., Jr., and Fortin, M.-J. (1995) Neighborhood models of clonal growth in the white clover *Trifolium repens. American Naturalist* 145:888–917.

Cain, M. L., Subler, S., Evans, J. P., and Fortin, M.-J. (1999) Sampling spatial and temporal variation in soil nitrogen availability. *Oecologia* 118:397–404.

Callahan, H., and M. Pigliucci (2001) Shade-induced plasticity and its ecological significance in wild populations of *Arabidopsis thaliana. Ecology,* in press.

Callahan, H. S., Pigliucci, M., and Schlichting, C. D. (1997) Developmental phenotypic plasticity: where ecology and evolution meet molecular biology. *BioEssays* 19: 519–525.

Callahan, H. S., Wells, C. L., and Pigliucci, M. (1999) Light-sensitive plasticity genes in *Arabidopsis thaliana:* mutant analysis and ecological genetics. *Evolutionary Ecology Research* 1:731–751.

Camara, M., and Pigliucci, M. (1999) Mutational contributions to genetic variance/covariance matrices: an experimental approach using induced mutations in *Arabidopsis thaliana. Evolution* 53:1692–1703.

Camara, M. D., Ancell, C. A., and Pigliucci, M. (2000) Induced mutations: a novel tool to study phenotypic integration and evolutionary constraints in *Arabidopsis thaliana*. *Evolutionary Ecology Research* 2:1009–1029.

Carroll, R. L. (2000) Towards a new evolutionary synthesis. *Trends in Ecology and Evolution* 15:27–32.

Carroll, S. B., Weatherbee, S. D., and Langeland, J. A. (1995) Homeotic genes and the regulation and evolution of insect wing number. *Nature* 375:58–61.

Casal, J. J., and Smith, H. (1989) The function, action and adaptive significance of phytochrome in light-grown plants. *Plant, Cell and Environment* 12:855–862.

Cavalli-Sforza, L. L. (1974) The role of plasticity in biological and cultural evolution. *Annals of the New York Academy of Sciences* 231:43–59.

Cavicchi, S., Guerra, D., Giorgi, G., and Pezzoli, C. (1985) Temperature-related divergence in experimental populations of *Drosophila melanogaster*. I. Genetic and developmental basis of wing size and shape variation. *Genetics* 109:665–689.

Cavicchi, S., Guerra, D., Natali, V., Pezzoli, C., and Giorgi, G. (1989) Temperature-related divergence in experimental populations of *Drosophila melanogaster*. II. Correlation between fitness and body dimensions. *Journal of Evolutionary Biology* 2:235–251.

Chandler, P. M., and Robertson, M. (1994) Gene expression regulated by abscisic acid and its relation to stress tolerance. *Annual Review of Plant Physiology and Plant Molecular Biology* 45:113–141.

Charlesworth, B. (1990) Optimization models, quantitative genetics, and mutation. *Evolution* 44:520–538.

Cheplick, G. P. (1995a) Genotypic variation and plasticity of clonal growth in relation to nutrient availability in *Amphibromus scabrivalvis*. *Journal of Ecology* 83: 459–468.

Cheplick, G. P. (1995b) Plasticity of seed number, mass, and allocation in clones of the perennial grass *Amphibromus scabrivalvis*. *International Journal of Plant Sciences* 156:522–529.

Cheverud, J. M. (1984) Quantitative genetics and developmental constraints on evolution by selection. *Journal of Theoretical Biology* 110:155–171.

Cheverud, J. M. (1988) The evolution of genetic correlation and developmental constraints, pp. 94–101 in: *Population Genetics and Evolution,* G. de Jong (ed.). Berlin: Springer-Verlag.

Cheverud, J. M., and Moore, A. J. (1994) Quantitative genetics and the role of the environment provided by relatives in behavioral evolution, pp. 67–100 in: *Quantitative Genetic Studies of Behavioral Evolution,* C. R. B. Boake (ed.). Chicago: University of Chicago Press.

Cheverud, J. M., and Routman, E. J. (1995) Epistasis and its contribution to genetic variance components. *Genetics* 139:1455–1461.

Cheverud, J. M., Rutledge, J. J., and Atchley, W. R. (1983) Quantitative genetics of development: genetic correlations among age-specific trait values, and the evolution of ontogeny. *Evolution* 37:895–905.

Cheverud, J. M., Routman, E. J., and Irschick, D. J. (1997) Pleiotropic effects of individual gene loci on mandibular morphology. *Evolution* 51:2006–2016.

Chippindale, A. K., Leroi, A. M., Saing, H., Borash, D. J., and Rose, M. R. (1997) Phenotypic plasticity and selection in *Drosophila* life history evolution. 2. Diet, mates and the cost of reproduction. *Journal of Evolutionary Biology* 10:269–294.

Chittka, L., and Geiger, K. (1995) Honeybee long-distance orientation in a controlled environment. *Ethology* 99:117–126.

Chory, J. (1993) Out of darkness: mutants reveal pathways controlling light-regulated development in plants. *Trends in Genetics* 9:167–172.

Chuang, S.-E., Daniels, D. L., and Blattner, F. R. (1993) Global regulation of gene expression in *Escherichia coli. Journal of Bacteriology* 175:2026–2036.

Ciceran, M., Murray, A.-M., and Rowell, G. (1994) Natural variation in the temporal patterning of calling song structure in the field cricket *Gryllus pennsylvanicus:* effects of temperature, age, mass, time of day, and nearest neighbour. *Canadian Journal of Zoology* 72:38–42.

Clack, T., Mathews, S., and Sharrock, R. A. (1994) The phytochrome apoprotein family in *Arabidopsis* is encoded by five genes: the sequences and expression of PHYD and PHYE. *Plant Molecular Biology* 25:413–427.

Clark, C. (1993) Dynamic models of behavior—an extension of life-history theory. *Trends in Ecology and Evolution* 8:338.

Clausen, J., and Hiesey, W. M. (1960) The balance between coherence and variation in evolution. *Proceedings of the National Academy of Sciences of the USA* 46:494–506.

Clausen, J., Keck, D., and Hiesey, W. M. (1940) Experimental studies on the nature of plant species. I. Effect of varied environment on western north American plants. Washington, D.C.: Carnegie Institution.

Clauss, M. J., and Venable, D. L. (2000) Seed germination in desert annuals: an empirical test of adaptive bet hedging. *American Naturalist* 155:168–186.

Cohn, M. J., and Tickle, C. (1999) Developmental basis of limblessness and axial patterning in snakes. *Nature* 399:474–479.

Coleman, J. S., McConnaughay, K. D. M., and Ackerly, D. D. (1994) Interpreting phenotypic variation in plants. *Trends in Ecology and Evolution* 9:187–190.

Collett, T. S., and Rees, J. A. (1997) View-based navigation in Hymenoptera: multiple strategies of landmark guidance in the approach to a feeder. *Journal of Comparative Physiology* 181:47–58.

Conway, L. J., and Poethig, R. S. (1993) Heterochrony in plant development. *Developmental Biology* 4:65–72.

Cook, C. D. K. (1968) Phenotypic plasticity with particular reference to three amphibious plant species, pp. 97–111 in: *Modern Methods in Plant Taxonomy,* V. H. Heywood (ed.). London: Academic Press.

Cook, S. A., and Johnson, M. P. (1968) Adaptation to heterogeneous environments. I. Variation in heterophylly in *Ranunculus flammula* L. *Evolution* 22:496–516.

Cooper, R. M., and Zubek, J. P. (1958) Effects of enriched and restricted early environments on the learning ability of bright and dull rats. *Canadian Journal of Psychology* 12:159–164.

Coyne, J. A. (1992) Genetics and speciation. *Nature* 355:511–515.

Cremer, F., Walle, C. v. d., and Bernier, G. (1991) Changes in mRNA level rhythmicity in the leaves of *Sinapsis alba* during a lengthening of the photoperiod which induces flowering. *Plant Molecular Biology* 17:465–473.

Crespi, B. J., and Bookstein, F. L. (1989) A path-analytic model for the measurement of selection on morphology. *Evolution* 43:18–28.

Crews, D. (1994) Temperature, steroids and sex determination. *Journal of Endocrinology* 142:1–8.

Crews, D. (1996) Temperature-dependent sex determination: the interplay of steroid hormones and temperature. *Zoological Science* 13:1–13.

Crews, D., Bergeron, J. M., Bull, J. J., Flores, D., Tousignant, A., Skipper, J. K., and Wibbels, T. (1994) Temperature-dependent sex determination in reptiles: proximate mechanisms, ultimate outcomes, and practical applications. *Developmental Genetics* 15:297–312.

Cubo, J., Fouces, V., Gonzales-Martin, M., Pedrocchi, V., and Ruiz, X. (2000) Non-heterochronic development changes underlie morphological heterochrony in the evolution of the Ardeidae. *Journal of Evolutionary Biology* 13:269–276.

Daday, H., Binet, F. E., Grassia, A., and Peak, J. W. (1973) The effect of environment on heritability and predicted selection response in *Medicago sativa. Heredity* 31: 293–308.

Dahlhoff, E., and Somero, G. N. (1993) Effects of temperature on mitochrondria from Abalone (genus *Haliotis*): adaptive plasticity and its limits. *Journal of Experimental Biology* 185:151–168.

Darwin, C. (1910 [1859]) *The Origin of Species by Means of Natural Selection: Or the Preservation of Favored Races in the Struggle for Life.* New York: A. L. Burt.

David, J. R., Capy, P., and Gauthier, J. P. (1990) Abdominal pigmentation and growth temperature in *Drosophila melanogaster:* similarities and differences in the norms of reaction of successive segments. *Journal of Evolutionary Biology* 3:429–445.

Day, T., and McPhail, J. D. (1996) The effect of behavioural and morphological plasticity on foraging efficiency in the threespine stickleback (*Gasterosteus* sp.). *Oecologia* 108:380–388.

Day, T., Pritchard, J., and Schluter, D. (1994) Ecology and genetics of phenotypic plasticity: a comparison of two sticklebacks. *Evolution* 48:1723–1734.

Decker, K. L., and Pilson, D. (2000) Biased sex ratios in the dioecious annual *Croton texensis* (Euphorbiaceae) are not due to environmental sex determination. *American Journal of Botany* 87:221–229.

de Jong, G. (1990a) Genotype-by-environment interaction and the genetic covariance between environments: multilocus genetics. *Genetica* 81:171–177.

de Jong, G. (1990b) Quantitative genetics of reaction norms. *Journal of Evolutionary Biology* 3:447–468.

de Jong, G. (1994) The fitness of fitness concepts and the description of natural selection. *Quarterly Review of Biology* 69:3–29.

de Jong, G. (1995) Phenotypic plasticity as a product of selection in a variable environment. *American Naturalist* 145:493–512.

de Jong, G. (1999) Unpredictable selection in a structured population leads to local genetic differentiation in evolved reaction norms. *Journal of Evolutionary Biology* 12:839–851.

Delasalle, V. A., and Blum, S. (1994) Variation in germination and survival among families of *Sagittaria latifolia* in response to salinity and temperature. *International Journal of Plant Sciences* 155:187–195.

De Meester, L. (1993) Genotype, fish-mediated chemicals, and phototactic behavior in *Daphnia magna. Ecology* 74:1467–1474.

De Meester, L., Weider, L. J., and Tollrian, R. (1995) Alternative antipredator defences and genetic polymorphism in a pelagic predator-prey system. *Nature* 378:483–485.

den Boer, P. J. (1999) Natural selection, or the non-survival of the non-fit. *Acta Biotheoretica* 47:83–97.

Denver, R. J. (1997a) Environmental stress as a developmental cue: corticotropin-releasing hormone is a proximate mediator of adaptive phenotypic plasticity in amphibian metamorphosis. *Hormones and Behavior* 31:169–179.

Denver, R. J. (1997b) Proximate mechanisms of phenotypic plasticity in amphibian metamorphosis. *American Zoologist* 37:172–184.

Dequeiroz, A., and Wimberger, P. H. (1993) The usefulness of behavior for phylogeny estimation—levels of homoplasy in behavioral and morphological characters. *Evolution* 47:46–60.

DeRocher, A. E., Helm, K. W., Lauzon, L. M., and Vierling, E. (1991) Expression of a conserved family of cytoplasmic low molecular weight heat shock proteins during heat stress and recovery. *Plant Physiology* 96:1038–1047.

Deschamp, P. A., and Cooke, T. J. (1985) Leaf dimorphism in the aquatic Angiosperm *Callitriche heterophylla. American Journal of Botany* 72:1377–1387.

Devlin, P. F., Patel, S. R., and Whitelam, G. C. (1998) Phytochrome E influences internode elongation and flowering time in *Arabidopsis. Plant Cell* 10:1479–1487.

Devlin P. F., Robson, P. R. H., Patel, S. R., Goosey, L., Sharrock, R. A., and Whitelam, G. C. (1999) Phytochrome D acts in the shade-avoidance syndrome in *Arabidopsis* by controlling elongation growth and flowering time. *Plant Physiology* 119:909–915.

DeWitt, T. J. (1998) Costs and limits of phenotypic plasticity: tests with predator-induced morphology and life history in a freshwater snail. *Journal of Evolutionary Biology* 11:465–480.

DeWitt, T. J., Sih, A., and Wilson, D. S. (1998) Costs and limits of phenotypic plasticity. *Trends in Ecology and Evolution* 13:77–81.

Dickerson, G. E. (1962) Implications of genetic-environmental interaction in animal breeding. *Animal Products* 4:47–63.

Diggle, P. K. (1993) Developmental plasticity, genetic variation, and the evolution of andromonoecy in *Solanum hirtum* (Solanaceae). *American Journal of Botany* 80:967–973.

Diggle, P. K. (1994) The expression of andromonoecy in *Solanum hirtum* (Solanaceae): phenotypic plasticity and ontogenetic contingency. *American Journal of Botany* 81:1354–1365.

Diggle, P. K. (1997) Extreme preformation in alpine *Polygonum viviparum:* an architectural and developmental analysis. *American Journal of Botany* 84:154–169.

Dobzhansky, T., and Spassky, B. (1944) Genetics of natural populations. XI. Manifestation of genetic variants in *Drosophila pseudoobscura* in different environments. *Genetics* 29:270–290.

Dodson, S. (1989) Predator-induced reaction norms. *BioScience* 39:447–452.

Doebeli, M. (1996) Quantitative genetics and population dynamics. *Evolution* 50:532–546.

Dolferus, R., Ellis, M., Bruxelles, G. d., Trevaskis, B., Hoeren, F., Dennis, E. S., and Peacock, W. J. (1997) Strategies of gene action in *Arabidopsis* during hypoxia. *Annals of Botany* 79:21–31.

Dong, M., and Pierdominici, M. G. (1995) Morphology and growth of stolons and rhizomes in three clonal grasses, as affected by different light supply. *Vegetatio* 116:25–32.

Dorweiler, J., Stec, A., Kermicle, J., and Doebley, J. (1993) Teosinte glume architecture. 1. A genetic locus controlling a key step in maize evolution. *Science* 262:233–235.

Doughty, P. (1995) Testing the ecological correlates of phenotypically plastic traits within a phylogenetic framework. *Acta Oecologica* 16:519–524.

Downs, C. A., Heckathorn, S. A., Bryan, J. K., and Coleman, J. S. (1998) The methionine-rich low-molecular-weight chloroplast heat-shock protein: evolutionary

conservation and accumulation in relation to thermotolerance. *American Journal of Botany* 85:175–183.
Doyle, J. J. (1994) Evolution of a plant homeotic multigene family: toward connecting molecular systematics and molecular developmental genetics. *Systematic Biology* 43:307–328.
Dubcovsky, J., Galvez, A. F., and Dvorak, J. (1994) Comparison of the genetic organization of the early salt-stress-response gene system in salt-tolerant *Lophopyrum elongatum* and salt-sensitive wheat. *Theoretical and Applied Genetics* 87:957–964.
Dudley, S. A., and Schmitt, J. (1996) Testing the adaptive plasticity hypothesis: density-dependent selection on manipulated stem length in *Impatiens capensis. American Naturalist* 147:445–465.
Dunn, A. M., Adams, J., and Smith, J. E. (1993) Is intersexuality a cost of environmental sex determination in *Gammarus duebeni? Journal of Zoology* 231:383–389.
Ebert, D. (1998) Experimental evolution of parasites. *Science* 282:1432–1435.
Ebert, D., Yampolsky, L., and Stearns, S. C. (1993) Genetics of life history in *Daphnia magna.* I. Heritabilities at two food levels. *Heredity* 70:335–343.
Eisen, E. J., and Saxton, A. M. (1983) Genotype by environment interactions and genetic correlations involving two environmental factors. *Theoretical and Applied Genetics* 67:75–86.
Eldredge, N., and Gould, S. J. (1972) Punctuated equilibria: an alternative to phyletic gradualism, pp. 82–115 in: *Models in Paleobiology,* T. J. M. Schopf (ed.). San Francisco: Freeman, Cooper.
Elena, S. F., and Lenski, R. E. (1997) Long-term experimental evolution in *Escherichia coli.* VII. Mechanisms maintaining genetic variability within populations. *Evolution* 51:1058–1067.
Emery, R. J. N., Reid, D. M., and Chinnappa, C. C. (1994) Phenotypic plasticity of stem elongation in two ecotypes of *Stellaria longipes:* the role of ethylene and response to wind. *Plant, Cell and Environment* 17:691–700.
Endler, J. A. (1986) *Natural Selection in the Wild.* Princeton: Princeton University Press.
Erickson, R. O., and Michelini, F. J. (1957) The plastochron index. *American Journal of Botany* 44:297–305.
Falconer, D. S. (1952) The problem of environment and selection. *American Naturalist* 86:293–298.
Falconer, D. S. (1989) *Introduction to Quantitative Genetics.* New York: Longman.
Falconer, D. S. (1990) Selection in different environments: effects on environmental sensitivity (reaction norm) and on mean performance. *Genetical Research* 56:57–70.
Fang, X.-M., Wu, C.-Y., and Brennan, M. D. (1991) Complexity in evolved regulatory variation for alcohol dehydrogenase genes in Hawaiian Drosophila. *Journal of Molecular Evolution* 32:220–226.
Feldman, M. W., and Lewontin, R. C. (1990) The heritability hang-up. *Science* 190:1163–1168.
Felsenstein, J. (1985) Phylogenies and the comparative method. *American Naturalist* 125:1–15.
Fernando, R. L., Knights, S. A., and Gianola, D. (1984) On a method of estimating the genetic correlation between characters measured in different experimental units. *Theoretical and Applied Genetics* 67:175–178.
Fink, A., Dutuit, M., Thiellement, H., Greppin, H., and Tacchini, P. (1997) Leaf protein analysis during gibberellic and photoperiodic induction of flowering in *Arabidopsis thaliana. Plant Physiology and Biochemistry* 35:665–670.

Fisher, R. A. (1930) *The Genetical Theory of Natural Selection.* Oxford: Clarendon.

Forbes, L. S. (1991) Optimal size and number of offspring in a variable environment. *Journal of Theoretical Biology* 150:299–304.

Fox, C. W., and Savalli, U. M. (1998) Inheritance of environmental variation in body size: superparasitism of seeds affects progeny and grandprogeny body size via a non-genetic maternal effect. *Evolution* 52:172–182.

Fox, C. W., Thakar, M. S., and Mousseau, T. A. (1997) Egg size plasticity in a seed beetle: an adaptive maternal effect. *American Naturalist* 149:149–163.

Fox, C. W., Czesak, M. E., Mousseau, T. A., and Roff, D. A. (1999) The evolutionary genetics of an adaptive maternal effect: egg size plasticity in a seed beetle. *Evolution* 53:552–560.

Freeman, D. C., Graham, J. H., and Emlen, J. M. (1993) Developmental stability in plants: symmetries, stress and epigenesis. *Genetica* 89:97–119.

Friedman, W. E., and Carmichael, J. S. (1998) Heterochrony and developmental innovation: evolution of female gametophyte ontogeny in *Gnetum,* a highly apomorphic seed plant. *Evolution* 52:1016–1030.

Fry, J. D. (1993) The "general vigor" problem: can antagonistic pleiotropy be detected when genetic covariances are positive? *Evolution* 47:327–332.

Fuglevand, G., Jackson, J. A., and Jenkins, G. I. (1996) UV-B, UV-A, and blue light signal transduction pathways interact synergistically to regulate chalcone synthase gene expression in *Arabidopsis. Plant Cell* 8:2347–2357.

Futuyma, D. (1998) *Evolutionary Biology.* Sunderland, Mass.: Sinauer.

Galen, C. (2000) High and dry: drought stress, sex-allocation trade-offs, and selection on flower size in the alpine wildflower *Polemonium viscosum* (Polemoniaceae). *American Naturalist* 156:72–83.

Galen, C., Shore, J. S., and Deyoe, H. (1991) Ecotypic divergence in Alpine *Polemonium viscosum:* genetic structure, quantitative variation, and local adaptation. *Evolution* 45:1218–1228.

Galloway, L. F. (1995) Response to natural environmental heterogeneity: maternal effects and selection on life-history characters and plasticities in *Mimulus guttatus. Evolution* 49:1095–1107.

Ganfornina, M. D., and Sanchez, D. (1999) Generation of evolutionary novelty by functional shift. *BioEssays* 21:432–439.

Garcia-Vazquez, E., and Rubio, J. (1988) Canalization and phenotypic variation caused by changes in the temperature of development. *Journal of Heredity* 79:448–452.

Gatesy, S. M., and Dial, K. P. (1996) Locomotor modules and the evolution of avian flight. *Evolution* 50:331–340.

Gavrilets, S. (1986) An approach to modeling the evolution of populations with consideration of genotype-environment interaction. *Genetika* 22:36–44.

Gavrilets, S. (1988) Evolution of modificational variability in random environment. *Journal of General Biology* 49:271–276.

Gavrilets, S. (1997) Hybrid zones with Dobzhansky-type epistatic selection. *Evolution* 51:1027–1035.

Gavrilets, S., and de Jong, G. (1993) Pleiotropic models of polygenic variation, stabilizing selection, and epistasis. *Genetics* 134:609–625.

Gavrilets, S., and Hastings, A. (1994) A quantitative-genetic model for selection on developmental noise. *Evolution* 48:1478–1486.

Gavrilets, S., and Scheiner, S. M. (1993) The genetics of phenotypic plasticity. V. Evolution of reaction norm shape. *Journal of Evolutionary Biology* 6:31–48.

Gavrilets, S., Li, H., and Vose, M. D. (1998) Rapid parapatric speciation on holey adaptive landscapes. *Proceedings of the Royal Society of London* B265:1483–1489.

Gazzaniga, M. S. (2000) Cerebral specialization and interhemispheric communication: does the corpus callosum enable the human condition? *Brain* 123:1293–1326.

Gedroc, J. J., McConnaughay, K. D. M., and Coleman, J. S. (1996) Plasticity in root/shoot partitioning: optimal, ontogenetic, or both? *Functional Ecology* 10:44–50.

Gehring, J. L., and Monson, R. K. (1994) Sexual differences in gas exchange and response to environmental stress in dioecious *Silene latifolia* (Caryphyllaceae). *American Journal of Botany* 81:166–174.

Gibert, P., Moreteau, B., David, J. R., and Scheiner, S. M. (1998) Describing the evolution of reaction norm shape: body pigmentation in *Drosophila. Evolution* 52: 1501–1506.

Gilbert, S. F. (1991) Epigenetic landscaping: Waddington's use of cell fate bifurcation diagrams. *Biology and Philosophy* 6:135–154.

Gillespie, J. H., and Turelli, M. (1989) Genotype-environment interactions and the maintenance of polygenic variation. *Genetics* 121:129–138.

Gimelfarb, A. (1990) How much genetic variation can be maintained by genotype-environment interactions? *Genetics* 124:443–445.

Gimelfarb, A. (1994) Additive-multiplicative approximation of genotype-environment interaction. *Genetics* 138:1339–1349.

Girondot, M., Zaborski, P., Servan, J., and Pieau, C. (1994) Genetic contribution to sex determination in turtles with environmental sex determination. *Genetical Research* 63:117–127.

Gittleman, J. L., and Decker, D. M. (1994) The phylogeny of behaviour, pp. 80–105 in: *Behaviour and Evolution,* P. J. B. Slater and T. R. Halliday (eds.). Cambridge: Cambridge University Press.

Gittleman, J. L., and Kot, M. (1990) Adaptation: statistics and a null model for estimating phylogenetic effects. *Systematic Zoology* 39:227–241.

Gittleman, J. L., and Luh, H.-K. (1992) On comparing comparative methods. *Annual Review of Ecology and Systematics* 23:383–404.

Givnish, T. J. (1982) On the adaptive significance of leaf height in forest herbs. *American Naturalist* 120:353–381.

Goho, S., and Bell, G. (2000) The ecology and genetics of fitness in *Chlamydomonas.* IX. The rate of accumulation of variation of fitness under selection. *Evolution* 54:416–424.

Goldschmidt, R. (1940) *The Material Basis of Evolution.* New Haven, Conn.: Yale University Press.

Goldstein, D. B., and Holsinger, K. E. (1992) Maintenance of polygenic variation in spatially structured populations: roles for local mating and genetic redundancy. *Evolution* 46:412–429.

Goliber, T. E., and Feldman, L. J. (1990) Developmental analysis of leaf plasticity in the heterophyllous aquatic plant *Hippuris vulgaris. American Journal of Botany* 77:399–412.

Gombos, Z., Wada, H., Hideg, E., and Murata, N. (1994) The unsaturation of membrane lipids stabilizes photosynthesis against heat stress. *Plant Physiology* 104:563–567.

Gomulkiewicz, R., and Kirkpatrick, M. (1992) Quantitative genetics and the evolution of reaction norms. *Evolution* 46:390–411.

Goodwin, D., Bradshaw, J. W., and Wickens, S. M. (1997) Paedomorphosis affects agonistic visual signals of domestic dogs. *Animal Behavior* 53:297–304.

Goransson, M., Sonden, B., Nilsson, P., Dagberg, B., Forsman, K., Emanuelsson, K., and Uhlin, B. E. (1990) Transcriptional silencing and thermoregulation of gene expression in *Escherichia coli. Nature* 344:682–685.

Gotthard, K. (1998) Life history plasticity in the satyrine butterfly *Lasiommata petropolitana:* investigating an adaptive reaction norm. *Journal of Evolutionary Biology* 11:21–40.

Gotthard, K., and Nylin, S. (1995) Adaptive plasticity and plasticity as an adaptation: a selective review of plasticity in animal morphology and life history. *Oikos* 74:3–17.

Gould, S. J. (1977) *Ontogeny and Phylogeny.* Cambridge, Mass.: Harvard University Press.

Gould, S. J. (1980) The evolutionary biology of constraint. *Daedalus* 109:39–52.

Gould, S. J. (1984) Covariance sets and ordered geographic variation in *Cerion* from Aruba, Bonaire and Curacao: a way of studying nonadaptation. *Systematic Zoology* 33:217–237.

Gould, S. J. (1989) A developmental constraint in *Cerion,* with comments on the definition and interpretation of constraint in evolution. *Evolution* 43:516–539.

Gould, S. J. (1996) *The Mismeasure of Man.* New York: W. W. Norton.

Gould, S. J., and Lewontin, R. C. (1979) The spandrels of San Marco and the Panglossian paradigm: a critique of the adaptationist programme. *Proceedings of the Royal Society of London* B205:581–598.

Gould, S. J., and Vrba, E. S. (1982) Exaptation—a missing term in the science of form. *Paleobiology* 8:4–15.

Grafen, A. (1988) On the uses of data on lifetime reproductive success, pp. 454–471 in: *Reproductive Success,* T. H. Clutton-Brock (ed.). Chicago: University of Chicago Press.

Grant, V. (1994) Evolution of the species concept. *Biologisches Zentralblatt* 113: 401–415.

Greene, E. (1989) A diet-induced developmental polymorphism in a caterpillar. *Science* 243:643–646.

Grime, J. P., Crick, J. C., and Rincon, J. E. (1986) The ecological significance of plasticity, pp. 5–29 in: *Plasticity in Plants,* D. H. Jennings and A. J. Trewavas (eds.). North Yorkshire: Pindar Scarborough.

Gromko, M. H. (1995) Unpredictability of correlated response to selection: pleiotropy and sampling interact. *Evolution* 49:685–693.

Gross, D., and Parthier, B. (1994) Novel natural substances acting in plant regulation. *Journal of Plant Growth and Regulation* 13:93–114.

Guerrant, E. O. J. (1982) Neotenic evolution of *Delphinium nudicaule* (Ranunculaceae): a hummingbird-pollinated larkspur. *Evolution* 36:699–712.

Gupta, A. P., and Lewontin, R. C. (1982) A study of reaction norms in natural populations of *Drosophila pseudoobscura. Evolution* 36:934–948.

Gurevitch, J. (1992) Sources of variation in leaf shape among two populations of *Achillea lanulosa. Genetics* 130:385–394.

Ha, S. B., and An, G. (1988) Identification of upstream regulatory elements involved in the developmental expression of the *Arabidopsis thaliana* cab1 gene. *Proceedings of the National Academy of Sciences of the USA* 85:8017–8021.

Hader, D. P., and Hansel, A. (1991) Responses of *Dictyostelium discoideum* to multiple environmental stimuli. *Botanica Acta* 104:200–205.

Haldane, J. B. S. (1932) The time of action of genes, and its bearing on some evolutionary problems. *American Naturalist* 66:5–24.

Hall, B. K. (1992a) *Bauplane:* fundamental body plans, pp. 66–82 in: *Evolutionary Developmental Biology.* London: Chapman & Hall.

Hall, B. K. (1992b) Waddington's legacy in development and evolution. *American Zoologist* 32:113–122.

Halliday, K. J., Thomas, B., and Whitelam, G. C. (1997) Expression of heterologous phytochromes A, B or C in transgenic tobacco plants alters vegetative development and flowering time. *Plant Journal* 12:1079–1090.

Harrington, H. M., Dash, S., Dharmasiri, N., and Dharmasiri, S. (1994) Heat-shock proteins: a search for functions. *Australian Journal of Plant Physiology* 21:843–855.

Hartl, D. L., and Clark, A. G. (1989) *Principles of Population Genetics.* Sunderland, Mass.: Sinauer.

Harvey, P. H., and Pagel, M. D. (1991) *The Comparative Method in Evolutionary Biology.* Oxford: Oxford University Press.

Harvey, P. H., and Purvis, A. (1991) Comparative methods for explaining adaptations. *Nature* 351:619–624.

Hawthorne, D. J. (1997) Ecological history and evolution in a novel environment: habitat heterogeneity and insect adaptation to a new host plant. *Evolution* 51: 153–162.

Hazel, W., Ante, S., and Stringfellow, B. (1998) The evolution of environmentally-cued pupal colour in swallowtail butterflies: natural selection for pupation site and pupal colour. *Ecological Entomology* 23:41–44.

Hazel, W., Brandt, R., and Grantham, T. (1987) Genetic variability and phenotypic plasticity in pupal colour and its adaptive significance in the swallowtail butterfly *Papilio polyxenes. Heredity* 59:449–455.

Hebert, D., Faure, S., and Olivieri, I. (1994) Genetic, phenotypic, and environmental correlations in black medic, *Medicago lupulina* L., grown in three different environments. *Theoretical and Applied Genetics* 88:604–613.

Herrera, A., Delgado, J., and Paraguately, I. (1991) Occurrence of inducible Crassulacean acid metabolism in leaves of *Talinum triangulare* (Portulacaceae). *Journal of Experimental Botany* 42:493–499.

Herrnstein, R. J., and Murray, C. (1994) *The Bell Curve: Intelligence and Class Structure in American Life.* New York: Free Press.

Hey, J. (1999) The neutralist, the fly and the selectionist. *Trends in Ecology and Evolution* 14:35–37.

Hillesheim, E., and Stearns, S. C. (1991) The responses of *Drosophila melanogaster* to artificial selection on body weight and its phenotypic plasticity in two larval food environments. *Evolution* 45:1909–1923.

Hillesheim, E., and Stearns, S. C. (1992) Correlated response in life-history traits to artificial selection for body size in *Drosophila melanogaster. Evolution* 46: 745–752.

Hillis, D. M., Huelsenbeck, J. P., and Cunningham, C. W. (1994) Application and accuracy of molecular phylogenies. *Science* 264:671–677.

Hodin, J. (2000) Plasticity and constraints in development and evolution. *Journal of Experimental Zoology* 288:1–20.

Hoekstra, R. F. (1988) Theory of phenotypic evolution: genetic or non-genetic models? pp. 33–41 in: *Population Genetics and Evolution,* G. de Jong (ed.). Berlin: Springer-Verlag.

Hoffmann, A. A., and Merila, J. (1999) Heritable variation and evolution under favourable and unfavourable conditions. *Trends in Ecology and Evolution* 14:96–101.

Hoffmann, A. A., and Parsons, P. A. (1993) Direct and correlated responses to selection for desiccation resistance: a comparison of *Drosophila melanogaster* and *D. simulans*. *Journal of Evolutionary Biology* 6:643–657.

Hoffmann, A. A., and Schiffer, M. (1998) Changes in the heritability of five morphological traits under combined environmental stresses in *Drosophila melanogaster*. *Evolution* 52:1207–1212.

Holloway, G. J., and Brakefield, P. M. (1995) Artificial selection of reaction norms of wing pattern elements in *Bicyclus anynana*. *Heredity* 74:91–99.

Holloway, G. J., Brakefield, P. M., Jong, P. W. d., Ottenheim, M. M., Vos, H. d., Kesbeke, F., and Peynenburg, L. (1995) A quantitative genetic analysis of an aposematic colour pattern and its ecological implications. *Philosophical Transactions of the Royal Society of London* B348:373–379.

Holloway, G. J., Povey, S. R., and Sibly, R. M. (1990) The effect of new environment on adapted genetic architecture. *Heredity* 64:323–330.

Holtsford, T. P., and Ellstrand, N. C. (1992) Genetic and environmental variation in floral traits affecting outcrossing rate in *Clarkia tembloriensis* (Onagraceae). *Evolution* 46:216–225.

Hong, S.-W., and Vierling, E. (2000) Mutants of *Arabidopsis thaliana* defective in the acquisition of tolerance to high temperature stress. *Proceedings of the National Academy of Sciences of the USA* 97:4392–4397.

Houle, D. (1991) Genetic covariance of fitness correlates: what genetic correlations are made of and why it matters. *Evolution* 45:630–648.

Houle, D., Hughes, K. A., Hoffmaster, D. K., Ihara, J., Assimacopoulos, S., Canada, D., and Charlesworth, B. (1994) The effects of spontaneous mutation on quantitative traits. I. Variances and covariances of life history traits. *Genetics* 138:773–785.

Houston, A. I., and McNamara, J. M. (1992) Phenotypic plasticity as a state-dependent life-history decision. *Evolutionary Ecology* 6:243–253.

Huberman, B. A., and Hogg, T. (1986) Complexity and adaptation. *Physica* 22D: 376–384.

Huey, R. B., Partridge, L., and Fowler, K. (1991) Thermal sensitivity of *Drosophila melanogaster* responds rapidly to laboratory natural selection. *Evolution* 45:751–756.

Hume, D. (1956 [1748]) *Enquiry Concerning Human Understanding*. Chicago: Gateway Editions.

Hume, D. (1978 [1739–40]) *A Treatise of Human Nature*. New York: Clarendon Press.

Huynen, M. A., Stadler, P. F., and Fontana, W. (1996) Smoothness within ruggedness: the role of neutrality in adaptation. *Proceedings of the National Academy of Sciences of the USA* 93:397–401.

Imasheva, A. G., Loeschcke, V., Zhivotovsky, L. A., and Lazebny, O. E. (1997) Effects of extreme temperatures on phenotypic variation and developmental stability in *Drosophila melanogaster* and *Drosophila buzzatii*. *Biological Journal of the Linnean Society* 61:117–126.

Ingram, G. C., Goodrich, J., Wilkinson, M. D., Simon, R., Haughn, G. W., and Coen, E. S. (1995) Parallels between UNUSUAL *floral organs* and *fimbriata*, genes controlling flower development in *Arabidopsis* and *Antirrhinum*. *Plant Cell* 7:1501–1510.

Jackson, R. B., and Caldwell, M. M. (1996) Integrating resource heterogeneity and plant plasticity: modelling nitrate and phosphate uptake in a patchy soil environment. *Journal of Ecology* 84:891–903.

Jacob, F. (1977) Evolution and tinkering. *Science* 196:1161–1166.

Jagtap, V., and Bhargava, S. (1995) Variation in the antioxidant metabolism of drought tolerant and drought susceptible varieties of *Sorghum bicolor* (L.) Moench. exposed to high light, low water and high temperature stress. *Journal of Plant Physiology* 145:195–197.

Jain, S. K. (1978) Inheritance of phenotypic plasticity in soft chess, *Bromus mollis* L. (Gramineae). *Experientia* 34:835–836.

Janzen, F. J. (1995) Experimental evidence for the evolutionary significance of temperature-dependent sex determination. *Evolution* 49:864–873.

Jasienski, M., Ayala, F. J., and Bazzaz, F. A. (1997) Phenotypic plasticity and similarity of DNA among genotypes of an annual plant. *Heredity* 78:176–181.

Jernigan, R. W., Culver, D. C., and Fong, D. W. (1994) The dual role of selection and evolutionary history as reflected in genetic correlations. *Evolution* 48:587–596.

Johannsen, W. (1911) The genotype conception of heredity. *American Naturalist* 45: 129–159.

Johnson, N. A., and Wade, M. J. (1996) Genetic covariances within and between species: indirect selection for hybrid inviability. *Journal of Evolutionary Biology* 9:205–214.

Jones, C. S. (1992) Comparative ontogeny of a wild cucurbit and its derived cultivar. *Evolution* 46:1827–1847.

Joshi, A., and Thompson, J. N. (1996) Evolution of broad and specific competitive ability in novel versus familiar environments in *Drosophila* species. *Evolution* 50:188–194.

Kacser, H., and Burns, J. A. (1981) The molecular basis of dominance. *Genetics* 97:639–666.

Kalisz, S. (1986) Variable selection on the timing of germination in *Collinsia verna* (Scrophulariaceae). *Evolution* 40:479–491.

Kane, M. E., and Albert, L. S. (1982) Environmental and growth regulator effects on heterophylly and growth of *Proserpinaca intermedia* (Haloragaceae). *Aquatic Botany* 13:73–85.

Kaplan, J. M. (2000) *The Limits and Lies of Human Genetic Research: Dangers for Social Policy* (Reflective Bioethics). New York: Routledge.

Kaplan, J. M., and Pigliucci, M. (2001) Genes "for" phenotypes: a modern history view. *Biology and Philosophy* 16:189–213.

Kaplan, R. H., and Cooper, W. S. (1984) The evolution of developmental plasticity in reproductive characteristics: an application of the "adaptive coin-flipping" principle. *American Naturalist* 123:393–410.

Kareiva, P., Parker, I. M., and Pascual, M. (1996) Can we use experiments and models in predicting the invasiveness of genetically engineered organisms? *Ecology* 77:1670–1676.

Kauffman, S. A. (1993) *The Origins of Order*. New York: Oxford University Press.

Kawecki, T. J., and Stearns, S. C. (1993) The evolution of life histories in spatially heterogeneous environments: optimal reaction norms revisited. *Evolutionary Ecology* 7:155–174.

Keightley, P. D., and Kacser, H. (1987) Dominance, pleiotropy and metabolic structure. *Genetics* 117:319–329.

Keller, L., and Ross, K. G. (1993a) Phenotypic basis of reproductive success in a social insect: genetic and social determinants. *Science* 260:1107–1110.

Keller, L., and Ross, K. G. (1993b) Phenotypic plasticity and "cultural transmission" of alternative social organizations in the fire ant *Solenopsis invicta*. *Behavioral Ecology and Sociobiology* 33:121–129.

Keller, L., and Ross, K. G. (1995) Gene by environment interaction: effects of a single gene and social environment on reproductive phenotypes of fire ant queens. *Functional Ecology* 9:667–676.

Kellogg, E. A., and Shaffer, H. B. (1993) Model organisms in evolutionary studies. *Systematic Biology* 42:409–414.

Kenyon, C., Chang, J., Gensch, E., Rudner, A., and Tabtiang, R. (1993) A *C. elegans* mutant that lives twice as long as wild type. *Nature* 366:461–464.

Khan, M. A., Antonovics, J., and Bradshaw, A. D. (1976) Adaptation to heterogeneous environments. III. The inheritance of response to spacing in flax and linseed (*Linum usitatissimum*). *Australian Journal of Agricultural Research* 27:649–659.

Kieber, J. J. (1997) The ethylene response pathway in *Arabidopsis. Annual Review of Plant Physiology and Plant Molecular Biology* 48:277–296.

Kimura, M. (1954) Process leading to quasi-fixation of genes in natural populations due to random fluctuations of selection intensities. *Genetics* 39:280–295.

Kimura, M. (1983) *The Neutral Theory of Molecular Evolution.* Cambridge: Cambridge University Press.

King, R. C., and Stansfield, W. D. (1990) *A Dictionary of Genetics.* New York: Oxford University Press.

Kingsolver, J. G. (1995) Fitness consequences of seasonal polyphenism in western white butterflies. *Evolution* 49:942–954.

Kirkpatrick, M., and Lofsvold, D. (1992) Measuring selection and constraint in the evolution of growth. *Evolution* 46:954–971.

Koch, M., Haubold, B., and Mitchell-Olds, T. (2000) Comparative evolutionary analysis of chalcone synthase and alcohol dehydrogenase loci in *Arabidopsis, Arabis,* and related genera (Brassicaceae). *Molecular Biology and Evolution* 17:1483–1498.

Koch, P. B., Brakefield, P. M., and Kesbeke, F. (1996) Ecdysteroids control eyespot size and wing color pattern in the polyphenic butterfly *Bicyclus anynana* (Lepidoptera: Satyridae). *Journal of Insect Physiology* 42:223–230.

Koga-Ban, Y., Abe, M., and Kitagawa, Y. (1991) Alteration in gene expression during cold treatment of rice plant. *Plant Cell Physiology* 32:901–905.

Kohler, H.-R., Zanger, M., Eckwert, H., and Einfeldt, I. (2000) Selection favours low hsp70 levels in chronically metal-stressed soil arthropods. *Journal of Evolutionary Biology* 13:569–582.

Komeda, Y. (1993) The use of transgenic *Arabidopsis thaliana* for studies of the regulation of genes for heat-shock proteins. *Journal of Plant Research* 3:213–219.

Komers, P. E. (1997) Behavioural plasticity in variable environments. *Canadian Journal of Zoology* 75:161–169.

Kondo, T., Tsinoremas, N. F., Golden, S. S., Johnson, C. H., Kutsuna, S., and Ishiura, M. (1994) Circadian clock mutants of Cyanobacteria. *Science* 266:1233–1236.

Krebs, R. A., and Feder, M. E. (1997a) Deleterious consequences of Hsp70 overexpression in *Drosophila melanogaster* larvae. *Cell Stress & Chaperones* 2:60–71.

Krebs, R. A., and Feder, M. E. (1997b) Natural variation in the expression of the heat-shock protein HSP70 in a population of *Drosophila melanogaster* and its correlation with tolerance of ecologically relevant thermal stress. *Evolution* 51:173–179.

Krebs, R. A., and Loeschcke, V. (1994a) Costs and benefits of activation of the heat-shock response in *Drosophila melanogster. Functional Ecology* 8:730–737.

Krebs, R. A., and Loeschcke, V. (1994b) Effects of exposure to short-term heat stress on fitness components in *Drosophila melanogaster. Journal of Evolutionary Biology* 7:39–49.

Krebs, R. A., Feder, M. E., and Lee, J. (1998) Heritability of expression of the 70KD heat-shock protein in *Drosophila melanogaster* and its relevance to the evolution of thermotolerance. *Evolution* 52:841–847.

Kreps, J. A., and Simon, A. E. (1997) Environmental and genetic effects on circadian clock-regulated gene expression in *Arabidopsis*. *Plant Cell* 9:297–304.

Kuhn, T. (1970) *The Structure of Scientific Revolutions*. Chicago: University of Chicago Press.

Kuno, N., Muramatsu, T., Hamazato, F., and Furuya, M. (2000) Identification by large-scale screening of phytochrome-regulated genes in etiolated seedlings of *Arabidopsis* using a fluorescent differential display technique. *Plant Physiology* 122:15–24.

Lacey, E. P. (1996) Parental effects in *Plantago lanceolata* L. I.: a growth chamber experiment to examine pre- and post-zygotic temperature effects. *Evolution* 50: 865–878.

Lacey, E. P., Real, L., Antonovics, J., and Heckel, D. G. (1983) Variance models in the study of life histories. *American Naturalist* 122:114–131.

Lande, R. (1980) Genetic variation and phenotypic evolution during allopatric speciation. *American Naturalist* 116:463–479.

Lande, R., and Arnold, S. J. (1983) The measurement of selection on correlated characters. *Evolution* 37:1210–1226.

Lande, R., and Shannon, S. (1996) The role of genetic variation in adaptation and population persistence in a changing environment. *Evolution* 50:434–437.

Landry, L. G., Chapple, C. C. S., and Last, R. L. (1995) Arabidopsis mutants lacking phenolic sunscreens exhibit enhanced ultraviolet-B injury and oxidative damage. *Plant Physiology* 109:1159–1166.

Lang, V., Mantyla, E., Welin, B., Sundberg, B., and Palva, E. T. (1994) Alterations in water status, endogenous abscisic acid content, and expression of rab18 gene during the development of freezing tolerance in *Arabidopsis thaliana*. *Plant Physiology* 104:1341–1349.

Larson, A. (1989) The relationship between speciation and morphological evolution, pp. 579–598 in: *Speciation and Its Consequences*, D. Otte and J. A. Endler (eds.). Sunderland, Mass.: Sinauer.

Lechowicz, M. J., and Bell, G. (1991) The ecology and genetic of fitness in forest plants. II. Microspatial heterogeneity of the edaphic environment. *Journal of Ecology* 79:687–696.

Lechowicz, M. J., and Blais, P. A. (1988) Assessing the contributions of multiple interacting traits to plant reproductive success: environmental dependence. *Journal of Evolutionary Biology* 1:255–273.

Lechowicz, M. J., Schoen, D. J., and Bell, G. (1988) Environmental correlates of habitat distribution and fitness components in *Impatiens capensis* and *Impatiens pallida*. *Journal of Ecology* 76:1043–1054.

Leclaire, M., and Brandle, R. (1994) Phenotypic plasticity and nutrition in a phytophagous insect: consequences of colonizing a new host. *Oecologia* 100:379–385.

Lee, D. W., Oberbauer, S. F., Johnson, P., Krishnapilay, B., Mansor, M., Mohamad, H., and Yap, S. K. (2000) Effects of irradiance and spectral quality on leaf structure and function in seedlings of two Southeast Asian *Hopea* (Dipterocarpaceae) species. *American Journal of Botany* 87:447–455.

Lee, M. S. (1993) The origin of the turtle body plan: bridging a famous morphological gap. *Science* 261:1716–1720.

Leibold, M. A., Tessier, A. J., and West, C. T. (1994) Genetic, acclimatization, and ontogenetic effects on habitat selection behavior in *Daphnia pulicaria. Evolution* 48:1324–1332.

Leimar, O. (1996) Life history plasticity: influence of photoperiod on growth and development in the common blue butterfly. *Oikos* 76:228–234.

Leips, J., and Mackay, T. F. C. (2000) Quantitative trait loci for life span in *Drosophila melanogaster:* interactions with genetic background and larval density. *Genetics* 155:1773–1788.

Leon-Kloosterziel, K. M., Bunt, G. A. v. d., Zeevaart, J. A. D., and Koornneef, M. (1996) *Arabidopsis* mutants with a reduced seed dormancy. *Plant Physiology* 110:233–240.

Lerner, I. M. (1954) *Genetic Homeostasis.* New York: Dover.

Leroi, A. M., Lenski, R. E., and Bennett, A. F. (1994) Evolutionary adaptation to temperature. III. Adaptation of *Escherichia coli* to a temporally varying environment. *Evolution* 48:1222–1229.

Levene, H. (1953) Genetic equilibrium when more than one ecological niche is available. *American Naturalist* 87:331–333.

Levin, D. A. (1983) Polyploidy and novelty in flowering plants. *American Naturalist* 122:1–25.

Levin, D. A. (1988) Plasticity, canalization and evolutionary stasis in plants, pp. 35–45 in: *Plant Population Biology,* A. J. Davy, M. J. Hutchings, and A. R. Watkinson (eds.). Oxford: Blackwell.

Levins, R. (1963) Theory of fitness in a heterogeneous environment. II. Developmental flexibility and niche selection. *American Naturalist* 47:75–90.

Levins, R. (1968) *Evolution in Changing Environments.* Princeton, N.J.: Princeton University Press.

Levins, R., and Lewontin, R. C. (1985) *The Dialectical Biologist.* Cambridge, Mass.: Harvard University Press.

Lewontin, R. C. (1974) The analysis of variance and the analysis of causes. *American Journal of Human Genetics* 26:400–411.

Lewontin, R. C. (1978) Adaptation. *Scientific American* 239:213–230.

Lewontin, R. C. (1984) Detecting population differences in quantitative characters as opposed to gene frequencies. *American Naturalist* 123:115–124.

Lewontin, R. C. (1992) *Biology as Ideology: The Doctrine of DNA.* New York: Harper Perennial.

Lewontin, R. C. (1998) The evolution of cognition: questions we will never answer, pp. 107–132 in: *An Invitation to Cognitive Science,* 2d ed., D. N. Osherson (gen. ed.), Vol. 4: *Methods, Models, and Conceptual Issues,* D. Scarborough and S. Sternberg (eds.). Cambridge, Mass.: MIT Press.

Lewontin, R. C. (2000) *The Triple Helix: Gene, Organism, and Environment.* Cambridge, Mass.: Harvard University Press.

Lewontin, R. C., and Hubby, J. L. (1966) A molecular approach to the study of genic heterozygosity in natural populations. II. Amount of variation and degree of heterozygosity in natural populations. *Genetics* 54:595–609.

Lewontin, R. C., Rose, S., and Kamin, L. J. (1984) *Not in Our Genes.* New York: Pantheon.

Lin, B.-L., and Yang, W.-J. (1999) Blue light and abscisic acid independently induce heterophyllous switch in *Marsilea quadrifolia. Plant Physiology* 119:429–434.

Liu, J., Ishitani, M., Halfter, U., Kim, C. S., and Zhu, J.-K. (2000) The *Arabidopsis thaliana* SOS2 gene encodes a protein kinase that is required for salt tolerance. *Proceedings of the National Academy of Sciences of the USA* 97:3730–3734.

Lively, C. M. (1986a) Canalization versus developmental conversion in a spatially variable environment. *American Naturalist* 128:561–572.
Lively, C. M. (1986b) Competition, comparative life histories, and maintenance of shell dimorphism in a barnacle. *Ecology* 67:858–864.
Lively, C. M. (1986c) Predator-induced shell dimorphism in the acorn barnacle *Chthamalus anisopoma. Evolution* 40:232–242.
Lively, C. M. (1999) Developmental strategies in spatially variable environments: barnacle shell dimorphism and strategic models of selection, pp. 215–258 in: *The Ecology and Evolution of Inducible Defenses,* R. Tollrian and C. D. Harvell (eds.). Princeton, N.J.: Princeton University Press.
Lloyd, D. G. (1984) Variation strategies of plants in heterogeneous environments. *Biological Journal of the Linnean Society* 21:357–385.
Loeschcke, V., and Krebs, R. A. (1994) Genetic variation for resistance and acclimation to high temperature stress in *Drosophila buzzatii. Biological Journal of the Linnean Society* 52:83–92.
Loeschcke, V., and Krebs, R. A. (1996) Selection for heat-shock resistance in larval and in adult *Drosophila buzzatii:* comparing direct and indirect responses. *Evolution* 50:2354–2359.
Lois, R., and Buchanan, B. B. (1994) Severe sensitivity to ultraviolet radiation in an *Arabidopsis* mutant deficient in flavonoid accumulation. II. Mechanisms of UV-resistance in *Arabidopsis. Planta* 194:504–509.
Lorenzi, R., Zonta, L. A., and Jayakar, S. D. (1989) Quantitative traits and temporally variable selection: two-locus models. *Journal of Genetics* 68:29–42.
Losos, J. B., and Miles, D. B. (1994) Adaptation, constraint, and the comparative method: phylogenetic issues and methods, pp. 60–98 in: *Ecological Morphology: Integrative Organismal Biology,* P. C. Wainwright and S. M. Reilly (eds.). Chicago: University of Chicago Press.
Losos, J. B., Creer, D. A., Glossip, D., Goellner, R., Hampton, A., Roberts, G., Haskell, N., Taylor, P., and Ettling, J. (2000). Evolutionary implications of phenotypic plasticity in the hindlimb of the lizard *Anolis sagrei. Evolution* 54:301–305.
Lotscher, M., and Hay, M. J. M. (1997) Genotypic differences in physiological integration, morphological plasticity and utilization of phosphorous induced by variation in phosphate supply in *Trifolium repens. Journal of Ecology* 85:341–350.
Lynch, C. B. (1994) Evolutionary inferences from genetic analyses of cold adaptation in laboratory and wild populations of the house mouse, pp. 278–301 in: *Quantitative Genetic Studies of Behavioral Evolution,* C. R. B. Boake (ed.). Chicago: University of Chicago Press.
Lynch, M., and Gabriel, W. (1987) Environmental tolerance. *American Naturalist* 129:283–303.
Mackay, T. F. C. (1981) Genetic variation in varying environments. *Genetical Research* 37:79–93.
Manly, B. F. J. (1985) *The Statistics of Natural Selection.* London: Chapman & Hall.
Mantyla, E., Lang, V., and Palva, E. T. (1995) Role of abscisic acid in drought-induced freezing tolerance, cold acclimation, and accumulation of LTI78 and RAB18 proteins in *Arabidopsis thaliana. Plant Physiology* 107:141–148.
Maresca, B., Patriarca, E., Goldenberg, C., and Sacco, M. (1988) Heat shock and cold adaptation in Antarctic fishes: a molecular approach. *Comparative Biochemistry and Physiology* 90B:623–629.

Markwell, J., and Osterman, J. C. (1992) Occurrence of temperature-sensitive phenotypic plasticity in chlorophyll-deficient mutants of *Arabidopsis thaliana. Plant Physiology* 98:392–394.
Marler, P. (1970) Bird song and speech development: could there be parallels? *American Scientist* 58:669–673.
Marler, P., and Nelson, D. A. (1993) Action-based learning: a new form of developmental plasticity in bird song. *Netherlands Journal of Zoology* 43:91–103.
Marshall, D. L., Levin, D. A., and Fowler, N. L. (1986) Plasticity of yield components in response to stress in *Sesbania macrocarpa* and *Sesbania vesicaria* (Leguminosae). *American Naturalist* 127:508–521.
Martin-Mora, E., and James, F. C. (1995) Developmental plasticity in the shell of the queen conch *Strombus gigas. Ecology* 76:981–994.
Martins, E. P. (2000). Adaptation and the comparative method. *Trends in Ecology and Evolution* 15:296–299.
Martins, E. P., and Hansen, T. F. (1997) Phylogenies and the comparative method: a general approach to incorporating phylogenetic information into the analysis of interspecific data. *American Naturalist* 149:646–667.
Mathews, S., and Sharrock, R. A. (1997) Phytochrome gene diversity. *Plant, Cell and Environment* 20:666–671.
Mathews, S., Lavin, M., and Sharrock, R. A. (1995) Evolution of the phytochrome gene family and its utility for phylogenetic analyses of angiosperms. *Annals of the Missouri Botanical Gardens* 82:296–321.
Matsuda, R. (1982) The evolutionary process in talitrid amphipods and salamanders in changing environments, with a discussion of "genetic assimilation" and some other evolutionary concepts. *Canadian Journal of Zoology* 60:733–749.
Maxwell, D. P., Falk, S., Trick, C. G., and Huner, N. P. A. (1994) Growth at low temperature mimics high-light acclimation in *Chlorella vulgaris. Plant Physiology* 105:535–543.
Maynard Smith, J., Burian, R., Kauffman, S., Alberch, P., Campbell, J., Goodwin, B., Lande, R., Raup, D., and Wolpert, L. (1985) Developmental constraints and evolution. *Quarterly Review of Biology* 60:265–287.
Mayr, E., and Provine, W. B. (1980) *The Evolutionary Synthesis: Perspectives on the Unification of Biology.* Cambridge, Mass.: Harvard University Press.
Mazer, S. J., and Delasalle, V. A. (1996) Temporal instability of genetic components of floral trait variation: maternal family and population effects in *Spergularia marina* (Caryophyllaceae). *Evolution* 50:2509–2515.
Mazer, S. J., and Gorchov, D. L. (1996) Parental effects on progeny phenotype in plants: distinguishing genetic and environmental causes. *Evolution* 50:44–53.
Mazer, S. J., and Schick, C. T. (1991a) Constancy of population parameters for life history and floral traits in *Raphanus sativus* L. I. Norms of reaction and the nature of genotype by environment interactions. *Heredity* 67:143–156.
Mazer, S. J., and Schick, C. T. (1991b) Constancy of population parameters for life-history and floral traits in *Raphanus sativus* L. II. Effects of planting density on phenotype and heritability estimates. *Evolution* 45:1888–1907.
McCabe, J., and Dunn, A. M. (1997) Adaptive significance of environmental sex determination in an amphipod. *Journal of Evolutionary Biology* 10:515–528.
McCollum, S. A., and van Buskirk, J. (1996) Costs and benefits of a predator-induced polyphenism in the gray treefrog *Hyla chrysoscelis. Evolution* 50:583–593.

McCormac, A. C., Wagner, D., Boylan, M. T., Quail, P. H., Smith, H., and Whitelam, G. C. (1993) Photoresponses of transgenic *Arabidopsis* seedlings expressing introduced phytochrome B-encoding cDNAs: evidence that phytochrome A and phytochrome B have distinct photoregulatory functions. *Plant Journal* 4:19–27.

McDowell, J. M., An, Y.-Q., Huang, S., McKinney, E. C., and Meagher, R. B. (1996) The *Arabidopsis* ACT7 actin gene is expressed in rapidly developing tissues and responds to several external stimuli. *Plant Physiology* 111:699–711.

McKenzie, G. J., Harris, R. S., Lee, P. L., and Rosenberg, S. M. (2000) The SOS response regulates adaptive mutation. *Proceedings of the National Academy of Sciences of the USA* 97:6646–6651.

McKinney, M. L., and Gittleman, J. L. (1995) Ontogeny and phylogeny: tinkering with covariation in life history, morphology and behaviour, pp. 21–47 in: *Evolutionary Change and Heterochrony,* K. J. McNamara (ed.). New York: John Wiley and Sons.

McKinney, M. L., and McNamara, K. J. (1991) *Heterochrony: The Evolution of Ontogeny.* New York: Plenum Press.

McKitrick, M. C. (1993) Phylogenetic constraint in evolutionary theory: has it any explanatory power? *Annual Review of Ecology and Systematics* 24:307–330.

McNamara, K. J. (1997) *Shapes of Time.* Baltimore: Johns Hopkins University Press.

McNeill Alexander, R. (1990) Apparent adaptation and actual performance. *Evolutionary Biology* 24:357–373.

McShea, D. W. (1994) Mechanisms of large-scale evolutionary trends. *Evolution* 48:1747–1763.

Menu, F., Roebuck, J.-P., and Viala, M. (2000). Bet-hedging diapause strategies in stochastic environments. *American Naturalist* 155:724–734.

Merila, J. (1997) Expression of genetic variation in body size of the collared flycatcher under different environmental conditions. *Evolution* 51:526–536.

Merila, J., and Bjorklund, M. (1999) Population divergence and morphometric integration in the Greenfinch (*Carduelis chloris*)—evolution against the trajectory of least resistance? *Journal of Evolutionary Biology* 12:103–112.

Merila, J., Bjorklund, M., and Gustafsson, L. (1994) Evolution of morphological differences with moderate genetic correlations among traits as exemplified by two flycatcher species (Ficedula; Muscicapidae). *Biological Journal of the Linnean Society* 52:19–30.

Meyer, A. (1987) Phenotypic plasticity and heterochrony in *Cichlasoma managuense* (Pisces, Cichlidae) and their implications for speciation in cichlid fishes. *Evolution* 41:1357–1369.

Meyer, G. A. (2000). Interactive effects of soil fertility and herbivory on *Brassica nigra. Oikos* 88:433–441.

Mezey, J. G., Cheverud, J. M., and Wagner, G. P. (2000) Is the genotype-phenotype map modular? A statistical approach using mouse quantitative trait loci data. *Genetics* 156:305–311.

Mhiri, C., Morel, J.-B., Vernhettes, S., Casacuberta, J. M., Lucas, H., and Grandbastien, M.-A. (1997) The promoter of the tobacco Tnt1 retrotransposon is induced by wounding and by abiotic stress. *Plant Molecular Biology* 33:257–266.

Mitchell, R. S. (1976) Submergence experiments on nine species of semi-aquatic *Polygonum. American Journal of Botany* 63:1158–1165.

Mitchell-Olds, T. (1995) The molecular basis of quantitative genetic variation in natural populations. *Trends in Ecology and Evolution* 10:324–327.

Mitchell-Olds, T. (1996) Genetic constraints on life-history evolution: quantitative-trait loci influencing growth and flowering in *Arabidopsis thaliana. Evolution* 50:140–145.

Mitchell-Olds, T., and Bradley, D. (1996) Genetics of *Brassica rapa.* 3. Costs of disease resistance to three fungal pathogens. *Evolution* 50:1859–1865.

Mitchell-Olds, T., and Rutledge, J. J. (1986) Quantitative genetics in natural plant populations: a review of the theory. *American Naturalist* 127:379–402.

Mizoguchi, T., Irie, K., Hirayama, T., Hayashida, N., Yamaguchi-Shinozaki, K., Matsumoto, K., and Shinozaki, K. (1996) A gene encoding a mitogen-activated protein kinase kinase kinase is induced simultaneously with genes for a mitogen-activated protein kinase and an S6 ribosomal protein kinase by touch, cold, and water stress in *Arabidopsis thaliana. Proceedings of the National Academy of Sciences of the USA* 93:765–769.

Monteiro, A. F., Brakefield, P. M., and French, V. (1994) The evolutionary genetics and developmental basis of wing pattern variation in the butterfly *Bicyclus anynana. Evolution* 48:1147–1157.

Monteiro, A., Brakefield, P. M., and French, V. (1997a) Butterfly eyespots: the genetics and development of the color rings. *Evolution* 51:1207–1216.

Monteiro, A., Brakefield, P. M., and French, V. (1997b) The genetics and development of an eyespot pattern in the butterfly *Bicyclus anynana:* response to selection for eyespot shape. *Genetics* 146:287–294.

Monteiro, A., Brakefield, P. M., and French, V. (1997c) The relationship between eyespot shape and wing shape in the butterfly *Bicyclus anynana:* a genetic and morphometrical approach. *Journal of Evolutionary Biology* 10:787–802.

Moran, N. A. (1992) The evolutionary maintenance of alternative phenotypes. *American Naturalist* 139:971–989.

Morin, J. P., Moreteau, B., Petavy, G., Parkash, R., and David, J. R. (1997) Reaction norms of morphological traits in *Drosophila:* adaptive shape changes in a stenotherm circumtropical species? *Evolution* 51:1140–1148.

Morton, N. E. (1974) Analysis of family resemblance. I. Introduction. *American Journal of Human Genetics* 26:318–330.

Mozley, D., and Thomas, B. (1995) Developmental and photobiological factors affecting photoperiodic induction in *Arabidopsis thaliana* Heynh. Landsberg erecta. *Journal of Experimental Botany* 46:173–179.

Mueller, L. D., Guo, P. Z., and Ayala, F. J. (1991) Density-dependent natural selection and trade-offs in life-history traits. *Science* 253:433–435.

Murray, C. (1998) IQ will put you in your place, Web page, www.eugenics.net (accessed 11/7/00).

Neill, W. E. (1992) Population variation in the ontogeny of predator-induced vertical migration of copepods. *Nature* 356:54–57.

Neuffer, B., and Meyer-Walf, M. (1996) Ecotypic variation in relation to man made habitats in *Capsella:* field and trampling area. *Flora* 191:49–57.

Newman, R. A. (1992) Adaptive plasticity in amphibian metamorphosis. *BioScience* 42:671–678.

Newman, R. A. (1994) Genetic variation for phenotypic plasticity in the larval life history of spadefoot toads (*Scaphiopus couchii*). *Evolution* 48:1773–1785.

Neyfakh, A. A., and Hartl, D. L. (1993) Genetic control of the rate of embryonic development: selection for faster development at elevated temperatures. *Evolution* 47: 1625–1631.

Nijhout, H. F. (1994) Developmental perspectives on evolution of butterfly mimicry. *BioScience* 44:148–156.

Nijhout, H. F., Wray, G. A., Kremen, C., and Teragawa, C. K. (1986) Ontogeny, phylogeny and evolution of form: an algorithmic approach. *Systematic Zoology* 35:445–457.

Novoplansky, A., Cohen, D., and Sachs, T. (1994) Responses of an annual plant to temporal changes in light environment: an interplay between plasticity and determination. *Oikos* 69:437–446.

Nylin, S., and Gotthard, K. (1997) Plasticity in life-history traits. *Annual Review of Entomology* 43:63–83.

Ollason, J. G. (1991) What is this stuff called fitness? *Biology and Philosophy* 6:81–92.

Olvido, A. E., and Mousseau, T. A. (1995) Effect of rearing environment on calling-song plasticity in the striped ground cricket. *Evolution* 49:1271–1277.

O'Neil, P. (1997) Natural selection on genetically correlated phenological characters in *Lythrum salicaria* L. (Lythraceae). *Evolution* 51:267–274.

Orr, A. (1998) The population genetics of adaptation: the distribution of factors fixed during adaptive evolution. *Evolution* 52:935–949.

Oyama, S. (1993) Constraints and development. *Netherlands Journal of Zoology* 43:6–16.

Palva, E. T. (1994) Gene expression under low temperature stress, pp. 103–130 in: *Stress-Induced Gene Expression in Plants,* A. S. Basra (ed.). Newark, N.J.: Harwood.

Parejko, K., and Dodson, S. I. (1991) The evolutionary ecology of an antipredator reaction norm: *Daphnia pulex* and *Chaoborus americanus*. *Evolution* 45: 1665–1674.

Parsons, B. L., and Mattoo, A. K. (1991) Wound-regulated accumulation of specific transcripts in tomato fruit: interactions with fruit development, ethylene and light. *Plant Molecular Biology* 17:453–464.

Parsons, P. A. (1987) Evolutionary rates under environmental stress. *Evolutionary Theory* 21:311–347.

Partridge, L., and Fowler, K. (1993) Responses and correlated responses to artificial selection on thorax length in *Drosophila melanogaster*. *Evolution* 47:213–226.

Passera, L., Roncin, E., Kaufmann, B., and Keller, L. (1996) Increased soldier production in ant colonies exposed to intraspecific competition. *Nature* 379:630–631.

Paterson, A. H., Damon, S., Hewitt, J. D., Zamir, D., Rabinowitch, H. D., Lincoln, S. E., Lander, E. S., and Tanksley, S. D. (1991) Mendelian factors underlying quantitative traits in tomato: comparison across species, generations, and environments. *Genetics* 127:181–197.

Paterson, A. M., Wallis, G. P., and Gray, R. D. (1995) Penguins, petrels, and parsimony: does cladistic analysis of behavior reflect seabird phylogeny? *Evolution* 49:974–989.

Paulson, S. G., Roberts, C. W., and Staley, L. M. (1973) The effect of environment of body weight heritability estimates. *Poultry Science* 52:1557–1563.

Pederson, D. G. (1968) Environmental stress, heterozygote advantage and genotype-environment interaction in *Arabidopsis*. *Heredity* 23:127–138.

Perkins, J. M., and Jinks, J. L. (1966) Environmental and genotype-environmental components of variability. III. Multiple lines and crosses. *Heredity* 21:399–405.

Petrov, A. P., and Petrosov, V. A. (1981) Variability in some lines of *Arabidopsis thaliana* (L.) Heynh. under different lighting conditions. *Genetika* 16:1596–1602.

Pfennig, D. W. (1992) Proximate and functional causes of polyphenism in an anuran tadpole. *Functional Ecology* 6:167–174.

Phillips, P. C. (1998) Genetic constraints at the metamorphic boundary: morphological development in the wood frog, *Rana sylvatica. Journal of Evolutionary Biology* 11:453–464.

Phillips, P. C., and Arnold, S. J. (1989) Visualizing multivariate selection. *Evolution* 43:1209–1222.

Pickett, F. B., and Meeks-Wagner, D. R. (1995) Seeing double: appreciating genetic redundancy. *Plant Cell* 7:1347–1356.

Piepho, H.-P. (1994) Application of a generalized Grubbs' model in the analysis of genotype-environment interaction. *Heredity* 73:113–116.

Piersma, T., and Lindstrom, A. (1997) Rapid reversible changes in organ size as a component of adaptive behaviour. *Trends in Ecology and Evolution* 12:134–138.

Pigliucci, M. (1992) Modelling phenotypic plasticity. I. Linear and higher order effects of dominance, drift, environmental frequency and selection on one-locus, two-alleles. *Journal of Genetics* 71:135–150.

Pigliucci, M. (1996a) How organisms respond to environmental changes: from phenotypes to molecules (and vice versa). *Trends in Ecology and Evolution* 11:168–173.

Pigliucci, M. (1996b) Modelling phenotypic plasticity. II. Do genetic correlations matter? *Heredity* 77:453–460.

Pigliucci, M. (1997) Ontogenetic phenotypic plasticity during the reproductive phase in *Arabidopsis thaliana* (Brassicaceae). *American Journal of Botany* 84:887–895.

Pigliucci, M. (2001) Characters and environments, pp. 363–388 in: *The Character Concept in Evolutionary Biology,* G. P. Wagner (ed.). San Diego: Academic Press.

Pigliucci, M., and Byrd, N. (1998) Genetics and evolution of phenotypic plasticity to nutrient stress in Arabidopsis: drift, constraints or selection? *Biological Journal of the Linnean Society* 64:17–40.

Pigliucci, M., and Kaplan, J. (2000) The fall and rise of Dr. Pangloss: adaptationism and the Spandrels paper 20 years later. *Trends in Ecology and Evolution* 15:66–70.

Pigliucci, M., and Schlichting, C. D. (1995) Ontogenetic reaction norms in *Lobelia siphilitica* (Lobeliaceae): response to shading. *Ecology* 76:2134–2144.

Pigliucci, M., and Schlichting, C. D. (1996) Reaction norms of *Arabidopsis.* IV. Relationships between plasticity and fitness. *Heredity* 76:427–436.

Pigliucci, M., and Schlichting, C. D. (1997) On the limits of quantitative genetics for the study of phenotypic evolution. *Acta Biotheoretica* 45:143–160.

Pigliucci, M., and Schlichting, C. D. (1998) Reaction norms of arabidopsis. V. Flowering time controls phenotypic architecture in response to nutrient stress. *Journal of Evolutionary Biology* 11:285–301.

Pigliucci, M., and Schmitt, J. (1999) Genes affecting phenotypic plasticity in *Arabidopsis:* pleiotropic effects and reproductive fitness of photomorphogenic mutants. *Journal of Evolutionary Biology* 12:551–562.

Pigliucci, M., Whitton, J., and Schlichting, C. D. (1995) Reaction norms of *Arabidopsis.* I. Plasticity of characters and correlations across water, nutrient and light gradients. *Journal of Evolutionary Biology* 8:421–438.

Pigliucci, M., Schlichting, C. D., Jones, C. S., and Schwenk, K. (1996) Developmental reaction norms: the interactions among allometry, ontogeny and plasticity. *Plant Species Biology* 11:69–85.

Pigliucci, M., deIorio, P., and Schlichting, C. D. (1997) Phenotypic plasticity of growth trajectories in two species of *Lobelia* in response to nutrient availability. *Journal of Ecology* 85:265–276.

Pigliucci, M., Tyler, G. A., and Schlichting, C. D. (1998) Mutational effects on constraints on character evolution and phenotypic plasticity in *Arabidopsis thaliana. Journal of Genetics* 77:95–103.

Pigliucci, M., Cammell, K., and Schmitt, J. (1999) Evolution of phenotypic plasticity: a comparative approach in the phylogenetic neighborhood of *Arabidopsis thaliana. Journal of Evolutionary Biology* 12:779–791.

Pilson, D. (1996) Two herbivores and constraints on selection for resistance in *Brassica rapa. Evolution* 50:1492–1500.

Pinker, S. (1997) *How the Mind Works.* New York: W. W. Norton.

Pollard, H., Cruzan, M., and Pigliucci, M. (2001) Comparative studies of reaction norms in *Arabidopsis.* I. Evolution of response to daylength. *Evolutionary Ecology Research* 3:129–155.

Pooni, H. S., and Treharne, A. J. (1994) The role of epistasis and background genotype in the expression of heterosis. *Heredity* 72:628–635.

Porras, R., and Munoz, J. M. (2000) Cleistogamous capitulum in *Centaurea melitensis* (Asteraceae): heterochronic origin. *American Journal of Botany* 87:925–933.

Prandl, R., Kloske, E., and Schoffl, F. (1995) Developmental regulation and tissue-specific differences of heat shock gene expression in transgenic tobacco and *Arabidopsis* plants. *Plant Molecular Biology* 28:73–82.

Price, P. W. (1994) Phylogenetic constraints, adaptive syndromes, and emergent properties: from individuals to populations dynamics. *Researches on Population Ecology* 36:3–14.

Provine, W. B. (1971) *The Origins of Theoretical Population Genetics.* Chicago: University of Chicago Press.

Provine, W. B. (1981) Origins of the GNP series, pp. 5–83 in: *Dobzhansky's Genetics of Natural Populations,* R. C. Lewontin, J. A. Moore, and W. B. Provine (eds.). New York: Columbia University Press.

Purugganan, M. D. (1998) The molecular evolution of development. *BioEssays* 20:700–711.

Purugganan, M. D., and Suddith, J. I. (1998) Molecular population genetics of the *Arabidopsis* CAULIFLOWER regulatory gene: nonneutral evolution and naturally occurring variation in floral homeotic function. *Proceedings of the National Academy of Sciences of the USA* 95:8130–8134.

Purugganan, M. D., and Suddith, J. I. (1999) Molecular population genetics of floral homeotic loci: departures from the equilibrium-neutral model at the APETALA3 and PISTILLATA genes of *Arabidopsis thaliana. Genetics* 151:839–848.

Qin, M., Kuhn, R., Moran, S., and Quail, P. H. (1997) Overexpressed phytochrome C has similar photosensory specificity to phytochrome B but a distinctive capacity to enhance primary leaf expansion. *Plant Journal* 12:1163–1172.

Quail, P. H. (1997a) An emerging molecular map of the phytochromes. *Plant, Cell and Environment* 20:657–665.

Quail, P. H. (1997b) The phytochromes: a biochemical mechanism of signaling in sight? *BioEssays* 19:571–579.

Quesada, V., Ponce, M. R., and Micol, J. L. (2000) Genetic analysis of salt-tolerance mutants in *Arabidopsis thaliana. Genetics* 154:421–436.

Quinn, J. A. (1987) Complex patterns of genetic differentiation and phenotypic plasticity versus an outmoded ecotype terminology, pp. 95–113 in: *Differentiation Patterns in Higher Plants,* K. M. Urbanska (ed.). London: Academic Press.

Qvarnstrom, A., Part, T., and Sheldon, B. C. (2000) Adaptive plasticity in mate preference linked to differences in reproductive effort. *Nature* 405:344–347.

Raff, R. A. (1996) *The Shape of Life.* Chicago: University of Chicago Press.

Raff, R. A., and Wray, G. A. (1989) Heterochrony: developmental mechanisms and evolutionary results. *Journal of Evolutionary Biology* 2:409–434.

Raff, R. A., Marshall, C. R., and Turbeville, J. M. (1994) Using DNA sequences to unravel the Cambrian radiation of the animal phyla. *Annual Review of Ecology and Systematics* 25:351–375.

Rakitina, T. Y., Vlasov, P. V., Jalilova, F. K., and Kefeli, V. I. (1994) Absiscic acid and ethylene in mutants of *Arabidopsis thaliana* differing in their resistance to ultraviolet (UV-B) radiation stress. *Russian Journal of Plant Physiology* 41:599–603.

Ramachandran, V. S. (1996) The evolutionary biology of self-deception, laughter, dreaming and depression: some clues from anosognosia. *Medical Hypotheses* 47: 347–362.

Rao, D. C., Morton, N. E., and Yee, S. (1974) Analysis of family resemblance. II. A linear model for familial correlation. *American Journal of Human Genetics* 26: 331–359.

Rausher, M. D. (1992) The measurement of selection on quantitative traits: biases due to environmental covariances between traits and fitness. *Evolution* 46:616–626.

Real, L. A., and Ellner, S. (1992) Life history evolution in stochastic environments: a graphical mean-variance approach. *Ecology* 73:1227–1236.

Reboud, X., and Bell, G. (1997) Experimental evolution in *Chlamydomonas.* III. Evolution of specialist and generalist types in environments that vary in space and time. *Heredity* 78:507–514.

Reed, J. W., Nagatani, A., Elich, T. D., Fagan, M., and Chory, J. (1994) Phytochrome A and phytochrome B have overlapping but distinct functions in *Arabidopsis* development. *Plant Physiology* 104:1139–1149.

Reekie, J. Y. C., and Hiclenton, P. R. (1994) Effects of elevated CO_2 on time of flowering in four short-day and four long-day species. *Canadian Journal of Botany* 72:533–538.

Reid, J. B. (1993) Plant hormone mutants. *Journal of Plant Growth Regulation* 12:207–226.

Reilly, S. M. (1994) The ecological morphology of metamorphosis: heterochrony and the evolution of feeding mechanisms in salamanders, pp. 319–338 in: *Ecological Morphology: Integrative Organismal Biology,* P. C. Wainwright and S. M. Reilly (eds.). Chicago: University of Chicago Press.

Resnik, D. (1994) The rebirth of rational morphology: process structuralism's philosophy of biology. *Acta Biotheoretica* 42:1–14.

Reyment, R. A., and Kennedy, W. J. (1991) Phenotypic plasticity in a Cretaceous ammonite analyzed by multivariate statistical methods. *Evolutionary Biology* 25: 411–426.

Reznick, D., and Travis, J. (1996) The empirical study of adaptation in natural populations, pp. 243–289 in: *Adaptation,* M. R. Rose and G. V. Lauder (eds.). San Diego: Academic Press.

Rhen, T., and Lang, J. W. (1995) Phenotypic plasticity for growth in the common snapping turtle: effects of incubation temperature, clutch, and their interaction. *American Naturalist* 146:726–747.

Rice, S. H. (1998) The evolution of canalization and the breaking of von Baer's laws: modeling the evolution of development with epistasis. *Evolution* 52:647–656.

Rickey, T. M., and Belknap, W. R. (1991) Comparison of the expression of several stress-responsive genes in potato tubers. *Plant Molecular Biology* 16:1009–1018.

Riechert, S. E., and Hall, R. F. (2000) Local population success in heterogeneous habitats: reciprocal transplant experiments completed on a desert spider. *Journal of Evolutionary Biology* 13:541–550.

Riska, B., Prout, T., and Turelli, M. (1989) Laboratory estimates of heritabilities and genetic correlations in nature. *Genetics* 123:865–871.

Roach, D. A., and Wulff, R. D. (1987) Maternal effects in plants. *Annual Review of Ecology and Systematics* 18:209–235.

Robert, C., Foulley, J. L., and Ducrocq, V. (1995a) Genetic variation of traits measured in several environments. I. Estimation and testing of homogeneous genetic and intra-class correlations between environments. *Genetics Selection and Evolution* 27:111–123.

Robert, C., Foulley, J. L., and Ducrocq, V. (1995b) Genetic variation of traits measured in several environments. II. Inference on between-environment homogeneity of intra-class correlations. *Genetics Selection and Evolution* 27:125–134.

Robertson, A. (1959a) Experimental design in the evaluation of genetic parameters. *Biometrics* 15:219–226.

Robertson, A. (1959b) The sampling variance of the genetic correlation coefficient. *Biometrics* 15:469–485.

Robertson, F. W. (1964) The ecological genetics of growth in *Drosophila.* 7. The role of canalization in the stability of growth relations. *Genetical Research* 5:107–126.

Roff, D. A. (1994) The evolution of dimorphic traits: predicting the genetic correlation between environments. *Genetics* 136:395–401.

Roff, D. A. (1997) *Evolutionary Quantitative Genetics.* New York: Chapman & Hall.

Roff, D. A. (2000) The evolution of the **G** matrix: selection or drift? *Heredity* 84:135–142.

Roff, D. A., and Bradford, M. J. (2000) A quantitative genetic analysis of phenotypic plasticity of diapause induction in the cricket *Allonemobius socius. Heredity* 84:193–200.

Roff, D. A., and Mousseau, T. A. (1999) Does natural selection alter genetic architecture? An evaluation of quantitative genetic variation among populations of *Allenomobius socius* and *A. fasciatus. Journal of Evolutionary Biology* 12: 361–369.

Roff, D. A., Stirling, G., and Fairbairn, D. J. (1997) The evolution of threshold traits: a quantitative genetic analysis of the physiological and life-history correlates of wing dimorphism in the sand cricket. *Evolution* 51:1910–1919.

Roper, C., Pignatelli, P., and Partridge, L. (1996) Evolutionary responses of *Drosophila melanogaster* life history to differences in larval density. *Journal of Evolutionary Biology* 9:609–622.

Rose, M. R., and Lauder, G. V. (1996a) *Adaptation.* San Diego: Academic Press.

Rose, M. R., and Lauder, G. V. (1996b) Post-Spandrel adaptationism, pp. 1–8 in: *Adaptation,* M. R. Rose and G. V. Lauder (eds.). San Diego: Academic Press.

Rose, M. R., Nusbaum, T. J., and Chippindale, A. K. (1996) Laboratory evolution: the experimental Wonderland and the Cheshire cat syndrome, pp. 221–241 in: *Adaptation,* M. R. Rose and G. V. Lauder (eds.). San Diego: Academic Press.

Roskam, J. C., and Brakefield, P. M. (1996) A comparison of temperature-induced polyphenism in African *Bicyclus* butterflies from a seasonal savannah-rainforest ecotone. *Evolution* 50:2360–2372.

Rossiter, M. C. (1996) Incidence and consequences of inherited environmental effects. *Annual Review of Ecology and Systematics* 27:451–476.
Rutherford, S. L., and Lindquist, S. (1998) Hsp90 as a capacitor for morphological evolution. *Nature* 396:336–342.
Sall, T., and Pettersson, P. (1994) A model of photosynthetic acclimation as a special case of reaction norms. *Journal of Theoretical Biology* 166:1–8.
Sandoval, C. P. (1994) Plasticity in web design in the spider *Parawixia bistriata:* a response to variable prey type. *Functional Ecology* 8:701–707.
Scharloo, W. (1989) Developmental and physiological aspects of reaction norms. *BioScience* 39:465–471.
Scheiner, S. M. (1993a) Genetics and evolution of phenotypic plasticity. *Annual Review of Ecology and Systematics* 24:35–68.
Scheiner, S. M. (1993b) Plasticity as a selectable trait: reply to Via. *American Naturalist* 142:371–373.
Scheiner, S. M. (1998) The genetics of phenotypic plasticity. VII. Evolution in a spatially-structured environment. *Journal of Evolutionary Biology* 11:303–320.
Scheiner, S. M., and Berrigan, D. (1998) The genetics of phenotypic plasticity. VIII. The cost of plasticity in *Daphnia pulex. Evolution* 52:368–378.
Scheiner, S. M., and Goodnight, C. J. (1984) The comparison of phenotypic plasticity and genetic variation in populations of the grass *Danthonia spicata. Evolution* 38:845–855.
Scheiner, S. M., and Lyman, R. F. (1989) The genetics of phenotypic plasticity. I. Heritability. *Journal of Evolutionary Biology* 2:95–107.
Scheiner, S. M., and Lyman, R. F. (1991) The genetics of phenotypic plasticity. II. Response to selection. *Journal of Evolutionary Biology* 4:23–50.
Scheiner, S. M., Caplan, R. L., and Lyman, R. F. (1991) The genetics of phenotypic plasticity. III. Genetic correlations and fluctuating asymmetries. *Journal of Evolutionary Biology* 4:51–68.
Schichnes, D. E., and Freeling, M. (1998) Lax midrib1-O, a systemic, heterochronic mutant of maize. *American Journal of Botany* 85:481–491.
Schlichting, C. D. (1986) The evolution of phenotypic plasticity in plants. *Annual Review of Ecology and Systematics* 17:667–693.
Schlichting, C. D. (1989a) Phenotypic integration and environmental change. *BioScience* 39:460–464.
Schlichting, C. D. (1989b) Phenotypic plasticity in *Phlox.* II. Plasticity of character correlations. *Oecologia* 78:496–501.
Schlichting, C. D., and Levin, D. A. (1984) Phenotypic plasticity of annual *Phlox:* tests of some hypotheses. *American Journal of Botany* 71:252–260.
Schlichting, C. D., and Levin, D. A. (1986a) Effects of inbreeding on phenotypic plasticity in cultivated *Phlox. Theoretical and Applied Genetics* 72:114–119.
Schlichting, C. D., and Levin, D. A. (1986b) Phenotypic plasticity: an evolving plant character. *Biological Journal of the Linnean Society* 29:37–47.
Schlichting, C. D., and Levin, D. A. (1988) Phenotypic plasticity in *Phlox.* I. Wild and cultivated populations of *P. drummondii. American Journal of Botany* 75: 161–169.
Schlichting, C. D., and Levin, D. A. (1990) Phenotypic plasticity in *Phlox.* III. Variation among natural populations of *Phlox drummondii. Journal of Evolutionary Biology* 3:411–428.
Schlichting, C. D., and Pigliucci, M. (1993) Evolution of phenotypic plasticity via regulatory genes. *American Naturalist* 142:366–370.

Schlichting, C. D., and Pigliucci, M. (1995) Gene regulation, quantitative genetics and the evolution of reaction norms. *Evolutionary Ecology* 9:154–168.

Schlichting, C. D., and Pigliucci, M. (1998) *Phenotypic Evolution: A Reaction Norm Perspective.* Sunderland, Mass.: Sinauer.

Schluter, D., and Nagel, L. M. (1995) Parallel speciation by natural selection. *American Naturalist* 146:292–301.

Schluter, D., and Nychka, D. (1994) Exploring fitness surfaces. *American Naturalist* 143:597–616.

Schmalhausen, I. I. (1949) *Factors of Evolution: The Theory of Stabilizing Selection.* Chicago: University of Chicago Press.

Schmitt, J. (1995) Genotype-environment interaction, parental effects, and the evolution of plant reproductive traits, pp. 1–16 in: *Experimental and Molecular Approaches to Plant Biosystematics,* P. Hoch (ed.). St. Louis: Missouri Botanical Gardens.

Schmitt, J. (1997) Is photomorphogenic shade avoidance adaptive? Perspectives from population biology. *Plant, Cell and Environment* 20:826–830.

Schmitt, J., and Wulff, R. D. (1993) Light spectral quality, phytochrome and plant competition. *Trends in Ecology and Evolution* 8:47–50.

Schmitt, J., Niles, J., and Wulff, R. D. (1992) Norms of reaction of seed traits to maternal environments in *Plantago lanceolata. American Naturalist* 139:451–466.

Schmitt, J., McCormac, A. C., and Smith, H. (1995) A test of the adaptive plasticity hypothesis using transgenic and mutant plants disabled in phytochrome-mediated elongation responses to neighbors. *American Naturalist* 146:937–953.

Schmitt, J., Dudley, S., and Pigliucci, M. (1999) Manipulative approaches to testing adaptive plasticity: phytochrome-mediated shade avoidance responses in plants. *American Naturalist* 154:S43–S54.

Schrag, S. J., Ndifon, G. T., and Read, A. F. (1994) Temperature-determined outcrossing ability in wild populations of a simultaneous hermaphrodite snail. *Ecology* 75:2066–2077.

Schwaegerle, K. E., McIntyre, H., and Swingley, C. (2000) Quantitative genetics and the persistence of environmental effects in clonally propagated organisms. *Evolution* 54:452–461.

Secor, E. M., and Diamond, J. (1995) Adaptive responses to feeding in Burmese pythons: Pay before pumping. *Journal of Experimental Biology* 198:1313–1325.

Seger, J., and Stubblefield, J. W. (1996) Optimization and adaptation, pp. 93–123 in: *Adaptation,* M. R. Rose and G. V. Lauder (eds.). San Diego: Academic Press.

Semlitsch, R. D. (1987) Paedomorphosis in *Ambystoma talpoideum:* effects of density, food, and pond drying. *Ecology* 68:994–1102.

Semlitsch, R. D., and Wilbur, H. M. (1989) Artificial selection for paedomorphosis in the salamander *Ambystoma talpoideum. Evolution* 43:105–112.

Service, P. M., and Rose, M. R. (1985) Genetic covariation among life-history components: the effect of novel environments. *Evolution* 39:943–945.

Sgrò, C. M., and Hoffmann, A. A. (1998) Heritable variation for fecundity in field-collected *Drosophila melanogaster* and their offspring reared under different environmental temperatures. *Evolution* 52:134–143.

Sharrock, R. A., and Quail, P. H. (1989) Novel phytochrome sequences in *Arabidopsis thaliana:* structure, evolution, and differential expression of a plant regulatory photoreceptor family. *Genes & Development* 3:1745–1757.

Shaw, A. J., Weir, B. S., and Shaw, F. H. (1997) The occurrence and significance of epistatic variance for quantitative characters and its measurement in haploids. *Evolution* 51:348–353.

Shaw, F. H., Shaw, R. G., Wilkinson, G. S., and Turelli, M. (1995) Changes in genetic variances and covariances: **G** whiz! *Evolution* 49:1260–1267.

Sheldon, C. C., Rouse, D. T., Peacock, W. J., and Dennis, E. S. (2000) The molecular basis of vernalization: the central role of *flowering locus c* (FLC). *Proceedings of the National Academy of Sciences of the USA* 97:3753–3758.

Shimizu, T., and Masaki, S. (1993) Genetic variability of the wing-form response to photoperiod in a subtropical population of the ground cricket, *Dianemobius fascipes. Zoological Science* 10:935–944.

Shine, R. (1995) A new hypothesis for the evolution of viviparity in reptiles. *American Naturalist* 145:809–823.

Shine, R. (1999) Why is sex determined by nest temperature in many reptiles? *Trends in Ecology and Evolution* 14:186–189.

Shinomura, T., Nagatani, A., Chory, J., and Furuya, M. (1994) The induction of seed germination in *Arabidopsis thaliana* is regulated principally by phytochrome B and secondarily by phytochrome A. *Plant Physiology* 104:363–371.

Sibly, R. M. (1995) Life-history evolution in spatially heterogeneous environments, with and without phenotypic plasticity. *Evolutionary Ecology* 9:242–257.

Sih, A., and Gleeson, S. K. (1995) A limits-oriented approach to evolutionary ecology. *Trends in Ecology and Evolution* 10:378–382.

Sih, A., and Moore, R. D. (1993) Delayed hatching of salamander eggs in response to enhanced larval predation risk. *American Naturalist* 142:947–960.

Sills, G. R., and Nienhuis, J. (1995) Maternal phenotypic effects due to soil nutrient levels and sink removal in *Arabidopsis thaliana* (Brassicaceae). *American Journal of Botany* 82:491–495.

Simms, E. L., and Triplett, J. (1994) Costs and benefits of plant responses to disease: resistance and tolerance. *Evolution* 48:1973–1985.

Simons, A. M., and Roff, D. A. (1994) The effect of environmental variability on the heritabilities of traits of a field cricket. *Evolution* 48:1637–1649.

Simons, A. M., and Roff, D. A. (1996) The effect of a variable environment on the genetic correlation structure in a field cricket. *Evolution* 50:267–275.

Sinervo, B., and Doughty, P. (1996) Interactive effects of offspring size and timing of reproduction on offspring reproduction: experimental, maternal, and quantitative genetic aspects. *Evolution* 50:1314–1327.

Sinervo, B., and Svensson, E. (1998) Mechanistic and selective causes of life history trade-offs and plasticity. *Oikos* 83:432–442.

Slatkin, M. (1987) Quantitative genetics of heterochrony. *Evolution* 41:799–811.

Slijper, E. J. (1942) Biologic-anatomical investigations on the bipedal gait and upright posture in mammals, with special reference to a little goat, born without forelegs. *Proc. Koninkl. Ned. Akad. Wetensch.* 45:288–295, 407–415.

Smith, H. (1990) Signal perception, differential expression within multigene families and the molecular basis of phenotypic plasticity. *Plant, Cell and Environment* 13:585–594.

Smith, H. (1995) Physiological and ecological function within the phytochrome family. *Annual Review of Plant Physiology and Plant Molecular Biology* 46:289–315.

Smith, H., and Whitelam, G. C. (1997) The shade avoidance syndrome: multiple responses mediated by multiple phytochromes. *Plant, Cell and Environment* 20:840–844.

Smith-Gill, S. J. (1983) Developmental plasticity: developmental conversion versus phenotypic modulation. *American Zoologist* 23:47–55.

Sober, E. (1984) *The Nature of Selection: Evolutionary Theory in Philosophical Focus.* Cambridge, Mass.: MIT Press.

Spencer, W. E., and Wetzel, R. G. (1993) Acclimation of photosynthesis and dark respiration of a submersed Angiosperm beneath ice in a temperate lake. *Plant Physiology* 101:985–991.

Stanley, S. M. (1975) A theory of evolution above the species level. *Proceedings of the National Academy of Sciences of the USA* 72:646–650.

Stanton, M. L., Roy, B. A., and Thiede, D. A. (2000) Evolution in stressful environments. I. Phenotypic variability, phenotypic selection, and response to selection in five distinct environmental stresses. *Evolution* 54:93–111.

Starck, Z., and Witek-Czuprynska, B. (1993) Diverse response of tomato fruit explants to high temperature. *Acta Societatis Botanicorum Poloniae* 62:165–169.

Stearns, S., de Jong, G., and Newman, B. (1991) The effects of phenotypic plasticity on genetic correlations. *Trends in Ecology and Evolution* 6:122–126.

Stearns, S. C., and Kawecki, T. J. (1994) Fitness sensitivity and the canalization of life-history traits. *Evolution* 48:1438–1450.

Stearns, S. C., and Koella, J. C. (1986) The evolution of phenotypic plasticity in life-history traits: predictions of reaction norms for age and size at maturity. *Evolution* 40:893–913.

Stebbins, G. L., and Ayala, F. J. (1981) Is a new evolutionary synthesis necessary? *Science* 213:967–971.

Stern, C., and Sherwood, E. R. (1966) *The Origin of Genetics: A Mendel Source Book.* New York: W. H. Freeman.

Stewart, C. N. J., and Nilsen, E. T. (1995) Phenotypic plasticity and genetic variation of *Vaccinium macrocarpon,* the American cranberry. I. Reaction norms of clones from central and marginal populations in a common garden. *International Journal of Plant Science* 156:687–697.

Stowe, K. A. (1998) Experimental evolution of resistance in *Brassica rapa:* correlated response of tolerance in lines selected for glucosinolate content. *Evolution* 52:703–712.

Strathmann, R. R., Fenaux, L., and Strathmann, M. F. (1992) Heterochronic developmental plasticity in larval sea urchins and its implications for evolution of nonfeeding larvae. *Evolution* 46:972–986.

Stratton, D. A. (1994) Genotype-by-environment interactions for fitness of *Erigeron annuus* show fine-scale selective heterogeneity. *Evolution* 48:1607–1618.

Stratton, D. A., and Bennington, C. C. (1996) Measuring spatial variation in natural selection using randomly-sown seeds of *Arabidopsis thaliana. Journal of Evolutionary Biology* 9:215–228.

Stratton, D. A., and Bennington, C. C. (1998) Fine-grained spatial and temporal variation in selection does not maintain genetic variation in *Erigeron annuus. Evolution* 52:678–691.

Sturtevant, A. H. (1965) *A History of Genetics.* New York: Harper & Row.

Sultan, S. E. (1987) Evolutionary implications of phenotypic plasticity in plants. *Evolutionary Biology* 21:127–178.

Sultan, S. E. (1992) What has survived of Darwin's theory? Phenotypic plasticity and the neo-Darwinian legacy. *Evolutionary Trends in Plants* 6:61–71.

Sultan, S. E. (1995) Phenotypic plasticity and plant adaptation. *Acta Botanica Neederlandica* 44:363–383.

Sultan, S. E., and Bazzaz, F. A. (1993) Phenotypic plasticity in *Polygonum persicaria.* I. Diversity and uniformity in genotypic norms of reaction to light. *Evolution* 47:1009–1031.

Suzuki, D. T., Griffiths, A. J. F., Miller, J. H., and Lewontin, R. C. (1989) *An Introduction to Genetic Analysis.* New York: W. H. Freeman.

Swalla, B. J., and Jeffery, W. R. (1996) Requirement of the Manx gene for expression of chordate features in a tailless ascidian larva. *Science* 274:1205–1208.

Takahashi, N. (1975) Adaptive behaviors of rice plants in seed germination and seedling growth, pp. 147–153 in: *Adaptability in Plants,* T. Matsuo (ed.). Tokyo: University of Tokyo Press.

Tarasjev, A. (1995) Relationship between phenotypic plasticity and developmental instability in *Iris pumila* L. *Russian Journal of Genetics* 31:1409–1416.

Tauber, C. A., and Tauber, M. J. (1992) Phenotypic plasticity in *Chrysoperla:* genetic variation in the sensory mechanism and in correlated reproductive traits. *Evolution* 46:1754–1773.

Temte, J. L. (1994) Photoperiod control of birth timing in the harbour seal (*Phoca vitulina*). *Journal of Zoology* 233:369–384.

Thewissen, J. G. M., Hussain, S. T., and Arif, M. (1994) Fossil evidence for the origin of aquatic locomotion in *Archeocete* whales. *Science* 263:210–212.

Thomas, H., and Stoddart, J. L. (1995) Temperature sensitivities of *Festuca arundinacea* Schreb., and *Dactylis glomerata* L. ecotypes. *New Phytologist* 130:125–134.

Thomas, H., and Villiers, L. d. (1996) Gene expression in leaves of *Arabidopsis thaliana* induced to senescence by nutrient deprivation. *Journal of Experimental Botany* 47:1845–1852.

Thomas, R. D. K., and Reif, W. E. (1993) The skeleton space: a finite set of organic designs. *Evolution* 47:341–360.

Thompson, D. B. (1999) Genotype-environment interaction and the ontogeny of diet-induced phenotypic plasticity in size and shape of *Melanoplus femurrubrum* (Orthoptera: Acrididae). *Journal of Evolutionary Biology* 12:38–48.

Thompson, J. D. (1991) Phenotypic plasticity as a component of evolutionary change. *Trends in Ecology and Evolution* 6:246–249.

Thomson, K. S. (1992) Macroevolution: the morphological problem. *American Zoologist* 32:106–112.

Tollrian, R. (1993) Neckteeth formation in *Daphnia pulex* as an example of continuous phenotypic plasticity: morphological effects of *Chaoborus* kairomone concentration and their quantification. *Journal of Plankton Research* 15:1309–1318.

Tollrian, R. (1995) Predator-induced morphological defenses: costs, life history shifts, and maternal effects in *Daphnia pulex. Ecology* 76:1691–1705.

Trainor, F. R., and Egan, P. F. (1991) Discovering the various ecomorphs of *Scenedesmus:* the end of a taxonomic era. *Archiv für Protistenkunde* 139:125–132.

Travisano, M., Mongold, J. A., Bennett, A. F., and Lenski, R. E. (1995a) Experimental tests of the roles of adaptation, chance, and history in evolution. *Science* 267:87–90.

Travisano, M., Vasi, F., and Lenski, R. E. (1995b) Long-term experimental evolution in *Escherichia coli.* III. Variation among replicate populations in correlated responses to novel environments. *Evolution* 49:189–200.

Trussell, G. C. (1996) Phenotypic plasticity in an intertidal snail: the role of a common crab predator. *Evolution* 50:448–454.

Tufto, J. (2000) The evolution of plasticity and nonplastic spatial and temporal adaptations in the presence of imperfect environmental cues. *American Naturalist* 156: 121–130.

Turelli, M. (1988) Phenotypic evolution, constant covariances, and the maintenance of additive variance. *Evolution* 42:1342–1347.

Turelli, M., and Barton, N. H. (1994) Genetic and statistical analyses of strong selection on polygenic traits: what, me normal? *Genetics* 138:913–941.

Turesson, G. (1922) The genotypical response of the plant species to the habitat. *Hereditas* 3:211–350.

Turkington, R. (1983) Plasticity in growth and patterns of dry matter distribution of two genotypes of *Trifolium repens* grown in different environments of neighbours. *Canadian Journal of Botany* 61:2186–2194.

van Buskirk, J., and McCollum, S. A. (2000) Functional mechanisms of an inducible defence in tadpoles: morphology and behaviour influence mortality risk from predation. *Journal of Evolutionary Biology* 13:336–347.

van Buskirk, J., and Relyea, R. A. (1998) Selection for phenotypic plasticity in *Rana sylvatica* tadpoles. *Biological Journal of the Linnean Society* 65:301–328.

van Dam, N. M., Hadwich, K., and Baldwin, I. T. (2000) Induced responses in *Nicotiana attenuata* affect behavior and growth of the specialist herbivore *Manduca sexta. Oecologia* 122:371–379.

van der Weele, C. (1993) Metaphors and the privileging of causes: the place of environmental influences in explanations of development. *Acta Biotheoretica* 41:315–327.

van Klausen, M., Fischer, M., and Schmid, B. (2000) Costs of plasticity in foraging characteristics of the clonal plant *Ranunculus reptans. Evolution* 54:1947–1955.

van Noordwijk, A. J., and de Jong, G. (1986) Acquisition and allocation of resources: their influence on variation in life history tactics. *American Naturalist* 128: 137–142.

van Tienderen, P. H. (1990) Morphological variation in *Plantago lanceolata:* limits of plasticity. *Evolutionary Trends in Plants* 44:35–43.

van Tienderen, P. H. (1991) Evolution of generalists and specialists in spatially heterogeneous environments. *Evolution* 45:1317–1331.

van Tienderen, P. H., and Antonovics, J. (1994) Constraints in evolution: on the baby and the bath water. *Functional Ecology* 8:139–140.

van Tienderen, P. H., and Koelewijn, H. P. (1994) Selection on reaction norms, genetic correlations and constraints. *Genetical Research* 64:115–125.

van Tienderen, P. H., Hammad, I., and Zwaal, F. C. (1996) Pleiotropic effects of flowering time genes in the annual crucifer *Arabidopsis thaliana* (Brassicaceae). *American Journal of Botany* 83:169–174.

Vela-Cardenas, M., and Frey, K. J. (1972) Optimum environment for maximizing heritability and genetic gain from selection. *Iowa State Journal of Science* 46:381–394.

Via, S. (1984a) The quantitative genetics of polyphagy in an insect herbivore. I. Genotype-environment interaction in larval performance on different host plant species. *Evolution* 38:881–895.

Via, S. (1984b) The quantitative genetics of polyphagy in an insect herbivore. II. Genetic correlations in larval performance within and among host plants. *Evolution* 38:896–905.

Via, S. (1987) Genetic constraints on the evolution of phenotypic plasticity, pp. 47–71 in: *Genetic Constraints on Adaptive Evolution,* V. Loeschcke (ed.). Berlin: Springer-Verlag.

Via, S. (1993) Adaptive phenotypic plasticity: target or by-product of selection in a variable environment? *American Naturalist* 142:352–365.

Via, S., and Lande, R. (1985) Genotype-environment interaction and the evolution of phenotypic plasticity. *Evolution* 39:505–522.

Via, S., and Lande, R. (1987) Evolution of genetic variability in a spatially heterogeneous environment: effects of genotype-environment interaction. *Genetical Research* 49:147–156.

Via, S., Gomulkiewicz, R., de Jong, G., Scheiner, S. M., Schlichting, C. D., and Tienderen, P. H. V. (1995) Adaptive phenotypic plasticity: consensus and controversy. *Trends in Ecology and Evolution* 10:212–216.

Voesenek, L. A. C. J., and van der Veen, R. (1994) The role of phytohormones in plant stress: too much or too little water. *Acta Botanica Neederlandica* 43:91–127.

Vogl, C., and Rienesl, J. (1991) Testing for developmental constraints: carpal fusions in *Urodeles. Evolution* 45:1516–1519.

Waddington, C. H. (1942) Canalization of development and the inheritance of acquired characters. *Nature* 150:563–565.

Waddington, C. H. (1952) Selection of the genetic basis for an acquired character. *Nature* 169:278.

Waddington, C. H. (1953) Genetic assimilation of an acquired character. *Evolution* 7:118–126.

Waddington, C. H. (1960) Experiments on canalizing selection. *Genetical Research* 1:140–150.

Waddington, C. H. (1961) Genetic assimilation. *Advances in Genetics* 10:257–290.

Wagner, A. (1996) Does evolutionary plasticity evolve? *Evolution* 50:1008–1023.

Wagner, A., Wagner, G. P., and Similion, P. (1994) Epistasis can facilitate the evolution of reproductive isolation by peak shifts: a two-locus two-allele model. *Genetics* 138:533–545.

Wagner, G. P. (1988) The influence of variation and of developmental constraints on the rate of multivariate phenotypic evolution. *Journal of Evolutionary Biology* 1:45–66.

Wagner, G. P. (1989) The origin of morphological characters and the biological basis of homology. *Evolution* 43:1157–1171.

Wagner, G. P. (1995) Adaptation and the modular design of organisms, pp. 317–328 in: *Advances in Artificial Life,* F. Moran, A. Moreno, J. J. Merelo, and P. Chacon (eds.). Berlin: Springer-Verlag.

Wagner, G. P. (ed.) (2001) *The Character Concept in Evolutionary Biology.* San Diego: Academic Press.

Wagner, G. P., and Altenberg, L. (1996) Complex adaptations and the evolution of evolvability. *Evolution* 50:967–976.

Waitt, D. E., and Levin, D. A. (1993) Phenotypic integration and plastic correlations in *Phlox drummondii* (Polemoniaceae). *American Journal of Botany* 80:1224–1233.

Wake, D. B. (1991) Homoplasy: the result of natural selection, or evidence of design limitations? *American Naturalist* 138:543–567.

Wang, L.-W., Showalter, A. M., and Ungar, I. A. (1997) Effect of salinity on growth, ion content, and cell wall chemistry in *Atriplex prostrata* (Chenopodiaceae). *American Journal of Botany* 84:1247–1255.

Ward, P. J. (1994) Parent-offspring regression and extreme environments. *Heredity* 72:574–581.

Watson, M. J. O., and Hoffmann, A. A. (1996) Acclimation, cross-generation effects, and the response to selection for increased cold resistance in *Drosophila. Evolution* 50:1182–1192.

Weber, S. L., and Scheiner, S. M. (1991) The genetics of phenotypic plasticity. IV. Chromosomal localization. *Journal of Evolutionary Biology* 2:109–120.

Weinig, C. (2000a) Plasticity versus canalization: population differences in the timing of shade-avoidance responses. *Evolution* 54:441–451.

Weinig, C. (2000b) Differing selection in alternative competitive environments: shade-avoidance responses and germination timing. *Evolution* 54:124–136.

Weis, A. E., and Gorman, W. L. (1990) Measuring selection on reaction norms: an exploration of the *Eurosta-Solidago* system. *Evolution* 44:820–831.

Welin, B. V., Olson, A., Nylander, M., and Palva, E. T. (1994) Characterization and differential expression of dhn/lea/rab-like genes during cold acclimation and drought stress in *Arabidopsis thaliana. Plant Molecular Biology* 26:131–144.

Wells, C., and Pigliucci, M. (2000) Heterophylly in aquatic plants: considering the evidence for adaptive plasticity. *Perspectives in Plant Ecology Evolution and Systematics* 3:1–18.

Wells, M. M., and Henry, C. (1992) The role of courtship songs in reproductive isolation among populations of green lacewings of the genus *Chrysoperla* (Neuroptera: Chrysopidae). *Evolution* 46:31–42.

Westcott, B. (1986) Some methods of analysing genotype-environment interaction. *Heredity* 56:243–253.

West-Eberhard, M. J. (1989) Phenotypic plasticity and the origins of diversity. *Annual Review of Ecology and Systematics* 20:249–278.

West-Eberhard, M. J. (1992) Behavior and evolution, pp. 57–75 in: *Molds, Molecules and Metazoa,* P. R. Grant and H. S. Horn (eds.). Princeton, N.J.: Princeton University Press.

Whiteman, H. H. (1994) Evolution of facultative paedomorphosis in salamanders. *Quarterly Review of Biology* 69:205–221.

Whiteman, H. H., Wissinger, S. A., and Brown, W. S. (1996) Growth and foraging consequences of facultative paedomorphosis in the tiger salamander, *Ambystoma tigrinum nebulosum. Evolutionary Ecology* 10:433–446.

Whitlock, M. C. (1996) The Red Queen beats the Jack-of-all-Trades: the limitations on the evolution of phenotypic plasticity and niche breadth. *American Naturalist* 148:S65–S77.

Wilkinson, G. S., Fowler, K., and Partridge, L. (1990) Resistance of genetic correlation structure to directional selection in *Drosophila melanogaster. Evolution* 44:1990–2003.

Williams, G. C. (1992) Optimization and related concepts, pp. 56–71 in: *Natural Selection: Domains, Levels and Challenges.* New York: Cambridge University Press.

Williams, J., Bulman, M., Huttly, A., Phillips, A., and Neill, S. (1994a) Characterization of a cDNA from *Arabidopsis thaliana* encoding a potential thiol protease whose expression is induced independently by wilting and abscisic acid. *Plant Molecular Biology* 25:259–270.

Williams, J., Bulman, M. P., and Neill, S. J. (1994b) Wilt-induced ABA biosynthesis, gene expression and down-regulation of rbcS mRNA levels in *Arabidopsis thaliana. Physiologia Plantarum* 91:177–182.

Willis, J. H. (1996) Measures of phenotypic selection are biased by partial inbreeding. *Evolution* 50:1501–1511.

Willis, J. H., and Orr, H. A. (1993) Increased heritable variation following population bottlenecks: the role of dominance. *Evolution* 47:949–956.

Wilson, E. O. (1978) *On Human Nature.* Cambridge, Mass.: Harvard University Press.

Wimberger, P. H. (1994) Trophic polymorphisms, plasticity, and speciation in vertebrates, pp. 19–43 in: *Theory and Application of Fish Marine Science,* D. J. Stouder, K. L. Fresh, and R. J. Feller (eds.). Columbia: University of South Carolina Press.

Windig, J. J. (1994) Genetic correlations and reaction norms in wing pattern of the tropical butterfly *Bicyclus anynana. Heredity* 73:459–470.

Winicov, I. (1994) Gene expression in relation to salt tolerance, pp. 61–85 in: *Stress-induced gene expression in plants.* A. S. Basra (ed.). Chur, Switzerland: Harwood.

Winn, A. A. (1996a) Adaptation to fine-grained environmental variation: an analysis of within-individual leaf variation in an annual plant. *Evolution* 50:1111–1118.

Winn, A. A. (1996b) The contributions of programmed developmental change and phenotypic plasticity to within-individual variation in leaf traits in *Dicerandra linearifolia. Journal of Evolutionary Biology* 9:737–752.

Winn, A. A. (1999) Is seasonal variation in leaf traits adaptive for the annual plant *Dicerandra linearifolia? Journal of Evolutionary Biology* 12:306–313.

Witelson, S. F., Kigar, D. L., and Harvey, T. (1999) The exceptional brain of Albert Einstein. *Lancet* 353:2149–2153.

Witte, F., Barel, C. D. N., and Hoogerhoud, R. J. C. (1990) Phenotypic plasticity of anatomical structures and its ecomorphological significance. *Netherlands Journal of Zoology* 40:278–298.

Woltereck, R. (1909) Weitere experimentelle Untersuchungen über Artveränderung, speziell über das Wesen quantitativer Artunterschiede bei Daphniden. *Versuche Deutsche Zoologische Geselleschaft* 19:110–172.

Wray, G. A., and McClay, D. R. (1989) Molecular heterochronies amd heterotopies in early echinoid development. *Evolution* 43:803–813.

Wright, S. (1932) Evolution in Mendelian populations. *Genetics* 16:97–159.

Wu, R. (1998) The detection of plasticity genes in heterogeneous environments. *Evolution* 52:967–977.

Wu, R., Bradshaw, H. D., and Stettler, R. F. (1997) Molecular genetics of growth and development in *Populus* (Salicaceae). V. Mapping quantitative trait loci affecting leaf variation. *American Journal of Botany* 84:143–153.

Wu, S.-H., and Lagarias, J. C. (1997) The phytochrome photoreceptor in the green alga *Mesotaenium caldariorum:* implication for a conserved mechanism of phytochrome action. *Plant, Cell and Environment* 20:691–699.

Wu, S.-H., Wang, C., Chen, J., and Lin, B.-L. (1994) Isolation of a cDNA encoding a 70 kDa heat-shock cognate protein expressed in vegetative tissues of *Arabidopsis thaliana. Plant Molecular Biology* 25:577–583.

Yamada, U. (1962) Genotype by environment interaction and genetic correlation of the same trait under different environments. *Japanese Journal of Genetics* 37:498–509.

Yampolsky, L. Y., and Scheiner, S. M. (1994) Developmental noise, phenotypic plasticity, and allozyme heterozygosity in *Daphnia. Evolution* 48:1715–1722.

Yanovsky, M. J., Casal, J. J., and Whitelam, G. C. (1995) Phytochrome A, phytochrome B and HY4 are involved in hypocotyl growth responses to natural radiation in *Arabidopsis:* weak de-etiolation of the phyA mutant under dense canopies. *Plant, Cell and Environment* 18:788–794.

Yates, F., and Cochran, W. G. (1938) The analysis of groups of experiments. *Journal of Agricultural Science* 28:556.

Ylonen, H., and Ronkainen, H. (1994) Breeding suppression in the bank vole as anti-predatory adaptation in a predictable environment. *Evolutionary Ecology* 8:658–666.

Young, H. J., Stanton, M. L., Ellstrand, N. C., and Clegg, J. M. (1994) Temporal and spatial variation in heritability and genetic correlations among floral traits in *Raphanus sativus,* wild radish. *Heredity* 73:298–308.

Young, K. A., and Schmitt, J. S. (1995) Genetic variation and phenotypic plasticity of pollen release and capture height in *Plantago lanceolata. Functional Ecology* 9:725–733.

Zelditch, M. L., Bookstein, F. L., and Lundrigan, B. L. (1993) The ontogenetic complexity of developmental constraints. *Journal of Evolutionary Biology* 6:621–641.

Zhikang, L., Pinson, S. R. M., Stansel, J. W., and Park, W. D. (1995) Identification of quantitative trait loci (QTLs) for heading date and plant height in cultivated rice (*Oryza sativa* L.). *Theoretical and Applied Genetics* 91:374–381.

Zhivotovsky, L. A., Feldman, M. W., and Bergman, A. (1996) On the evolution of phenotypic plasticity in a spatially heterogeneous environment. *Evolution* 50:547–558.

Zhong, R., Taylor, J. J., and Ye, Z.-H. (1999) Transformation of the collateral vascular bundles into amphivasal vascular bundles in an *Arabidopsis* mutant. *Plant Physiology* 120:53–64.

Zolman, J. F. (1993) *Biostatistics: Experimental Design and Statistical Inference.* Oxford: Oxford University Press.

Zonta, L., and Jayakar, S. D. (1988) Models of fluctuating selection for a quantitative trait, pp. 102–108 in: *Population Genetics and Evolution,* G. de Jong (ed.). Berlin: Springer-Verlag.

Zrsavy, J., and Stys, P. (1997) The basic body plan of arthropods: insights from evolutionary morphology and developmental biology. *Journal of Evolutionary Biology* 10:353–368.

— Author Index —

— Subject Index —

Page numbers followed by letters *f, n,* and *t* refer to material presented in figures, notes, and tables, respectively.

Milton Keynes UK
Ingram Content Group UK Ltd.
UKHW010605010923
427853UK00003B/21/J